Lecture Notes in Computer Science 7239

Commenced Publication in 1973
Founding and Former Series Editors:
Gerhard Goos, Juris Hartmanis, and Jan van Leeuwen

Sang-goo Lee   Zhiyong Peng
Xiaofang Zhou   Yang-Sae Moon
Rainer Unland   Jaesoo Yoo (Eds.)

Database Systems for Advanced Applications

17th International Conference, DASFAA 2012
Busan, South Korea, April 15-18, 2012
Proceedings, Part II

 Springer

Volume Editors

Sang-goo Lee
Seoul National University, Seoul 151747, South Korea
E-mail: sglee@snu.ac.kr

Zhiyong Peng
Wuhan University, Wuhan 430081, Hubei Province, China
E-mail: peng@whu.edu.cn

Xiaofang Zhou
University of Queensland, Brisbane, QLD 4072, Australia
E-mail: zxf@itee.uq.edu.au

Yang-Sae Moon
Kangwon National University, Chuncheon 200701, Kangwon, South Korea
E-mail: ysmoon@kangwon.ac.kr

Rainer Unland
University of Duisburg-Essen, 45117 Essen, Germany
E-mail: rainer.unland@icb.uni-due.de

Jaesoo Yoo
Chungbuk National University, Cheongju 361-763, Chungbuk, South Korea
E-mail: yjs@chungbuk.ac.kr

ISSN 0302-9743   e-ISSN 1611-3349
ISBN 978-3-642-29034-3   e-ISBN 978-3-642-29035-0
DOI 10.1007/978-3-642-29035-0
Springer Heidelberg Dordrecht London New York

Library of Congress Control Number: 2012933786

CR Subject Classification (1998): H.2-5, C.2, J.1, J.3

LNCS Sublibrary: SL 3 – Information Systems and Application, incl. Internet/Web
and HCI

Typesetting: Camera-ready by author, data conversion by Scientific Publishing Services, Chennai, India

Printed on acid-free paper

Springer is part of Springer Science+Business Media (www.springer.com)

Preface

It is our great pleasure to welcome you to the proceedings of the 17th International Conference on Database Systems for Advanced Applications (DASFAA 2012), which was held in Busan, Korea, in April, 2012. DASFAA is an international conference which provides a forum for technical presentations and discussions among database researchers, developers and users from academia, business and industry, in the areas of databases, large-scale data management, data mining, search and recommendation, and the Web.

The call for papers attracted 159 research submissions from 24 countries (based on the affiliation of the first author). Among them, the Program Committee selected, through a comprehensive review process, 44 regular papers and 8 short papers for presentation. The Industrial Committee, chaired by Won Suk Lee, Mukesh Mohania and Jeffrey Yu, selected 8 industrial papers for presentation. The conference program also included 8 demo presentations selected from 17 submissions by the Demo Committee chaired by Wolf-Tilo Balke and Seung-Won Hwang.

This volume also includes extended abstracts of the two invited keynote lectures by Divesh Srivastava (AT&T Research) and Sang Kyun Cha (Seoul National University and SAP Labs Korea), whose topics were on "Enabling Real-Time Data Analysis" and "A New Paradigm of Thinking and Architecture for Real-Time Information Processing at Fingertips," respectively. The Tutorial Chair, Wook-Shin Han, organized four tutorials by leading experts on topics ranging from probabilistic databases to detecting clones and reuse on the Web. A stimulating panel was organized by the Panel Chair, Kyuseok Shim. This rich and attractive conference program boasts conference proceedings that span two volumes of Springer's *Lecture Notes in Computer Science* series.

Beyond the main conference Hwanjo Yu, Yu Ge and Wynne Hsu, who chaired the Workshop Committee, put together five workshops that catered to specific interests of the conference participants. The workshop papers are included in a separate volume of proceedings also published by Springer in its *Lecture Notes in Computer Science* series.

DASFAA 2012 was jointly organized by Pusan National University and the Database Society of Korea. It received in-cooperation sponsorship from the Korea Institute of Information Scientists and Engineers, the Database Society of Japan, the China Computer Federation Database Technical Committee, and the Korea Database Agency. We are grateful to the industry and institutional sponsors who contributed generously to making DASFAA 2012 successful.

The conference would not have been possible without the support and hard work of many colleagues. We would like to express our special thanks to Honorary Conference Chair, Kyu-Young Whang, for his valuable advice on all aspects of organizing the conference. We thank the DASFAA Steering Committee for their leaderships and encouragement. We thank the General Co-chairs, Yoon Joon Lee and Kazutoshi Sumiya, Organizing Committee Chair, Bonghee Hong, Publicity Co-chairs, Eenjun Hwang, Jae-Gil Lee and YunChan Chang, Local Arrangements Committee Co-chairs, Joonho Kwon and Ok-Ran Jeong, Finance Chair, Min-Su Lee, Web Co-chairs, Ha-Joo Song and Young-Koo Lee, Demo Award Committee Co-chairs, Young-Kuk Kim, Takahiro Hara and Kyoung-Gu Woo, Best Paper Committee Co-chairs, SangKeun Lee, Hiroyuki Kitagawa and Xiaofeng Meng, Sponsor Co-chairs, Yunmook Nah and Kyu-Chul Lee, Registration Chair, Sanghyun Park, Steering Committee Liaison, Byeong-Soo Jeong, APWEB Liaison, Wookey Lee, and EDB Liason, Jinho Kim.

Finally, our thanks go to all the committee members and other individuals involved in putting this all together, and to all authors who submitted their papers to this conference.

April 2012

Sang-goo Lee
Zhiyong Peng
Xiaofang Zhou
Yang-Sae Moon
Rainer Unland
Jaesoo Yoo

Organization

Honorary Conference Chair

Kyu-Young Whang   KAIST, South Korea

Conference General Co-chairs

Yoon Joon Lee   KAIST, South Korea
Kazutoshi Sumiya   University of Hyogo, Japan

Program Committee Co-chairs

Sang-goo Lee   Seoul National University, South Korea
Zhiyong Peng   Wuhan University, China
Xiaofang Zhou   University of Queensland, Australia

Organizing Committee Chair

Bonghee Hong   Pusan National University, South Korea

Workshop Co-chairs

Hwanjo Yu   POSTECH, South Korea
Yu Ge   Northeastern University, China
Wynne Hsu   National University of Singapore, Singapore

Industrial Co-chairs

Won Suk Lee   Yonsei University, South Korea
Mukesh K. Mohania   IBM Research, India
Jeffrey Xu Yu   Chinese University of Hong Kong, China

Tutorial Chair

Wook-Shin Han   Kyungbook National University, South Korea

Panel Chair

Kyuseok Shim   Seoul National University, South Korea

Demo Co-chairs

Wolf-Tilo Balke   TU-Braunschweig, Germany
Seung-Won Hwang   POSTECH, South Korea

Publicity Co-chairs

Eenjun Hwang   Korea University, South Korea
Jae-Gil Lee   KAIST, South Korea
YunChan Chang   Victoria University, Australia

Local Arrangements Co-chairs

Joonho Kwon   Pusan National University, South Korea
Ok-Ran Jeong   Gachon University, South Korea

Finance Chair

Min-Su Lee   Ewha Womans University, South Korea

Publication Co-chairs

Rainer Unland   University of Duisburg-Essen, Germany
Jaesoo Yoo   Chungbuk National University, South Korea
Yang-Sae Moon   Kangwon National University, South Korea

Web Co-chairs

Ha-Joo Song   Pukyong National University, South Korea
Young-Koo Lee   Kyung Hee University, South Korea

Demo Award Committee Co-chairs

Young-Kuk Kim   Chungnam National University, South Korea
Takahiro Hara   Osaka University, Japan
Kyoung-Gu Woo   Samsung Electronics, South Korea

Best Paper Committee Co-chairs

SangKeun Lee   Korea University, South Korea
Hiroyuki Kitagawa   University of Tsukuba, Japan
Xiaofeng Meng   Renmin University of China, China

Steering Committee Liaison

Byeong-Soo Jeong   Kyung Hee University, South Korea

Sponsor Co-chairs

Yunmook Nah   Dankook University, South Korea
Kyu-Chul Lee   Chungnam National University, South Korea

Registration Chair

Sanghyun Park   Yonsei University, South Korea

APWEB Liaison

Wookey Lee   Inha University, South Korea

EDB (International Conference on Emerging Databases) Liaison

Jinho Kim   Kangwon National University, South Korea

DASFAA Steering Committee

Ramamohanarao
  Kotagiri (Chair)   University of Melbourne, Australia
Jianzhong Li (Vice Chair)   Harbin Institute of Technology, China
Katsumi Tanaka (Advisor)   Kyoto University, Japan
Kazutoshi Sumiya (Treasurer) University of Hyogo, Japan
Qing Li (Secretary)   City University of Hong Kong, China
Masaru Kitsuregawa   University of Tokyo, Japan
Mukesh K. Mohania   IBM Research, India
Byeong-Soo Jeong   Kyung Hee University, South Korea
Ming-Syan Chen   National Taiwan University, Taiwan
Eui Kyeong Hong   University of Seoul, South Korea
Hiroyuki Kitagawa   University of Tsukuba, Japan
Li-Zhu Zhou   Tsinghua University, China
Stephane Bressan   National University of Singapore, Singapore
BongHee Hong   Pusan National University, South Korea

Program Committees

Research Track

Toshiyuki Amagasa	University of Tsukuba, Japan
Masayoshi Aritsugi	Kumamoto University, Japan
Zhifeng Bao	National University of Singapore, Singapore
Ladjel Bellatreche	Poitiers University, France
Boualem Benatallah	University of New South Wales, Australia
Sourav Bhowmick	Nanyang Technological University, Singapore
Cui Bin	Peking University, China
Athman Bouguettaya	RMIT, Australia
Jinseok Chae	University of Incheon, South Korea
Chee Yong Chan	National University of Singapore, Singapore
Jae Woo Chang	Chonbuk National University, South Korea
Jae-young Chang	Hansung University, South Korea
Lei Chen	HKUST, China
Ming-Syan Chen	National Taiwan University, Taiwan
Yi Chen	Arizona State University, USA
Hong Cheng	Chinese University of Hong Kong, China
James Cheng	Nanyang Technological University, Singapore
Reynold Cheng	University of Hong Kong, China
Jae-heon Cheong	Shingu University, South Korea
Byron Choi	Hong Kong Baptist University, China
Yon Dohn Chung	Korea University, South Korea
Gao Cong	Nanyang Technological University, Singapore
Alfredo Cuzzocrea	ICAR-CNR / University of Calabria, Italy
Gill Dobbie	University of Auckland, New Zealand
Xiaoyong Du	Renmin University of China, China
Jianhua Feng	Tsinghua University, China
Ling Feng	Tsinghua University, China
Yunjun Gao	Zhejiang University, China
Yu Ge	Northeastern University, China
Stephane Grumbach	INRIA, France
Takahiro Hara	Osaka University, Japan
Bingsheng He	Nanyang Technological University, Singapore
Wynne Hsu	National University of Singapore, Singapore
Haibo Hu	Hong Kong Baptist University, China
Ming Hua	Facebook, USA
Dong-Hyuk Im	Seoul National University, South Korea
Yoshiharu Ishikawa	Nagoya University, Japan
Adam Jatowt	Kyoto University, Japan
Ruoming Jin	Kent State University, USA
Sungwon Jung	Sogang University, South Korea

Norio Katayama   National Institute of Informatics, Japan
Yiping Ke   Chinese University of Hong Kong, China
Chulyon Kim   Kyungwon University, South Korea
Dongkyu Kim   Georgetown University, USA
Han-joon Kim   University of Seoul, South Korea
Jinho Kim   Kangwon National University, South Korea
Sang-Wook Kim   Hanyang University, South Korea
Markus Kirchberg   HP Labs Singapore, Singapore
Hiroyuki Kitagawa   University of Tsukuba, Japan
Ig-hoon Lee   Seoul National University, South Korea
Mong Li Lee   National University of Singapore, Singapore
Sang-Won Lee   Sungkyunkwan University, South Korea
Wang-Chien Lee   Pennsylvania State University, USA
Cuiping Li   Renmin University of China, China
Jianzhong Li   Harbin Institute of Technology, China
Xuemin Lin   University of New South Wales, Australia
Chengfei Liu   Swinburne University of Technology, Australia
Eric Lo   Hong Kong Polytechnic University, China
Jiaheng Lu   Renmin University of China, China
Nikos Mamoulis   University of Hong Kong, China
Weiyi Meng   Binghamton University, USA
Xiaofeng Meng   Renmin University of China, China
Jun-Ki Min   Korea University of Technology and Education,
                        South Korea
Jun Miyazaki   Nara Advanced Institute of Science and
                        Technology, Japan
Bongki Moon   University of Arizona, USA
Yang-Sae Moon   Kangwon National University, South Korea
Yasuhiko Morimoto   Hiroshima University, Japan
Atsuyuki Morishima   University of Tsukuba, Japan
Miyuki Nakano   University of Tokyo, Japan
Tadashi Ohmori   University of Electro-Communications, Japan
Makoto Onizuka   NTT Corporation, Japan
Hyoungmin Park   University of Brithish Columbia, Canada
Min Sik Park   Korea Database Agency, South Korea
Sanghyun Park   Yonsei University, South Korea
Young-Ho Park   Sookmyung Women's University, South Korea
Jian Pei   Simon Fraser University, Canada
Wen-Chih Peng   National Chiao Tung University, Taiwan
Lu Qin   Chinese University of Hong Kong, China
Keun Ho Ryu   Chungbuk National University, South Korea
Simonas Saltenis   Aalborg University, Denmark
Markus Schneider   University of Florida, USA
Jialie Shen   Singapore Management University, Singapore

Junho Shim	Sookmyung Women's University, South Korea
Hyoseop Shin	Konkuk University, South Korea
Jung Hyeon Sin	INET-Hosting, South Korea
Atsuhiro Takasu	National Institute of Informatics, Japan
David Taniar	Monash University, Australia
Vincent Tseng	National Cheng Kung University, Taiwan
Haixun Wang	Microsoft Research Asia, China
Jianyong Wang	Tsinghua University, China
John Wang	Griffith University, Australia
Wei Wang	University of New South Wales, Australia
Raymond Wong	HKUST, China
Xiaokui Xiao	Nanyang Technological University, Singapore
Jianliang Xu	Hong Kong Baptist University, China
Ke Yi	HKUST, China
Man Lung Yiu	Hong Kong Polytechnic University, China
Haruo Yokota	Tokyo Institute of Technology, Japan
Jaesoo Yoo	Chungbuk National University, South Korea
Rui Zhang	University of Melbourne, Australia
Wenjie Zhang	University of New South Wales, Australia
Baihua Zheng	Singapore Management University, Singapore
Bin Zhou	University of Maryland Baltimore County, USA

Industrial Track

Haibo Hu	Hong Kong Baptist University, China
Weining Qian	Fudan University, China
Bingsheng HE	Nanyang Technological University, Singapore
Marek Kowalkiewicz	SAP, Australia
Jilei Tian	Nokia Research China, China
Unil Yun	Chungbuk National University, South Korea
Yang-Sae Moon	Kangwon National University, South Korea
Byungjoo Chung	Cubrid, South Korea

Demo Track

Ilaria Bartolini	University of Bologna, Italy
Changkyu Kim	Intel Labs, USA
Jiaheng Lu	Renmin University of China, China
Yaokai Feng	Kyushu University, Japan
Young-In Song	Microsoft Research Asia, China
Yoonkyong Lee	Samsung Electronics, South Korea
Georgia Koutrika	IBM Research, USA
Christoph Lofi	TU-Braunschweig, Germany

External Reviewers

Brian Ackerman
Kamel Boukhalfa
Panagiotis Bouros
Yulei Fan
Wei Feng
Shen Ge
Reza Hemayati
Hai Huang
Zheng Huo
Stéphane Jean
Yu Jiang
Selma Khouri
Chungrim Kim
Young-kook Kim

Ryan Ko
Erwin Leonardi
Jing Li
Jianxin Li
Wenxin Liang
Pan Lin
Lin Liu
Yunzhong Liu
Cheng Long
Youzhong Ma
Luo Min
Jaehui Park
Peng Peng
Yu Peng

Zhenhua Song
Yu Shyang Tan
Yongxin Tong
Guan Wang
Hao Wang
Yousuke Watanabe
Kefeng Xuan
MingxuanYuan
Geng Zhao
Jinzeng Zhang
Rui Zhou
Wei Zhang
Xiaojian Zhang

Table of Contents – Part II

Demo Papers I: Social Data

Demo Papers II: Data Mining

Panel

Tutorials

Table of Contents – Part I

XML and Semi-structured Data II

Data Mining and Knowledge Discovery I

Data Mining and Knowledge Discovery II

Data Mining and Knowledge Discovery III

Privacy and Anonymity

Data Management in the Web

Graphs and Data Mining Applications

Temporal and Spatial Data I

Temporal and Spatial Data II

Top-k Best Probability Queries
on Probabilistic Data

Trieu Minh Nhut Le and Jinli Cao

Department of Computer Science and Computer Engineering,
La Trobe University Victoria 3086 Australia
nhuttrieu@gmail.com, j.cao@latrobe.edu.au

Abstract. There has been much interest in answering top-k queries on probabilistic data in various applications such as market analysis, personalised services, and decision making. In relation to probabilistic data, the most common problem in answering top-k queries is selecting the semantics of results according to their scores and top-k probabilities. In this paper, we propose a novel top-k best probability query to obtain results which are not only the best top-k scores but also the best top-k probabilities. We also introduce an efficient algorithm for top-k best probability queries without requiring the user's defined threshold. Then, the top-k best probability answer is analysed, which satisfies the semantic ranking properties of queries [3, 18] on uncertain data. The experimental studies are tested with both the real data to verify the effectiveness of the top-k best probability queries and the efficiency of our algorithm.

Keywords: Top-k Query, Query Processing, Uncertain data.

1 Introduction

Uncertain data has arisen in some important applications such as personalized services, market analysis and decision making, because data sources of these applications are collected from data integration, data analysis, data statistics, and results prediction. These data are usually inconsistent or likelihood information. Thus, selecting the best choice from various alternatives of uncertain data is an important challenge facing these applications. The top-k queries that return the k best answers according to a user's function score are essential for exploring uncertain data on these applications [7]. Uncertain data have been used extensively in many research areas such as modelling uncertain data [15, 5], managing uncertain data [22], and mining uncertain data [1, 12].

In business, investors often make decisions about their products based on analysis and statistical data [12], which provide predictions relating to successful and unsuccessful projects.

1.1 Motivation

As an example, assume that the data in Table 1 has been collected and analysed statistically, according to historical data resources. Each tuple represents

S.-g. Lee et al. (Eds.): DASFAA 2012, Part II, LNCS 7239, pp. 1–16, 2012.
© Springer-Verlag Berlin Heidelberg 2012

an investment project of USD $100 to produce a specific product (Product ID). Investing a profit (Profit) on the product will receive with the probability estimates of success on this investment (Probability) e.g. in tuple t_1, a businessman invests USD $100 on product A, and it has a 29% chance of obtaining a profit of USD $25.

Table 1. Predicted profit of USD $100 investment

Tuple	Product ID	Profit	Probability
t_1	A ·	25	0.29
t_2	B1	18	0.3
t_3	E1	17	0.8
t_4	B2	13	0.4
t_5	C	12	1.0
t_6	E2	11	0.2

In probabilistic data, almost tuples are independent of each other. However, in the real world, when analysing historical data, some special tuples are mutually exclusive. e.g. in tuple t_2, product B1 has 0.3 probability of making a profit of USD $18, and in tuple t_4, product B2 has 0.4 probability of making a profit of USD $13. In this case, if the prediction for B1 is true, then the prediction for B2 will not be true. In probabilistic data model, these mutually exclusive tuples can be controlled by a set of generation rules. In this example, probabilistic data in Table 1 are restricted by the generation rules $R_1 = t_2 \oplus t_4$ and $R_2 = t_3 \oplus t_6$.

In previous research [6, 16, 15, 19], to find the top-k answers to queries, the probabilistic data can be represented by listing all possible worlds. A possible world contains a number of tuples in probabilistic data set. Each possible world has a non zero probability for its existence. Table 2 lists three dimensions: the possible world, the probability of existence, and the top-2 tuples.

Table 2. All the possible worlds

Possible world	Pro. of existence	Top-2
$W_1 = \{t_1, t_2, t_3, t_5\}$	0.0696	t_1, t_2
$W_2 = \{t_1, t_2, t_5, t_6\}$	0.0174	t_1, t_2
$W_3 = \{t_1, t_3, t_4, t_5\}$	0.0928	t_1, t_3
$W_4 = \{t_1, t_4, t_5, t_6\}$	0.0232	t_1, t_4
$W_5 = \{t_1, t_3, t_5\}$	0.0696	t_1, t_3
$W_6 = \{t_1, t_5, t_6\}$	0.0174	t_1, t_5
$W_7 = \{t_2, t_3, t_5\}$	0.1704	t_2, t_3
$W_8 = \{t_2, t_5, t_6\}$	0.0426	t_2, t_5
$W_9 = \{t_3, t_4, t_5\}$	0.2272	t_3, t_4
$W_{10} = \{t_4, t_5, t_6\}$	0.0568	t_4, t_5
$W_{11} = \{t_3, t_5\}$	0.01704	t_3, t_5
$W_{12} = \{t_5, t_6\}$	0.0426	t_5, t_6

Table 3. Top-2 probabilities

Tuple	Profit	Probability	Top-2 probability
t_1	25	0.29	0.29
t_2	18	0.3	0.3
t_3	17	0.8	0.7304
t_4	13	0.4	0.3072
t_5	12	1.0	0.3298
t_6	11	0.2	0.0426

The top-2 probability of a tuple is aggregated by the sum of its probabilities of existence in the top-2 in Table 2.

In previous research by [6], the top-k answers are found using the probability threshold approach called PT-k. The PT-k queries return a set of tuples with top-k probabilities greater than the users' threshold value. In Table 3, the answer to the PT-2 queries with threshold 0.3 is the set containing 4 tuples $\{t_2, t_3, t_4, t_5\}$. This result of PT-k will be discussed based on both profit and top-2 probability attributes of the top-2 tuples in the following.

- The PT-k queries may lose some important results. According to PT-2 queries, tuple $t_1(25, 0.29)$ is eliminated by the PT-2 algorithm because its top-k probability is less than the threshold 0.3. In this case, we recommend that tuple t_1 should be in the result, the reason being that tuple $t_1(25, 0.29)$ is not worse than tuple $t_4(13, 0.3072)$, when comparing both attributes of profit and top-2 probability. That is, t_1.profit(25) is greater than t_4.profit(13) and t_1.top-2 probability(0.29) is less than t_2.top-2 probability(0.3072). Investors may like to choose t_1 because they can earn nearly double the profit from t_1 compared to t_4, while t_1 is only slightly riskier than t_4 with a top-2 probability 0.0172. Therefore, t_1 should be acceptable in the top-k answers.
- The PT-k answer may contain some redundant tuples which should be eliminated earlier in the top-k result. Referring to Table 3 again, tuples t_4 and t_5 should be eliminated immediately from the answer because the values of both attributes profit and top-2 probability in $t_4(13, 0.3072)$ and $t_5(12, 0.3298)$ are less than those in $t_3(18, 0.7304)$. It is obvious that the investor will choose t_3 which is more dominant than t_4 and t_5 in both profit and top-k probability.
- The threshold of PT-k is an unclear value for users. The threshold is a crucial factor used for efficiency and effectiveness in PT-k queries [6]. A threshold is required from users who may not know much about the probability. Therefore, they may assign any random threshold value several times until they can obtain the best answers with the PT-k algorithm, which may waste time.

With the aforementioned observations, there is a need to study the top-k query answers for a better solution. Thus, it is necessary to take both *top-k profit* and *top-k probability* into account to select the best top-k answer. In our recommendation, the top-2 results should include tuples $\{t_1, t_2, t_3\}$, for the following properties:

1. Inclusion of important tuples: The answer to the top-k queries should be *"the k highest scoring tuples"*. i.e. in Table 1, tuples t_1 and t_2 the top-2 highest profit should be in the top-2 answer.
2. Elimination of redundant tuples: The dominating concept on *score* and *top-k probability* of tuples are used to process non-top-k score tuples. These tuples will have *"best top-k probabilities"*. i.e. tuple t_3 also is added into the result because it has greater top-k probability than a tuple in the k highest ranking tuples. Then, the tuples t_4, t_5, t_6 in the data set, which are dominated by t_3 on both *score* and *top-k probability*, are eliminated.

3. Removal of the threshold: The combination of the two previous properties
will remove the need of the threshold.

Therefore, the answer set contains tuples $\{t_1, t_2, t_3\}$ for the top-2 query on probabilistic data. This solution provides *not only the best top-2 ranking scores but also the top-2 highest probabilities* to users.

The above example demonstrates and analyses the PT-k approach with our proposed approach. This paper will discuss the novel approach for top-k best probability queries. The proposed algorithm for new top-k queries is created to illustrate its effectiveness and efficiency.

1.2 Contributions

Our contributions are summarized as follows:

- The new definition: "Top-k best probability" answer to the top-k queries on probabilistic data is proposed based on common top-k answer and a dominating concept.
- Some pruning rules are introduced by formal formulas. Their correctness has been mathematically proven. These rules will be used to improve the effectiveness of the proposed algorithm by reducing the computation cost of top-k probabilities of tuples.
- The proposed algorithm demonstrates that the answers to top-k best probability queries are more efficient than the answers to PT-k under various probabilistic data sources. Moreover, the proposed approach is proven to meet the semantic top-k ranking properties
- A real data set is used in our extensive experimental study to evaluate the proposed approach. The PT-k method is compared with our method in terms of effectiveness and efficiency.

The rest of this paper is organized as follows. Section 2 presents the preliminaries to support our approach. In Section 3, we illustrate several new concepts and the proposed algorithm. Section 4 evaluates and discusses the proof of the proposed approach. Extensive experiments are conducted to show the significance of this study in Section 5. The final section briefly concludes our contribution and outlines future research.

2 Preliminary

In this section, several previous definitions on probabilistic data are presented formally, which are similar to papers [4, 6, 22, 11, 1, 21, 2, 19]. Generally, the probabilistic data $D = \{t_1, ..., t_n\}$ is a finite set of probabilistic tuples, where each probabilistic tuple is a tuple being associated with probability to denote the uncertainty of the data.

Definition 1. *Probability of tuple is the likelihood of a tuple appearing in the data set* $\mathcal{D}$*. The probability* $p(t_i)$ *of tuple* t_i *can be presented as a new dimension in the data with values of* $0 < p(t_i) \leq 1$*.*

In Table 1, tuple t_1 has a probability 0.29 of obtaining a profit.

Definition 2. *A possible world represents semantics of probabilistic data. Each possible world $W_j = \{t_1, ..., t_k\}$ contains a number of tuples which are members of probabilistic data. Each possible world is associated with a probability to indicate its existence. The probability of existence of possible world W_j is calculated by multiplying the probabilities of the tuples, which are the likelihood and unlikelihood of tuples in W_j. All possible worlds $\mathcal{W} = \{W_1, ..., W_m\}$ is called the possible world space.*

Table 2 is a possible world space that lists all possible worlds of Table 1.

Definition 3. *In probabilistic data, a generation rule is a set of special tuples, which are mutually exclusive in a possible world space. The generation rule is in a form of $R_h = t_{h_1} \oplus t_{h_2} \oplus ... \oplus t_{h_q}$, where $\oplus$ is an exclusive operator, and $t_{h_1}, t_{h_2}, ..., t_{h_q}$ are members of probabilistic data. The sum of all probabilities of tuples in a same generation rule must be less than or equal to 1, $\sum_{i=1}^{q} p(t_{h_i}) \leq 1$.*

In Table 1, data includes 2 generation rules $R_1 = t_2 \oplus t_4$ and $R_2 = t_3 \oplus t_6$.

2.1   Calculation Top-k Probability

In this section, we represent the calculation of top-k probabilities in a possible world space. The top-k probability of a tuple $pr_{top}^{k}(t_i)$ is the sum of the probabilities of existence of possible worlds in which the tuple is in the top-k [6]. In the probabilistic data set, when the number of tuples is increasing, it is impossible to list all the possible worlds and calculate every probability of existence of a possible world at a limited time, because the number of all the possible worlds is 2^n and the computation cost of all probabilities of its existence is very expensive. Then, the formulas to calculate the top-k probability of tuples are required to avoid listing and calculating for effective algorithms.

Let a probabilistic data set $\mathcal{D}$ be a ranked sequence $(t_1, t_2, ..., t_n)$ with $t_1.score \preceq t_2.score \preceq ... \preceq t_n.score$, and $S_{t_i} = (t_1, t_2, ..., t_i)$ $(1 \leq i \leq n)$ be a subsequence from t_1 to t_i.

Theorem 1. *Given a ranked sequence $S_{t_i} = (t_1, t_2, ..., t_i)$. $pr(S_{t_i}, j)$ is a probability of any j tuples $(n > i \geq j > 0)$ appearing in the set S_{t_i}. $pr(S_{t_i}, j)$ is calculated as follows:*

$$pr(S_{t_i}, j) = pr(S_{t_{i-1}}, j-1) \times p(t_i) + pr(S_{t_{i-1}}, j) \times (1 - p(t_i))$$

In special cases
- $pr(\phi, 0) = 1$
- $pr(\phi, j) = 0$
- $pr(S_{t_{i-1}}, 0) = \prod_{j=1}^{i} (1 - p(t_j))$

This Poisson binomial recurrence has been proved in [10].

Example: Given the data set in Table 1

$$pr(S_{t_1}, 1) = pr(\phi, 0) \times p(t_1) + pr(\phi, 1) \times (1 - p(t_1)) = 0.29$$
$$pr(S_{t_2}, 1) = pr(S_{t_1}, 0) \times p(t_2) + pr(S_{t_1}, 1) \times (1 - p(t_2)) = 0.416$$

Theorem 2. *Suppose that tuple t_i and the subsequence $S_{t_{i-1}}$. $pr(t_i, j)$ is the probability of tuple t_i which is ranked at the exact j^{th} position ($n > i \geq j > 0$). $pr(t_i, j)$ is calculated as follows:*

$$pr(t_i, j) = p(t_i) \times pr(S_{t_{i-1}}, j-1)$$

This Poisson binomial recurrence has been proved in [10].

Example: The 1st and 2nd rank probabilities of tuple t_3 are computed as follows:

$$pr(t_3, 1) = p(t_3) \times pr(S_{t_2}, 0) = 0.3976$$
$$pr(t_3, 2) = p(t_3) \times pr(S_{t_2}, 1) = 0.3328$$

Theorem 3. *Given a tuple t_i, $pr_{top}^k(t_i)$ is the top-k probability of t_i in the possible world space, then $pr_{top}^k(t_i)$ is calculated as follows:*

$$pr_{top}^k(t_i) = \sum_{j=1}^{k} pr(t_i, j) = p(t_i) \times \sum_{j=1}^{k} pr(S_{t_{i-1}}, j-1)$$

If $i \leq k$ then $pr_{top}^k(t_i) = p(t_i)$

Example: The top-2 probability of tuple t_3 is computed as follows:

$$pr_{top}^2(t_3) = pr(t_3, 1) + pr(t_3, 2) = 0.3976 + 0.3328 = 0.7304$$

2.2   Calculation of Top-k Probability with Generation Rules

Generally, the above theorems can be used to calculate the top-k probability of each tuple. However, real probabilistic data involves mutually exclusive rules. Therefore, the calculations of top-k probabilities have to take those rules into account. We follow paper [6] for calculating the top-k probabilities with added generation rules as follows:

Tuples in the same generation rule are mutually exclusive. Therefore, the rule can produce as a tuple. Then, the previous formulas are used to calculate the top-k probability.

Let $t_1.score \preceq t_2.score \preceq ... \preceq t_n.score$ be ranked as the sequence $\mathcal{D} = (t_1, t_2, ..., t_n)$. To compute the top-$k$ probability $pr_{top}^k(t_i)$ of tuple $t_i \in \mathcal{D}(1 \leq i \leq n)$, t_i divides the generation rule $R_h = t_{h_1} \oplus ... \oplus t_{h_m}$ into two parts $R_{hLeft} = t_{h_1} \oplus ... \oplus t_{h_j}$ and $R_{hRight} = t_{h_{j+1}} \oplus ... \oplus t_{h_m}$. The tuples involved in R_{hLeft} are ranked higher than or equal to t_i. R_{hRight} contains the ranked tuples lower than tuple t_i. According to this division, the following cases are demonstrated for reducing the tuples.

- Case 1: $R_{hLeft} = \phi$, i.e. all the tuples in rule R_h are ranked lower than or equal to tuple t_i. Therefore, all tuples in R_h are not considered in calculating the top-k probability of tuple t_i. Consequently, all tuples in R_h are ignored.
- Case 2: $R_{hLeft} \neq \phi$, i.e. all tuples in R_{hRight} can be ignored, and the tuples in R_{hLeft} have been changed for calculating the top-k probability of t_i. There are two sub-cases for these changes.

+ Sub-case 1: $t_i \in R_{hLeft}$, i.e. t_i has already appeared in the possible world space, the other tuples in R_{hLeft} will be removed from subsequence $S_{t_{i-1}}$ when calculating $pr_{top}^k(t_i)$.

+ Sub-case 2: $t_i \notin R_{hLeft}$, all tuples in R_{hLeft} will be produced and considered as a tuple $t_{R_{hLeft}}$ with probability $p(t_{R_{hLeft}}) = \sum_{R_{hLeft}} p(t_{hLeft})$.

After all the generation rules are produced, the formulas for calculating top-k probability in the previously mentioned theorems are used normally.

Example: In Table 1, $pr_{top}^2(t_6)$ will be calculated by applying two generation rules $R_1 = t_2 \oplus t_4$ and $R_2 = t_3 \oplus t_6$. The subsequence S_{t_6} containing the tuples with their probabilities is presented in Table 4.

Table 4. The subsequence S_{t_6}

t_1	t_2	t_3	t_4	t_5	t_6
0.29	0.3	0.8	0.4	1.0	0.2

Table 5. The produced subsequence S_{t_6}

t_1	$t_{2\oplus4}$	t_5	t_6
0.29	0.7	1.0	0.2

The generation rules are produced in subsequence S_{t_6}.

- In the generation rule R_2, t_3 is removed because t_3 and t_6 are in the same generation rule (Sub-case 1)

- In the generation rule R_1, $t_2 \oplus t_4$ are produced as $t_{2\oplus4}$. The probability of $t_{2\oplus4}$ is 0.7 (Sub-case 2)

The subsequence S_{t_6} is produced by the generation rules with their probabilities as presented in Table 5.

$pr_{top}^2(t_6)$ with the set $S_{t_5}\{t_1, t_{2\oplus4}, t_5\}$ is calculated as follows:

$$pr_{top}^2(t_6) = p(t_6) \times (pr(S_{t_5}, 0) + pr(S_{t_5}, 1)) = 0.2 \times (0 + 0.213) = 0.0426$$

3   The Top-k Best Probability Queries

Tuples which are the result of probabilistic top-k queries must consider not only the ranking score but also the top-k probability [13, 21]. In the area of probabilistic data, significant research is being conducted on semantics of top-k queries. However, the semantics between high scoring tuples and high probability of tuples is interpreted differently by various researchers. Our ranking approach considers both dimensions of ranking score and top-k probability independently, in which, the ranking score cannot be considered more important than the top-k probability and vice versa.

3.1   Definition of the Top-k Best Probability

To answer a probabilistic query, every tuple has two associated dimensions a top-k probability and its ranking score. These two dimensions are crucial for choosing the answer to the top-k queries on probabilistic data. They are also independent in real world applications. In this research, we introduce the concept of dominating tuples for selecting the full meaning top-k tuples which are non-dominated tuples. This concept is widely used for multiple independent dimensions for skyline queries in many papers [2, 14, 8].

Definition 4. *(domination of top-k probability tuples) For any tuples t_i and t_j in probabilistic data, t_i dominates t_j $(t_i \prec t_j)$, if and only if t_i has been ranked higher than t_j and the top-k probability of t_i is greater than the top-k probability of t_j $(pr^k_{top}(t_i) > pr^k_{top}(t_j))$, else t_i does not dominate t_j $(t_i \nprec t_j)$.*

Example: in Table 3, tuple t_3 dominates t_5, because t_3 has a higher rank than t_5 $(rank(t_3.score) = 3, rank(t_5.score) = 5)$ and top-2 probability (0.7304) of t_3 is greater than top-2 probability (0.3072) of t_5.

Now, we introduce Definition 5 to select the best tuples using the domination concept to improve the quality of the top-k answers. That is, we are looking at tuples which are in top-k best ranking scores and best top-k probabilities.

Definition 5. *The answer L to the top-k best probability query consists of two sets L_{score} and L_{pro} where L_{score} contains the top-k ranking score tuples in the data without considering the probabilities, and L_{pro} contains non-dominated tuples on top-k probability in the set $\{\{\mathcal{D} \setminus L_{score}\} \cup \{t_{pmin}\}\}$, where t_{pmin} is the tuple with the lowest top-k probability in L_{score}.*

The conditions to get the final answer to top-k best probability queries are described in Definition 5. With the example of Table 1, the answer set to the top-2 best probability query is $L = \{t_1, t_2, t_3\}$. In this answer, L_{score} is $\{t_1, t_2\}$ the top-2 highest ranked profit tuples (the common result), and L_{pro} is $\{t_1, t_3\}$ where tuples t_1 and t_3 are non-dominated tuples on top-k probability in $\{t_1, t_3, t_4, t_5, t_6\}$, while t_1 is t_{pmin}.

3.2   Finding Top-k Best Probability and Pruning Rules

We now describe a technique for selecting the top-k best probability answers, and present effective pruning rules for the top-k best probability algorithm.

a. Selecting the Answer to the Top-k Best Probability

Suppose that the data set has been ranked by score, we have divided n tuples of data set $\mathcal{D}$ into two sets $L_{score} = \{t_1, t_2, ..., t_k\}$ and $\mathcal{D} \setminus L_{score} = \{t_{k+1}, t_{k+2}, ..., t_n\}$. L_{score} contains the first k highest ranking score tuples.

To select non-dominated tuples for L_{pro}, we first pick up the tuple t_{pmin} which has the lowest top-k probability in L_{score}. This lowest top-k probability tuple is also the first non-dominated tuple in L_{pro}, because t_{pmin} has higher rank than all other tuples in $\mathcal{D} \setminus L_{score}$. The rest non-dominated tuples are selected only considering their top-k probabilities due to their ranking order. The initial value of $bestPr$ is $bestPr = \min^k_{i=1}(pr^k_{t_i})$. The non-dominated tuples from $\{t_{k+1}, t_{k+2}, ..., t_n\}$, the top-$k$ probability of each tuple from t_{k+1} to t_n is calculated in succession. In each tuple, the top-k probability will be compared to $bestPr$. If it is greater than $bestPr$, the tuple will be inserted into L_{pro}, and $bestPr$ assigns the greater top-k probability. The inserted tuple is the non-dominated in L_{pro} because its top-k probability is greater than all top-k probabilities of tuples in L_{pro}. At last, all the non-dominated tuples of L_{pro} are found when all tuples are executed. The answer set $L = L_{score} \cup L_{pro}$ is returned.

The value of $bestPr$ will be increased during selecting the top-k higher probability to the answer L_{pro}, therefore $bestPr$ is called "best top-k probability". The $bestPr$ is also the key value to improve the effectiveness and efficiency of our proposed algorithm because it is used to eliminate all tuples with low top-k probabilities without conducting calculation by following the pruning rules.

b. Pruning Rules

In this section, several theorems on pruning rules are created for our proposal.

Theorem 4. *Given any* t_i *and* $bestPr$, *if* $bestPr \geq p(t_i)$ *then* $bestPr \geq pr_{top}^k(t_i)$
The proof of theorem 4 is based on theorem 3.

The top-k probabilities of tuples do not need to be calculated when their probabilities are less than $bestPr$. Moreover, the following theorem is used to stop the algorithm before it calculates the top-k probabilities of lower ranking score tuples.

Theorem 5. *For any* t_i *and* $bestPr$, *if* $bestPr \geq \sum_{j=1}^{k} pr(S_{t_i} \setminus \{t_{L_{max}}\}, j-1)$
then $bestPr \geq pr_{top}^k(t_{i+m})$ *for any* $m \geq 1$ *where*
- $S_{t_i} = (t_1, t_2, ..., t_i)$ *is the ranked sequence.*
- $t_{L_{max}}$ *is the produced tuple of the generation rule which has the highest produced probability than the other produced probabilities in generation rules.*

Proof is omitted due to page limitations.

3.3 The Top-k Best Probability Algorithm

Input: probabilistic data $\mathcal{D}$ in ranking order, generation rules $\mathcal{R}$.
Output: L answer set to top-k best probability query
1 **foreach** *tuple* t_i *{i=1 to n}* **do**
2   **if** $i \leq k$ **then**
3          $pr_{top}^k(t_i) \leftarrow p(t_i)$ *(Theorem 3)*;
4          $L \leftarrow L \cup (t_i, pr_{top}^k(t_i))$;
5          $bestPr \leftarrow min\{pr_{top}^k(t_i)\}$;
6   **else**
7   **if** $p(t_i) > bestPr$ *(Theorem 4)* **then**
8             $Computing\ pr_{top}^k(t_i)$;
9   **if** $pr_{top}^k(t_i) > bestPr$ **then**
10               $bestPr \leftarrow pr_{top}^k(t_i)$;
11               $L \leftarrow L \cup (t_i, pr_{top}^k(t_i))$;
12   **if** $bestPr$ *satisfies Theorem 5* **then**
13             $Exit$;

Algorithm 1. The top-k best probability

Algorithm 1 is to find the top-k best probability answer $L = L_{score} \cup L_{pro}$ in section 3.2. L_{score} is found from line 2 to line 5, and non-dominated tuples in L_{pro} are selected from line 8 to line 11. For effectiveness, the pruning rules are applied in line 7 and line 12. Finally, the answer L to the top-k best probability query is returned.

4   Significance of Top-k Best Probability Query

In this section, we further analyse the top-k best probability query matching the proposed requirements.

4.1   Dominating Concept for Semantic Answers

The top-k best probability is used to find the non-dominated tuples for the best top-k probability value. This is the key of our proposal which has solved almost all the limitations of the previous approaches. A number of papers have discussed about probabilistic data queries. The semantics of answering a query can have different meanings in various applications [1, 17, 20]. The U-top-k [16], C-Typical-top-k [4], and E-rank [3] were trying to create relations between the ranking scores and top-k probabilities on probabilistic data in different domains. These approaches are suitable in their specific situations, because these relations have different meanings semantically. However, the real semantics of top-k answers are mainly based on the users' decisions on both top-k ranking score and top-k probabilities. This is appropriate to use the *non-dominated concept* for selecting answers to users.

4.2   Threshold vs. BestPr

A threshold plays an important role in the PT-k algorithm in paper [6]. The authors used the static threshold to prune irrelevant answers. Hence, the PT-k algorithm of top-k query seems to be more effective and efficient for processing the top-k answers. However, it could be ineffective and inefficient for users, because they ingeniously gave the problem of selecting the threshold value to the users. The users randomly select the threshold from 0 to 1 several times until obtaining a satisfactory answer. If the users choose a lower threshold than the top-k probabilities of tuples, the computation cost of PT-k algorithm to the top-k queries will be expensive due to the need to calculate almost all the top-k probabilities of tuples. Otherwise, if the users choose a higher threshold than top-k probabilities, the result set of PT-k algorithm will be empty due to pruning all top-k probability tuples by the higher threshold. Moreover, users sometimes do not have enough information or do not know about threshold for probabilistic data. Therefore, it is not easy to choose a suitable threshold value as the threshold is an unclear value for the users'. Therefore, based on the observation our proposed algorithm does not use the threshold for pruning technique. Our pruning technique uses *bestPr* which is automatically initialized and updated during execution.

4.3   Semantics Ranking Properties

The previous work [3, 18] formally defined the key properties of ranking query semantics on probabilistic data. These properties are *Exact-k, Containment, Unique ranking, Value invariance*, and *Stability*. However, it is desirable that the users also expect the results containing the k-best ranking scores. We consider the capability of providing these scores as another property of ranking query semantics on probabilistic data. We name this property as *k-best ranking scores*. The following will present a comparison between our approach and the previous work in terms of these properties.

U-top-k [16]: The Uncertain-top-k approach returns a set of top-k tuples, namely, a tuple vector. This vector has the highest aggregated probabilities in all possible worlds. The k-tuples in the tuple vector appear restrictively together in the same possible worlds. Given the example of the probabilistic data in Table 1, the U-top-2 answer is the vector (t_3, t_4) which has the highest top-2 probability vector 0.2272 in the possible world space. The answer to the U-top-2 query does not contain the top-2 highest score tuples t_1 and t_2. As a results, the *k-best ranking scores* property of U-top-k query answer is fail (Fail).

E-rank [3]: A new expected score is produced by the expected ranking to select the top-k ranking query for probabilistic data. The authors used the ranking to calculate the new expected score for their proposal to remove the magnitude of normal expected score limitations. The magnitude of the normal expected score is a tuple having low top-k probability and a high score, giving it the highest expected score. If the score has been adjusted to just greater than the next highest score, it will fall down the ranking. Applying E-rank for probabilistic data in Table 1, the final expected score is ranked $(t_1, t_3, t_2, t_5, t_4, t_6)$. The E-rank top-2 query answer is (t_1, t_3) which does not contain the top-2 highest score tuple t_2. Hence, the E-rank query answer fail to reach the *k-best ranking scores* property.

M. Hue and J. Pei [6] proposed a PT-k query on probabilistic data using the threshold value to cut off irrelevant candidate tuples having lower top-k probability. This was analysed in section I. The answer to the PT-2 query also does not contain the top-2 highest score tuple t_1 when the threshold is assigned 0.3 by users. The PT-2 query answer can contain the top-2 highest score tuples when the threshold is assigned less than or equal to 0.29. Therefore, the *k-best ranking scores'* property is weakly acceptable (Ok).

To analyse these properties for our top-k best probability queries, the four properties *Containment, Unique ranking, Value invariance*, and *Stability* are proved similar to the PT-k method in paper [3]. The *Exact-k* property is defined as "the top-k answer contains exactly k tuples". The top-k best probability query is fairly satisfied with this property, because this query always returns at *least k tuples as the answer*. The answer to the top-k best probability query contains top-k highest ranked score tuples L_{score} and non-dominated tuples L_{pro}. In the worst case, L_{pro} can contain only one tuple t_{pmin} in L_{score}. Hence, the *Exact-k* property also is acceptable (Good) in the top-k best probability approach.

Table 6. Summary of Top-k Methods

Method	k-best ranking score	Exact-k	Containment	Unique ranking	Value invariance	Stability
U-top-k	Fail	Fail	Fail	Good	Good	Good
PT-k	Ok	Fail	Ok	Good	Good	Good
E-Rank	Fail	Good	Good	Good	Good	Good
Top-k bestPr	Good	Ok	Ok	Good	Good	Good

The last property *k-best ranking score* is also obtained by top-k best probability result because of containing L_{score} the k highest ranking tuples.

Overall, Table 6 illustrates that our proposed approach covers six properties of semantic ranking while the previous studies all fail in at least one of these properties. Hence, the answers to the top-k best probability queries overcome the other answers to earlier top-k queries on probabilistic data in terms of semantic ranking properties.

5   Experimental Study

In this section, we report an extensive empirical study over a real data set to examine which is more effective and efficient than the ranking queries with the threshold approach [6]. All the experiments were conducted on a PC with a 2.33GHz Intel Core 2 Duo, 3 GB RAM, and 350GB HDD, running Windows XP Professional Operating system. All algorithms were implemented in C++.

We use the *International Ice Patrol Iceberg Sighting dataset* for our real experiment[1]. This data set was used in previous work on ranking and skyline queries in probabilistic data [6, 11, 9]. The IIP's mission is to survey, plot and predict iceberg drift to prevent icebergs threatening ships' travel. The data in the IIP collection is structured in ASCII text files. The data contains a set of iceberg sightings as tuples, each tuple including the important attributes (Sighting Source, Position, Date/Time, Iceberg Description, Drifted days). These are very important for detecting the travel of icebergs. Making the top-k queries using the number of days of iceberg drift as a ranking score is considered an important tool for predicting the icebergs' travel. Moreover, there are six types of sighting sources which are R/V (Radar and Visual), VIS (VISual only), RAD (RADar only), SAT-LOW (LOW earth orbit SATellite), SAT-MED (MEDium earth orbit SATellite), and SAT-HIGH (HIGH earth orbit SATellite). The differences in confidence-levels of these sighting sources are classified and presented by

---

[1] ftp://sidads.colorado.edu/pub/DATASETS/NOAA/G00807

probabilistic values which are R/V(0.7), VIS(0.6), RAD(0.5), SAT-LOW(0.4), SAT-MED (0.3), and SAT-HIGH(0.2). These numbers are also considered as probabilities of independent tuples in IIP data.

In the IIP data, some sighting tuples from different sighting sources could detect the same iceberg at the same time. In this situation, if the distance of these sighting tuples is less than 0.1 mile, they are considered to be sightings of one iceberg collected from different sources. Therefore, of these sighting tuples, only one tuple can be correct. These tuples are recorded on the IIP data which is controlled by generation rules $R_r = t_{r_1} \oplus t_{r_2} \oplus ... \oplus t_{r_q}$. The probabilities of these records are adjusted by the following formula:

$$p(t_{r_j}) = \frac{p(t_{r_j})}{\sum_{i=1}^{m} p(t_{r_i})} \times max(p(t_{r_1}), ..., p(t_{r_m}))$$

where $p(t_{r_j})$ is the probability of sighting sources of tuple t_{r_j}.

After reprocessing all sighting tuples in IIP data 2009, the IIP probabilistic data contains 13,039 tuples and 2,137 generation rules. One of the generation rules is $R_r = t_7 \oplus t_8 \oplus t_9 \oplus t_{10}$. The proposed algorithm is executed on this IIP probabilistic data for the top-10 best probability query, the answer to this query being $(t_1, t_2, t_3, t_4, t_5, t_6, t_7, t_8, t_9, t_{10}, t_{11})$ as shown in Table 7. Tuple t_1 has the highest value in the *Drifted Days* attribute, t_2 is the other tuple which has the same value or the second highest value in the *Drifted Days* attribute, and so on. The answer to the top-10 best probability query contains the 10 highest scoring tuples in the data set $(t_1, t_2, t_3, t_4, t_5, t_6, t_7, t_8, t_9, t_{10})$. Tuple t_{11} is in the answer due to the fact that it has a top-10 probability better than a tuple in $(t_1, t_2, t_3, t_4, t_5, t_6, t_7, t_8, t_9, t_{10})$. t_{11} is a non-dominated tuple. Moreover, tuples $(t_{12}, t_{13}, t_{14}, t_{15}, ...)$ are not in the answer because they are dominated by t_{11} on both ranking score and top-k probability.

Table 7. Highest scores of tuples in IIP (2009)

Tuple	Drifted days	Pro. of tuples	Top-10 pro.
t_1	500.0	0.2917	0.2917
t_2	500.0	0.2333	0.2333
t_3	495.8	0.7	0.7
t_4	488.7	0.35	0.35
t_5	455.5	0.6	0.6
t_6	439.5	0.7	0.7
t_7	435.2	0.15	0.15
t_8	431.6	0.15	0.15
t_9	431.0	0.15	0.15
t_{10}	430.9	0.15	0.15
t_{11}	427.6	0.7	0.7
t_{12}	423.5	0.7	0.7
t_{13}	416.2	0.7	0.6
t_{14}	414.5	0.3	0.299
t_{15}	408.8	0.2	0.198

On this IIP probabilistic data, we also applied PT-10 queries [6] by setting different thresholds, as shown in Figure 1. This illustrates that the number of tuples in the answer to PT-10 are different to the different values of threshold. If the threshold is greater than or equal to 0.35, the number of tuples in the PT-10 queries is less than 10. This is surely due to the PT-10 answers missing some of tuples which have the top-10 highest scores in the data set, which is the reason for losing important tuples in PT-k approach. Therefore, threshold plays a main role in selecting an answer to the PT-k queries. It is not easy for the users to choose the threshold to obtain suitable answers. In our approach, the users do not need to choose the threshold. The best answer will be returned without losing any important tuples in terms of the non-dominated set, whereupon, the users can choose the best one in both ranking score and top-k probability.

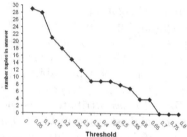

Fig. 1. The answer to the PT-10 vs. thresholds

Moreover, we compare the PT-k algorithm (PT-k), our proposed algorithm (top-k best probability), and the top-k normal algorithm (top-k normal) to evaluate which is the most effective calculation and the most efficient answers. The top-k normal algorithm simply returns the k top-k highest ranking scores in the data set. The top-k normal algorithm has been widely used on certain data, but has not been used on probabilistic data. We mention it as the axis for comparison of the PT-k results and top-k best probability results. The threshold of the PT-k algorithm is assigned 0.15 the minimum probabilities from t_1 to t_{15} of Table 7. This value is selected due to the fact that the PT-k satisfied the "k-best ranking score" attribute of Table 6. It means that all answers to the PT-k will contain the top-k highest scoring tuples when k is run from 1 to 15. For these settings, we execute programs to obtain the results in Figure 2 and 3.

The effectiveness of the proposed algorithm can be verified by counting the number of tuples which are accessed during the executing algorithm. The lower the number of tuples accessed, the more effective the algorithm. Figure 2 shows that the top-k best algorithm accesses fewer tuples than the PT-k algorithm. The top-k normal algorithm has the best performance in accessing the number of tuples. However, this algorithm has only been executed on certain data.

Figure 3 shows the number of tuples in the answers to the PT-k queries, the top-k best probability queries, and the normal queries. The users always expect that the answers to the top-k queries on probabilistic data are concise. Figure 3 shows that all the answers to the top-k best probability queries contain less tuples

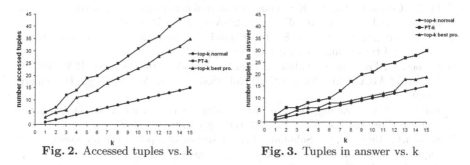

Fig. 2. Accessed tuples vs. k   **Fig. 3.** Tuples in answer vs. k

than the answers to PT-k. Also, they are closer to the top-k normal answers. This can explain why the PT-k answers can contain redundant tuples which do not have the top-k highest score and the best top-k probabilities. Hence, the answers to the top-k best probability queries are more concise than the answers to PT-k queries.

In real data, the results clearly show that the top-k best probability algorithm removed the unclear threshold value and reduced the number of accessed tuples compared to the PT-k algorithm. This makes the top-k best probability algorithm more effective. Moreover, the answers to the top-k best probability queries are concise.

6   Conclusions

In this paper, we proposed a novel top-k best probability for queries on probabilistic data, which selects the top-k best ranking score and the non-dominated tuples for users. Firstly, several concepts and theorems from previous studies were formally defined and discussed. Then, semantic ranking properties were introduced to answer top-k best probability queries, and an effective algorithm was created. This proposed approach has been demonstrated the improvement of previous state-of-the-art algorithms. The answer has not only the best top-k score but also the top-k best probability. Finally, the experimental results verified the efficiency and effectiveness of our approach on IIP real probabilistic data. The proposed approach has been proven outperformed PT-k.

In many real life domains, uncertain data is inherent in many applications and modern equipments. Therefore, discovering semantic answers to queries is a critical issue in relation to uncertain data. The proposed approach can be applied to modelling, managing, and analysing on uncertain data.

References

[1] Aggarwal, C.C., Yu, P.S.: A survey of uncertain data algorithms and applications. IEEE TKDE 21, 609–623 (2009)
[2] Atallah, M.J., Qi, Y.: Computing all skyline probabilities for uncertain data. In: PODS, pp. 279–287 (2009)

[3] Li, F., Cormode, G., Yi, K.: Semantics of ranking queries for probabilistic data and expected ranks. In: ICDE, March 29-April 2, pp. 305–316 (2009)

[4] Ge, T., Zdonik, S., Madden, S.: Top-k queries on uncertain data: on score distribution and typical answers. In: SIGMOD, pp. 375–388 (2009)

[5] Getoor, L.: Learning Probabilistic Relational Models. In: Choueiry, B.Y., Walsh, T. (eds.) SARA 2000. LNCS (LNAI), vol. 1864, pp. 322–323. Springer, Heidelberg (2000)

[6] Hua, M., Pei, J., Zhang, W., Lin, X.: Ranking queries on uncertain data: a probabilistic threshold approach. In: SIGMOD, pp. 673–686 (2008)

[7] Ilyas, I.F., Beskales, G., Soliman, M.A.: A survey of top-k query processing techniques in relational database systems. ACM 40, 1–58 (2008)

[8] Jan, C., Parke, G., Jarek, G., Dongming, L.: Skyline with presorting theory & optimizations. IIPWM 31, 595–604 (2005)

[9] Jin, C., Yi, K., Chen, L., Yu, J.X., Lin, X.: Sliding-window top-k queries on uncertain streams. In: VLDB, pp. 301–312 (2008)

[10] Lange, K.: Numerical analysis for statisticians. Springer, Heidelberg (1999)

[11] Li, J., Saha, B., Deshpande, A.: A unified approach to ranking in probabilistic databases. In: VLDB, pp. 502–513 (2009)

[12] Pang-Ning, T., Michael, S., Vipin, K.: Introduction to data mining. Library of Congress (2006)

[13] Papadias, D., Tao, Y., Fu, G., Seeger, B.: An optimal and progressive algorithm for skyline queries. In: SIGMOD, pp. 467–478 (2003)

[14] Pei, J., Jiang, B., Lin, X., Yuan, Y.: Probabilistic skylines on uncertain data. In: VLDB, pp. 15–26 (2007)

[15] Sarma, A.D., Benjelloun, O., Halevy, A., Widom, J.: Working models for uncertain data. In: ICDE (2006)

[16] Soliman, M.A., Ilyas, I.F., Chang, K.C.-C.: Top-k query processing in uncertain databases. In: ICDE, pp. 896–905 (2007)

[17] Soliman, M.A., Ilyas, I.F., Chang, K.C.-C.: Probabilistic top-k & ranking-aggregate queries. ACM Trans. Database Syst. 33, 13:1–13:54 (2008)

[18] Xi, Z., Jan, C.: Semantics and evaluation of top-k queries in probabilistic databases. Distributed Parallel Databases 26(1), 67–126 (2009)

[19] Yan, D., Ng, W.: Robust Ranking of Uncertain Data. In: Yu, J.X., Kim, M.H., Unland, R. (eds.) DASFAA 2011, Part I. LNCS, vol. 6587, pp. 254–268. Springer, Heidelberg (2011)

[20] Yi, K., Li, F., Kollios, G., Srivastava, D.: Efficient processing of top-k queries in uncertain databases with x-relations. TKDE 20, 1669–1682 (2008)

[21] Zhang, S., Zhang, C.: A probabilistic data model and its semantics. Journal of Research & Practice in Information Technology 35, 237–256 (2003)

[22] Zhang, W., Lin, X., Pei, J., Zhang, Y.: Managing uncertain data: probabilistic approaches. In: Web-Age Information Management (2008)

Probabilistic Reverse Skyline Query Processing over Uncertain Data Stream

Mei Bai, Junchang Xin, and Guoren Wang

College of Information Science & Engineering, Northeastern University, P.R. China
baimei1221@gmail.com, {xinjunchang,wangguoren}@ise.neu.edu.cn

Abstract. Reverse skyline plays an important role in market decision-making, environmental monitoring and market analysis. Now the flow property and uncertainty of data are more and more apparent, probabilistic reverse skyline query over uncertain data stream has become a new research topic. Firstly, a novel pruning technique is proposed to reduce the number of uncertain tuples reserved for processing continuous probabilistic reverse skyline query. Then some probability pruning techniques are proposed to reduce some redundant calculations. Next, an efficient algorithm, called Optimization Probabilistic Reverse Skyline (OPRS), is proposed to process continuous probabilistic reverse skyline queries. Finally, the performance of OPRS is verified through a large number of simulation experiments. The experimental results show that OPRS is an effective way to solve the problem of continuous probabilistic reverse skyline, and it could significantly reduce the executionx time of continuous probabilistic reverse skyline queries and meet the requirements of practical applications.

1 Introduction

With the development of information technology, the generation speed of mass data becomes faster and faster, and the flow property of data becomes more and more apparent. At the same time, the uncertainty of data has taken on, so uncertain data stream has become a new research topic. Traditional data processing technology cannot meet the requirements of uncertain data stream, hence it is necessary to study the processing technology of uncertain data stream.

The skyline operator [1] and its variations [2,3,5,6,7,8,9,4] play an important role in daily life. Given a set P of points, the *skyline* of P contains all the points which are not dominated by any other points in P. We can say a point x dominates y, if x is not worse than y for each dimension i, and x is better than y for at least one dimension j. In this paper, the smaller the value is, the better it is. Let $x.i$ denotes the value of x in dimension i, then x dominates y can be expressed by the formula, $x \prec y \Leftrightarrow (\forall i \in \{1, 2, ..., d\}, x.i \leq y.i) \land (\exists j \in \{1, 2, ..., d\}, x.j < y.j)$. Figure 1 illustrates an example of skyline in 2-D space, point a, c and e are the skyline points, because there are no other points can dominate them.

Given a data set P and a query point q, *Dynamic skyline* [5,6] of P contains all the points which are not dynamic dominated with respect to (w.r.t.) q by any

S.-g. Lee et al. (Eds.): DASFAA 2012, Part II, LNCS 7239, pp. 17–32, 2012.

other points in P. Tuple x dynamic dominates y, if x doesn't have longer distance to q than y for each dimension i, and x has shorter distance to q than y for at least one dimension j. Then x dynamic dominates y w.r.t. q can be expressed by the formula, $x \prec_q y \Leftrightarrow (\forall i \in \{1, 2, ...d\}, |x.i - q.i| \leq |y.i - q.i|) \wedge (\exists j \in \{1, 2, .., d\}, |x.j - q.j| < |y.j - q.j|)$. As illustrated in Figure 1, point c, d and f are the dynamic skyline points w.r.t. q.

Reverse skyline [6,7,8,9] is proposed based on the concept of dynamic skyline, given a data set P and a query point q, reverse skyline w.r.t. q of P contains all the points whose dynamic skyline contains q. As shown in Figure 1, q is the reverse skyline of c, since c is contained in the dynamic skyline w.r.t. q. Reverse skyline plays an important role in market decision-making.

With the development of economic and the progress of technological means, the flow property and uncertainty of data become more and more apparent. So reverse skyline query can be introduced to uncertain data streams which is very useful for many applications. For example, in the shopping site, lots f new product information can be modeled as uncertain date stream. If a shopkeeper want to take some new products to sell, it is desirable for the new product to attract the customers' attention. If a customer is interested in a product p, s/he might be also interested in the dynamic skyline of p. Therefore, the shopkeeper conduct a reverse skyline query over all the new products w.r.t. the best-selling product in the shopping site, then s/he can get the best purchase programme.

Although reverse skyline query has been studied in recent years [6,7,8,9,10,11], there has no efficient method to process reverse skyline query over uncertain data stream. In this paper, we propose an efficient algorithm, called *Optimization Probabilistic Reverse Skyline* (OPRS), to continuously answer probabilistic reverse skyline. The contributions of this paper are summarized as follows:

1. Through detailed and in-depth analysis of reverse skyline's properties over uncertain data stream, some pruning techniques are proposed.
2. Based the analysis of pruning techniques, a algorithm Optimization Probabilistic Reverse Skyline (OPRS) is proposed.
3. The experimental results show that OPRS can efficiently process probabilistic reverse skyline query over uncertain data stream.

The rest of the paper is organized as follows. Section 2 briefly review the related work. Section 3 introduce some related concepts and define the problem of probabilistic reverse skyline over uncertain data stream. Section 4 discuss the details of OPRS algorithm. Section 5 describe the experimental results, and the results are analyzed. Finally, we conclude the paper in Section 6.

2   Related Work

Since Borzsonyi et al. [1] introduced the concept of skyline to database in 2001, many efficient algorithms about skyline and its variations have been proposed. Papadias et al. [5] proposed the concept of dynamic skyline in 2003. There are many efficient algorithms for processing skyline query. For example, BBS [5] process skyline query based on R-tree, making use of the corner of MBR to prune some redundant tuples, which greatly improve the efficiency of the algorithm.

Based on dynamic skyline, Dellis et al. [6] proposed the concept of reverse skyline, they propose BBRS algorithm to process reverse skyline on the basis of BBS, BBRS make use of R-tree to compute *global skyline*, then do window query for all the global skyline points, get the final results. Wang et al. [10] proposed the concepts of semi-dominance and semi-skyline, then they propose an efficient algorithm to compute reverse skyline on sensor networks on basis of skyband.

Lin et al. [12] proposed CCS and PCS to compute skyline over data stream, CCS can real-timely respond to the change of data, and prune the data on data stream through dominance relationship transitivity. PCS is appropriate for the application environment which updates periodically, and it help save CPU resources. Zhang et al. [13] proposed probabilistic skyline query algorithm over sliding window, which reduce the number of data in the sliding window by dominance relationship transitivity, and then make use of the valid time of data in sliding window to prune. Zhu et al. [11] proposed DCRS to compute reverse skyline over data stream, based on the DC-tree and the concept of 2-skyline, DCRS can continuously compute the reverse skyline over data stream. There are many other algorithms [14, 15, 16, 17] proposed to process skyline query and its variations over data stream. In this paper, we propose a algorithm to process continuous probabilistic reverse skyline over uncertain data stream, which is different from all the studies above.

3   Problem Statement

First, we recall two important concepts, *full-dominate* [10] and *semi-dominate* [10]. All the full-dominance relationship and semi-dominance relationship in this paper are with respect to the query point q, later we omit "w.r.t. q" for shortness.

Definition 1 (full-dominate [10]). *Given a set P of tuples in a d-dimensional space D, q is the query point, x full-dominate y (denoted $x \precsim_q y$) if: 1)$\forall i \in D, |x.i - q.i| \leq |y.i - q.i|, |x.i - q.i||y.i - q.i| \geq 0$; 2)$\exists j \in D, |x.j - q.j| < |y.j - q.j|, |x.j - q.j||y.j - q.j| > 0$.*

Definition 2 (semi-dominate [10]). *Given a set P of tuples in a d-dimensional space D, q is the query point, x semi-dominate y (denoted $x \tilde{\precsim}_q y$) if: 1)$\forall i \in D, |x.i - q.i| \leq 2|y.i - q.i|, |x.i - q.i||y.i - q.i| \geq 0$; 2)$\exists j \in D, |x.j - q.j| < 2|y.j - q.j|, |x.j - q.j||y.j - q.j| > 0$.*

Figure 2 illustrates an example of full-dominance and semi-dominance, by Definition 1 and 2 we can see that, if two points have full-dominance or semi-dominance relationships, they must appear in the same region of the query point. For example, a can full-dominate b, f can full-dominate h, then a, b are in the first region of q, and f, h in the forth region. In Figure 2, point a, b, c are in the same region, b and c can semi-dominate a, and a, b cannot semi-dominate c, hence there are no points can semi-dominate c.

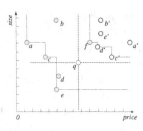
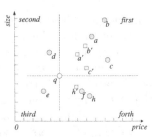

Fig. 1. Example of Skylines and Its Variations

Fig. 2. Example of Full-dominate and Semi-dominate

In paper [10], there has been an very useful theorem about semi-dominance, which can help to process reverse skyline query, as shown in Theorem 1. Due to the article length restrictions, proofs in this paper are omitted.

Theorem 1. *[10] Given a set P of tuples in d-dimensional space D, q is the query point, the reverse skyline of P w.r.t. q contains all the tuples which are not semi-dominated by any other tuples in P.*

Then we introduce the concept of probabilistic reverse skyline in uncertain database. In uncertain database U, every point u must have an *exist probability* to express the probability that u occurs, denote as $P(u)$. For uncertain tuple u, *reverse skyline probability* of u w.r.t. q means the probability that u become the reverse skyline w.r.t. q, denoted as $P_{RSky}(u, q)$. Through Theorem 1, we know that the reverse skyline probability of u equals to the product of u's exist probability and the probability that all the tuples which can semi-dominate u don't occur. So we can get Formula 1.

$$P_{RSky}(u, q) = \prod_{u' \in U, u' \preceq_q u} (1 - P(u')) \times P(u) \tag{1}$$

Definition 3 (probabilistic reverse skyline [8]). *Given a set U of uncertain tuples, the threshold t and the query point q, probabilistic reverse skyline of U w.r.t. q contains all the uncertain points in U whose reverse skyline probability is not smaller than t (denoted $PRSky(U, q) = \{u | P_{RSky}(u, q) \geq t\}$).*

Figure 3 illustrates an example of probabilistic reverse skyline. In Figure 3(a), only u_1 can semi-dominate u_2, so reverse skyline probability of u_2 w.r.t. q is $P_{RSky}(u_2, q) = (1 - P(u_1)) \times P(u_2) = 0.54$, then $P_{RSky}(u_1, q) = 0.04$, $P_{RSky}(u_3, q) = 0.052$, $P_{RSky}(u_4, q) = 0.1$, $P_{RSky}(u_5, q) = 0.026$. So in Figure 3(a), only u_2 is the probabilistic reverse skyline.

Finally, we introduce the probabilistic reverse skyline over uncertain data stream. Given an uncertain data stream UDS, every uncertain tuple u has a label $\kappa(u)$ to indicate its position in UDS. Supposed that in uncertain data stream, we can only process the recent N points, then use UDS_N to denote the most recent N uncertain points over the data stream. The probabilistic reverse skyline over uncertain data stream can be defined as Definition 4.

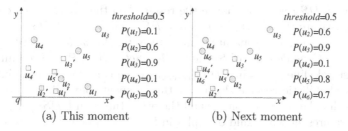

Fig. 3. Example of Probabilistic Reverse Skyline over Uncertain Data Stream

Definition 4 (probabilistic reverse skyline over uncertain data stream).
Given an uncertain data stream UDS, threshold t and query point q, probabilistic reverse skyline over UDS w.r.t. q continuously retrieves all the points in the most recent N points to find the probabilistic reverse skyline in UDS_N.

As illustrated in Figure 3, supposed that $N = 5$, Figure 3(a) contains the tuples of UDS_N at this moment, u_2 is the probabilistic reverse skyline. Next moment, u_6 enters into the data stream and u_1 is overdue, then UDS_N is shown as Figure 3(b), u_2 and u_6 are probabilistic reverse skyline at next moment.

4   Optimization Probabilistic Reverse Skyline

4.1   Preliminaries

First, we introduce some common symbols in this paper. UDS_N express the recent N uncertain tuples over data stream, q and t represent the query point and threshold respectively, and u, u' denote a uncertain tuple in UDS_N.

We know that there is sequence between tuples in UDS_N. For a tuple u, u' is u's *prefix tuple/postfix tuple* if u' enters into UDS_N before/after u. If u' is u's prefix/postfix tuple and u' can semi-dominate u, then we call u' is u's *pre-semidom tuple/post-semidom tuple*. If u' is u's postfix tuple and u' can full-dominate u, then u' is u's *post-fuldom tuple*. All the pre-semidom tuples of u in UDS_N constitute the set $PreSemiDom(u,q)$, all the post-semidom tuples of u in UDS_N constitute the set $PostSemiDom(u,q)$, while all the post-fuldom tuples of u in UDS_N constitute the set $PostFulDom(u,q)$. We use $P_{NpreSD}(u,q)$, $P_{NpostSD}(u,q)$ and $P_{NpostFD}(u,q)$ to denote the probability that all the tuples in $PreSemiDom(u,q)$, $PostSemiDom(u,q)$ and $PostFulDom(u,q)$ don't exist respectively. Then probability $P_{NpreSD}(u,q) = \prod_{u' \in PreSemiDom(u,q)} (1 - P(u'))$, $P_{NpostSD}(u,q) = \prod_{u' \in PostSemiDom(u,q)} (1 - P(u'))$, and then $P_{NpostFD}(u,q) = \prod_{u' \in PostFulDom(u,q)} (1 - P(u'))$. Formula 1 can be transformed as Formula 2.

$$P_{RSky}(u,q) = P_{NpreSD}(u,q) \times P(u) \times P_{NpostSD}(u,q) \tag{2}$$

Reserved Tuples of UDS_N. Through the detailed and in-depth analysis of reverse skyline's properties over uncertain data stream, we know that delete some tuples from UDS_N will not change the result. And the tuples which cannot be deleted constitute the *reserved tuples of* UDS_N (denoted as $UDS_{N,reserved}$).

$UDS_{N,reserved}$ can be separated into three sets, *Probabilistic Reverse Skyline Set* (PRS_N), *Candidate Set* (CS_N) and *Effect Set* (ES_N). PRS_N contains all the probabilistic skyline points in $UDS_{N,reserved}$, CS_N contains all the points which are not probabilistic reverse skyline at this moment, but will be probabilistic reverse skyline later, ES_N contains all the other tuples in $UDS_{N,reserved}$.

Next, we present what kind of tuples in UDS_N can be deleted. According to Formula 2, for a tuple u in UDS_N, we can know that $P_{NpreSD}(u,q)$ can't become smaller with the invalidation of old tuple, while $P_{NpostSD}(u,q)$ can't become larger with the insertion of new tuple. From Definition 3 and 4, if u wants to be a probabilistic reverse skyline point, then $P_{RSky}(u,q) \geq t$, so $P_{NpreSD}(u,q)$, $P(u)$ and $P_{NpostSD}(u,q)$ must not smaller than t. For a tuple u in UDS_N, u belongs to PRS_N if $P_{RSky}(u,q) \geq t$, u belongs to CS_N if $P(u) \times P_{NpostSD}(u,q) \geq t$ and $P_{RSky}(u,q) < t$. Then we consider the situation in which u can be deleted, there is a theorem between full-dominance and semi-dominance as Theorem 2.

Theorem 2. *[10] If tuple x full-dominate tuple y, and y semi-dominate tuple z, then x can semi-dominate z.*

According to Theorem 2, if any tuple u' can be semi-dominated by u, then u' can be semi-dominated by all the tuples in $PostFulDom(u,q)$. If the probability $P_{NpostFD}(u,q) < t$, u can't belong to PRS_N and CS_N, and u can be directly deleted without affecting continuously probabilistic reverse skyline query. Based on the delete condition, we can get $UDS_{N,reserved}$ as Formula 3.

$$UDS_{N,reserved} = \{u|u \in UDS_N \wedge P_{NpostFD}(u,q) \geq t\} \qquad (3)$$

Then we can prove that at any time, the set PRS_N of UDS_N and the PRS_N of $UDS_{N,reserved}$ are equal. In order to distinguish in theorems, we use superscript RE to denote corresponding value of $UDS_{N,reserved}$. For example, $P_{RSky}^{RE}(u,q)$ denote the probability $P_{RSky}(u,q)$ in $UDS_{N,reserved}$.

Theorem 3. *For any tuple u in UDS_N, if u belongs to PRS_N, then u must belong to PRS_N^{RE} of the same moment $UDS_{N,reserved}$.*

If $u \in PRS_N$, then $u \in UDS_{N,reserved}$, because all the tuples in $UDS_{N,reserved}$ must be in UDS_N, then we can get $P_{RSky}^{RE}(u,q) \geq t$, so $u \in PRS_N^{RE}$.

Theorem 4. *For any tuple u in UDS_N, if u doesn't belong to PRS_N, and u is in $UDS_{N,reserved}$, then u must not belong to PRS_N^{RE}.*

If $u \notin PRS_N$ and $u \in UDS_{N,reserved}$, then $P_{RSky}(u,q) < t$. For any tuple $u' \tilde{\succ} u$, if $u' \in UDS_{N,reserved}$, then $P_{RSky}^{RE}(u,q) < t$. If $u' \notin UDS_{N,reserved}$, every tuples in $PostFulDom(u',q)$ can semi-dominate u, also get $P_{RSky}^{RE}(u,q) < t$, so $u \notin PRS_N^{RE}$. Then easily get Theorem 5 from Theorem 3 and 4.

Theorem 5. *At any moment, the set PRS_N of UDS_N must equal to PRS_N^{RE} of $UDS_{N,reserved}$.*

According to Theorem 5, the problem of reverse skyline query over uncertain data stream can be simplified to computing probabilistic reverse skyline over $UDS_{N,reserved}$, and we can directly delete the tuples which not meet the condition in Formula 3. The following calculations are carried out in $UDS_{N,reserved}$, and we omit superscript ^{RE} later in this paper. The tuples in $UDS_{N,reserved}$ can be summarized as follows.

- **Probabilistic Reverse Skyline Set** (PRS_N): All tuples belongs to set PRS_N are called *result tuple(RT)*, whose probability $P_{RSky}(u, q) \geq t$.
- **Candidate Set** (CS_N): The tuples belongs to set CS_N are called *candidate tuple(CT)*, whose probability $P(u) \times P_{NpostSD}(u, q) \geq t$ and $P_{RSky}(u, q) < t$.
- **Effect Set** (ES_N): The tuples belongs to set ES_N are called *effect tuple* (ET), whose probability $P_{NpostFD}(u, q) \geq t$ and $P(u) \times P_{NpostSD}(u, q) < t$.

Probability Prunings. In this section we present some pruning strategies about probability. If tuple u's probability $P_{RSky}(u, q) < t$, we don't need to compute the exactly value of $P_{RSky}(u, q)$, only need to know that $P_{RSky}(u, q)$ is smaller than t. Then we can get some probability pruning strategies.

If tuple u's exist probability $P(u) < t$, we can know u can't be a RT or CT, so u can only be a ET before it is deleted. Then we can easily get Theorem 6.

Theorem 6. *For a tuple u, if the exist probability $P(u) < t$, u can be only an effect tuple before it is deleted.*

For a CT u, we know $P(u) \times P_{NpostSD}(u, q) \geq t$ and $P_{RSky}(u, q) < t$, when u can become a RT depends on probability $P_{NpreSD}(u, q)$. So we should keep the set $PreSemiDom(u, q)$ to determine the moment that u become a RT when one tuple in $PreSemiDom(u, q)$ is overdue. We don't need to keep the complete set $PreSemiDom(u, q)$ if $P_{NpreSD}(u, q) < t$. We use $PreShortSD(u, q)$ to denote the set we need keep. $P_{NpreShort}(u, q)$ denote the tuples in $PreShortSD(u, q)$ don't exist, it is $P_{NpreShort}(u, q) = \prod_{u' \in PreShortSD(u,q)} (1 - P(u'))$. If $P(u) \times P_{NpreSD}(u, q) \geq t$, $PreShortSD(u, q)$ are equal to $PreSemiDom(u, q)$, otherwise $PreShortSD(u, q)$ is the shortness of $PreSemiDom(u, q)$.

For a RT u, we must keep its set $PreSemiDom(u, q)$, because we should update its probability timely, and we can also use $PreShortSD(u, q)$ instead of $PreSemiDom(u, q)$. When we compute the set $PreShortSD(u, q)$, accessing the data stream by reverse order can improve the calculation efficiency.

In Figure 3(a), u_5 is a CT and u_2 is a RT, keep $PreShortSD(u_2, q) = \{u_1\}$ and $PreShortSD(u_5, q) = \{u_3, u_4\}$. For u_5, its $PreSemiDom(u_5, q) = \{u_1, u_2, u_3, u_4\}$. Because $P(u_5) \times P_{NpreShort}(u_5, q) = 0.072 < t$ and $P(u_5) \times (1 - P(u_4)) = 0.72 > t$, we don't need further calculation and get $PreShortSD(u_5, q) = \{u_3, u_4\}$. Only if u_3 is overdue, u_5 will have a chance to become a RT.

Sliding Window Structure. We use sliding window to model data stream, each tuple is stored in the corresponding window w_i, besides there is also a *set* in each window w_i (denoted as $w_i.set$). For a tuple u_i whose label $\kappa(u_i) = i$, then u_i is stored in w_i whose label $\kappa(w_i) = i$. $w_i.set$ stores some tuples u, the first tuple in $PreShortSD(u, q)$ have the same label with w_i. Only RT and CT need to keep $PreShortSD(u, q)$, so in window set there are only RT and CT.

It is known that if probability $P_{NpostFD}(u_i, q) < t$, u_i can be deleted. If $w_i.set$ is not empty, we should reserve $P(u_i)$, otherwise u_i can be deleted completely. Figure 4 illustrates the sliding window model of Figure 3(a).

w_1	w_2	w_3	w_4	w_5

$PRS=\{u_2\}$
$CS=\{u_5\}$
$ES=\{u_1,u_4\}$
$PreShortSD(u_2,q)=\{u_1\}$
$PreShortSD(u_5,q)=\{u_3,u_4\}$

w_i	$w_i.tuple$	$w_i.set$
w_1	u_1	u_2
w_2	u_2	$\emptyset$
w_3	$P(u_3)$	u_5
w_4	u_4	$\emptyset$
w_5	u_5	$\emptyset$

Fig. 4. The Sliding Window Model of Figure 3(a)

In Figure 3(a), only u_2 and u_5 are RT or CT, so in Figure 4, window sets stores u_2 and u_5. The first tuple in $PreShortSD(u_5, q)$ is u_3, so u_5 stores in $w_3.set$. It is known u_3 can be deleted, but $u_3.set$ is not empty, so we keep $P(u_3)$.

4.2 OPRS Algorithm

Data Structure. According to Definition 1 and 2, full-dominance relationship and semi-dominance relationship only exist between two tuples in the same region w.r.t. q, so the whole data space can be divided into 2^d different regions. When a new tuple u arrives, we need to find the tuples in $UDS_{N,reserved}$ which has full-dominance relationship or semi-dominance relationship with u, and we just need to scan all the regions that u belongs to (if u has the same values as q in some dimensions, u belongs to more than one region). For simplicity, the tuples in different regions can be stored independently, and each region need to maintain its sets PRS_N, CS_N and ES_N. R-tree index can help to calculation, hence each region maintains an independent R-tree, as showed in Figure 5.

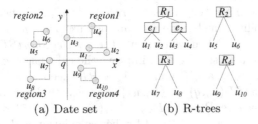

(a) Date set   (b) R-trees

Fig. 5. Example of Data Structure

In Figure 5(a), u_3 belongs to *region1* and *region2*. There is a R-tree for each region, as illustrated in Figure 5(b). A tuple can be stored in only one R-tree, but must be computed in all the corresponding R-trees of its regions. For example, when u_3 arrives, u_3 belongs to region1 and region2, then we compute the full-dominance and semi-dominance relationships between u_3 and the tuples in corresponding R-tree R_1 and R_2, finally u_3 is stored in only one R-tree R_1.

For simplicity, we reserve a label l_{max} for every entry of R-tree, the label $e.l_{max}$ represents the largest label of the tuples in entry e. For example, in R_1 of Figure 5(b), $e_1.l_{max} = 2$ and $e_2.l_{max} = 4$. Since our techniques are based on R-tree, we present the relationships between u and R-tree entry e, $e.min$ and $e.max$ denote the lower-left and the upper-right corner of e. The main relationships are illustrated in Figure 6.

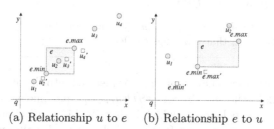

(a) Relationship u to e (b) Relationship e to u

Fig. 6. Example of Relationships between Tuples and R-tree Entry

Figure 6(a) illustrates the semi-dominance and full-dominance relationships from tuples to e. u_1 (u_2) can full-dominate (semi-dominate) $e.min$, then all the tuples in e can be full-dominated (semi-dominated) by u_1 (u_2), no further calculation. While u_3 (u_4) can't full-dominate (semi-dominate) $e.max$, then u_3 (u_4) can full-dominate (semi-dominate) no tuples in e, and we can ignore e.

Figure 6(b) illustrates the semi-dominance relationships from e to tuples. $e.min$ can't semi-dominate u_1, then all the tuples in e can't semi-dominate u_1, e can be ignored. While $e.max$ can semi-dominate u_2, then all the tuples in e can semi-dominate u_2, don't need further calculation.

OPRS Description. When a tuple u_{new} enters into the data stream, if its label $\kappa(u_{new}) > N$, the old window w_{old} whose label $\kappa(w_{old}) = \kappa(u_{new}) - N$ should be cleared, then deal with u_{new}. We can get the frame of OPRS as Algorithm 1.

Algorithm 1. Frame of OPRS

---

input : Sliding window size N, threshold t and query point q.
output: Continuously maintain probabilistic reverse skyline set PRS_N
1 **while** *a new tuple u_{new} arrives* **do**
2 | **if** $\kappa(u_{new}) > N$ **then**
3 | | Clearwindow(w_{old});
4 | Insert(u_{new});

Algorithm 2 describes how to clear a window w_{old}. If the old tuple u_{old} which should be in w_{old} has been expired, only need to dispose $w_{old}.set$. Otherwise u_{old} hasn't been expired, after dealing with $w_{old}.set$, u_{old} should be deleted. For every tuple $u' \in w_{old}.set$, its $P_{NpreSD}(u', q) = P_{NpreSD}(u', q)/(1 - P(u_{old}))$. If $P_{RSky}(u', q) \geq t$, u' can be added into result set PRS_N.

Algorithm 2. Clearwindow(w_{old})

---

    input : Threshold t , query point q and u_{old}'s probability $P(u_{old})$ in w_{old}.
    output: Changed probabilistic reverse skyline set PRS_N

1 **while** $w_{old}.set$ *is not empty* **do**
2   remove top tuple u' from $w_{old}.set$; $P_{NpreSD}(u', q)/ = (1 - P(u_{old}))$;
3   remove first tuple u_{old} from $PreShortSD(u', q)$;
4   add u' to $PreShortSD(u', q)$'s first tuple's window set;
5   **if** u' *is used to be a CT and* $P_{RSky}(u', q) \geq t$ **then**
6   add u' to result set PRS_N;

7 **if** *the tuple* u_{old} *should be in* w_{old} *hadn't been expired* **then**
8   find the R-tree R_i that u belongs to; delete u from R_i;

Algorithm 3 describes a new tuple u_{new} is inserted into the data stream. First, we should calculate the set $PreShortSD(u_{new}, q)$ and find the corresponding window set that u_{new} belongs to, which is described in CalPreShortSD(u_{new}, q). Then u_{new} is stored in any one R-tree it belongs to. Finally, we should consider the impact from u_{new} to the tuples in data stream, which is described in CalNewRtree(u_{new}, q).

Algorithm 3. Insert(u_{new})

---

    input : Threshold t , query point q and the new tuple u_{new}.
    output: Changed R-trees, changed sets PRS_N, CS_N and ES_N

1 find the R-tree set RS that u_{new} belongs to;
2 **if** $P(u_{new}) \geq t$ **then**
3   **for** *each* R_i *in RS* **do**
4   R_i.CalPreShortSD(u_{new}, q);
5   add u_{new} to $PreShortSD(u_{new}, q)$'s first tuple's window set;
6   **if** $P_{RSky}(u_{new}, q) \geq t$ **then**
7   add u_{new} to set PRS_N;

8 add u_{new} to any one R-tree it belongs to;
9 **while** RS *is not empty* **do**
10   remove the top R-tree R_i from RS; R_i.CalNewRtree(u_{new}, q);
11   **for** *each* u_i *in set TD* **do**
12   delete u_i from R-tree R_i; find the window w_i of u_i;
13   **if** $w_i.set$ *is not empty* **then**
14   reserve probability $P(u_i)$ in w_i
15   delete u_i from w_i;
16   **for** *each* u_i *in set TC* **do**
17   delete u_i from PRS_N;
18   **for** *each* u_i *in TE* **do**
19   save u_i as a effect tuple;

In Algorithm 3, we discuss the impact from u_{new} to every tuples in UDS_N. For a tuple $u_i \in UDS_N$, if u_i can be deleted directly under the influence of u_{new}, we can completely delete u_i if u_i's window set $w_i.set$ is empty. Otherwise we can delete u_i but have to reserve the probability $P(u_i)$ if there are some tuples in $w_i.set$. If u_i becomes a candidate tuple from a result tuple, we should delete u_i from result set PRS_N. If u_i becomes a effect tuple from a candidate tuple (or result tuple), we should reserve u_i as the format of a effect tuple.

Algorithm 4 describes the details of the function CalPreShortSD(u, q), which can calculate the set $PreShort(u, q)$ of the tuple u, then we can find u's corresponding window set. We can calculate $PreShort(u, q)$ through the semi-dominance relationship from the tuples in UDS_N to u. And we should access the tuples in UDS_N by reverse order to improve efficiency.

Algorithm 4. CalPreShortSD(u, q)

---

    input   : Threshold t, query point q, tuple u and R-tree R_i.
    output: $PreShortSD(u, q)$, $P_{RSky}(u, q)$ and $P_{NpreShort}(u, q)$.
1 insert all entries in $R_i.root$ into a heap H;
2 **while** H *is not empty* **do**
3   remove the top entry e from H;
4   take first tuple u' in $PreShortSD(u, q)$;
5   **if** $P_{NpreShort}(u, q) \times P(u) < t$ *and* $e.l_{max} < \kappa(u')$ **then**
6   break;
7   **if** e *is a intermediate entry* **then**
8   **if** $e.min$ *can semi-dominate* u **then**
9   **for** *each child* e_i *of* e **do**
10   **if** $e_i.min$ *can semi-dominate* u **then**
11   add e_i into H by descending order of l_{max};
12   **else if** e *can semi-dominate* u **then**
13   add e to $PreShortSD(u, q)$ by ascending order of $\kappa(e)$;
14 **for** *each* u_i *in* $PreShortSD(u, q)$ **do**
15   remove u_i from $PreShortSD(u, q)$;
16   **if** $P_{NpreShort}(u, q) \times P(u) \geq t$ **then**
17   add u_i to $PreShortSD(u, q)$;
18   break;
19 **return**;

Algorithm 5 describes the details of the function CalNewTree(u, q), which can calculate the set TD, TC and TE. TD contains all the tuples which should be deleted under the influence of u, every tuple u' in TD are full-dominated by u whose probability $P_{NpostFD}(u', q) < t$. TC contains all the tuples which should be removed to candidate set from probabilistic reverse skyline set under the influence of u, every tuple u' in TC are semi-dominated by u whose probability $P_{RSky}(u', q) < t$ and $P_{NpostSD}(u', q) \times P(u') \geq t$. TE contains all the tuples which change to be effect tuple, every tuple u' in TE are semi-dominated by u whose probability $P_{NpostSD}(u', q) \times P(u') < t$.

Algorithm $1 - 5$ can continuously maintain the probabilistic reverse skyline queries efficiently, which can prune some redundant data, and minimize the number of tuples that kept in the sliding window, the algorithms reduce the time and space cost greatly.

Algorithm 5. CalNewTree(u, q)

---

    input : Threshold t, query point q, tuple u and R-tree R_i.
    output: The set TD, TC and TE.
1 insert all entries in R_i.root into a heap H by descending order of l_{max};
2 **while** H *is not empty* **do**
3   remove the top entry e from H;
4   **if** e *is a intermediate entry* **then**
5   **if** u *can semi-dominate* $e.max$ **then**
6   **for** *each child* e_i *of* e **do**
7   **if** u *can semi-dominate* $e_i.max$ **then**
8   add e_i into H;

9   **else**
10   **if** u *can full-dominate* e **then**
11            $P_{NpostFD}(e, q) \times = 1 - P(u)$;
12   **if** $P_{NpostFD}(e, q) < t$ **then**
13   add e into TD; continue;

14   **if** u *can semi-dominate* e **then**
15            $P_{NpostSD}(e, q) \times = 1 - P(u)$;
16   **if** $P_{NpostSD}(e, q) \times P(e) < t$ *and* e *was not a effect tuple* **then**
17   add e into TE;
18   **if** $P_{RSky}(e, q) < t$ *and* e *was a result tuple* **then**
19   add e into TC;

20 **return**;

---

5   Experimental Evaluation

In this section, we have developed a simulator to evaluate the performance of our proposed OPRS approach with C++ programming language. All the experiments are run on a PC with Pentium IV 2.4GHz CPU, 512MB DDR memory, 80GB hard disk and Windows XP operating system.

We use real data set and synthetic data sets to verify the performance of our proposed OPRS algorithm. Real data set apply forest environmental monitoring data obtained by a sensor network, randomly generate a existing probability for each tuple, and the query point is generated randomly. In forest environmental monitoring data, there is 30000 tuples and we assume the sliding window size

$N = 15000$, each tuple has four attributes, including humidity, temperature, light and voltage, we take two, three, four dimensions to calculate continuously probabilistic reverse skyline as Figure 7.

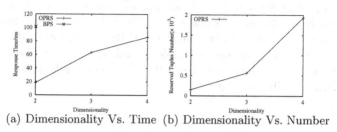

(a) Dimensionality Vs. Time (b) Dimensionality Vs. Number

Fig. 7. The Results of Real Data Set

In Figure 7(a), BPS is the basic approach to calculate continuously probabilistic reverse skyline, because the running time of BPS is too long, so we just record the response time of BPS in two-dimensions to compare with OPRS, in Figure 7(b), we record the max number of reserved tuples in OPRS.

Synthetic data sets apply two different distribution data, one is uniform distribution, the other one is cluster distribution, and the exist probability of tuples is uniform distribution. In Table 1, we give the parameters in simulation.

Table 1. Parameters in Simulation

Parameter	Default	Range
dimensionality	2	2, 3, 4, 5
sliding window size N	100K	100K,200K,300K,400K,500K
threshold t	0.5	0.3, 0.4, 0.5, 0.6, 0.7

First, we investigate the influence of dimensionality in OPRS. Figure 8 shows OPRS's changing performance from two-dimensions to five-dimensions when the window size $N = 100K$ and the threshold $t = 0.5$. Figure 8(a) records the average response time of each tuple in OPRS for two data distributions. Figure 8(b) records the max reserved tuples' number in OPRS for two data distributions,

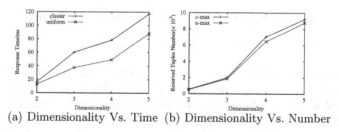

(a) Dimensionality Vs. Time (b) Dimensionality Vs. Number

Fig. 8. The Influence of Dimensionality

c-max represents the max reversed tuples' number in cluster distribution data set, while u-max represents that in uniform distribution data set.

Next, we investigate the influence of the sliding window size. Figure 9 shows OPRS's changing performance of the sliding window size from 100K to 500K when data set is two-dimensions and the threshold $t = 0.5$. Figure 9(a) records the average response time of each tuple in OPRS for two data distributions. Figure 9(b) records the max reserved tuples' number, c-max represents the max reserved tuples' number in cluster distribution data set, while u-max represents that in uniform distribution data set.

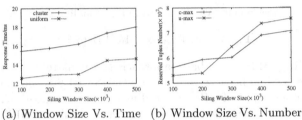

(a) Window Size Vs. Time   (b) Window Size Vs. Number

Fig. 9. The Influence of Sliding Window Size

Finally, we discuss the influence of threshold t. Figure 10 shows OPRS's changing performance of the threshold from 0.3 to 0.7 when data set is two-dimension and sliding window size $N = 100$K. Figure 10(a) records the average response time of each tuple in OPRS for two data distributions. Figure 10(b) records the max reserved tuples' number, c-max represents the max reserved tuples' number in cluster distribution data set, while u-max represents that in uniform distribution data set.

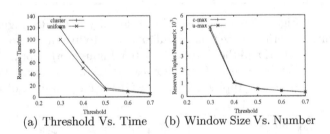

(a) Threshold Vs. Time   (b) Window Size Vs. Number

Fig. 10. The Influence of Threshold

Through the experiments above, we can conclude that no matter what kind of data distribution, average response time of OPRS become longer and max reserved tuples' number become more with the increase of data dimensionality and sliding window size, while the response time become shorter and max reserved tuples' number become fewer with the increase of the threshold. The experimental results accord with the actual situation.

6 Conclusions

Many applications can be modeled by uncertain data stream, reverse skyline plays an important role in market decision-making, environmental monitoring and market analysis. In this paper, we focus on the problem of continuously computing probabilistic reverse skyline over uncertain data stream. Through detailed and in-depth analysis of probabilistic reverse skyline's properties over uncertain data stream, we present some probability pruning strategies, which minimize the number of tuples that kept in the sliding window and reduce a lot of redundant calculations. Then an efficient approach, called Optimization Probabilistic Reverse Skyline (OPRS), is proposed for processing continuous probabilistic reverse skyline queries. Finally, a large number of experiments verify that our proposed approach OPRS is an efficient algorithm which can meet the practical requirements. The next step, we will focus on the problem of reverse skyline's properties over distributed uncertain data stream, designing an efficient approach to process continuous reveres skyline queries over distributed uncertain data stream is our future direction.

Acknowledgement. This research is supported by the State Key Program of National Natural Science of China (Grant No. 60933001), the National Science Foundation for Distinguished Young Scholars of China (Grant No. 61025007), the National Natural Science Foundation of China (Grant No. 60973020, 61073063), and the National Natural Science Foundation for Young Scientists of China (Grant No. 61100022).

References

1. Borzsonyi, S., Stocker, K., Kossmann, D.: The skyline operator. In: ICDE, Heidelberg, Germany, April 2-6, pp. 421–430 (2001)
2. Pei, J., Jiang, B., Lin, X., Yuan, Y.: Probabilistic Skylines on Uncertain Data. In: VLDB, Vienna, Austria, September 23-27, pp. 15–26 (2007)
3. Wu, X., Tao, Y., Wong, R.C.W., Ding, L., Yu, J.X.: Finding the Influence Set through Skylines. In: EDBT, Saint-Petersburg, Russia, March 23-26, pp. 1030–1041 (2009)
4. Deng, K., Zhou, X., Shen, H.T.: Multi-source skyline query processing in road networks. In: SIGMOD, Beijing, China, June 12-14, pp. 796–805 (2007)
5. Papadias, D., Tao, Y., Fu, G., Seeger, B.: An optimal and progressive algorithm for skyline queries. In: SIGMOD, San Diego, California, June 9-12, pp. 467–472 (2003)
6. Dellis, E., Seeger, B.: Effcient computation of reverse skyline queries. In: VLDB, Vienna, Austria, September 23-27, pp. 291–302 (2007)
7. Prasad, M.D., Deepak, P.: Efficient Reverse Skyline Retrieval with Arbitrary Non-Metric Similarity Measures. In: EDBT, pp. 319–330 (2011)
8. Lian, X., Chen, L.: Monochromatic and bichromatic reverse skyline search over uncertain databases. In: SIGMOD, Vancouver, BC, Canada, June 10-12, pp. 213–226 (2008)
9. Lian, X., Chen, L.: Reverse skyline search in uncertain databases. ACM Transactions on Database Systems 35(1), Article 3, 1–49 (2010)

10. Wang, G., Xin, J., Chen, L., Liu, Y.: Energy-efficient reverse skyline query processing over wireless sensor networks. IEEE TKDE,
    http://doi.ieeecomputersociety.org/10.1109/TKDE.2011.64
11. Zhu, L., Li, C., Chen, H.: Efficient computation of reverse skyline on data stream. In: CSO, Sanya, Hainan, China, April 24-26, pp. 735–739 (2009)
12. Lin, J., Lin, Q.: Algorithms for skyline query on data streams. Journal of FuZhou University 35(4), 526–531 (2007)
13. Zhang, W., Zhang, Y., Yu, J.X.: Probabilistic skyline operators over sliding windows windows. In: ICDE, Shanghai, China, March 29-April 2, pp. 1060–1071 (2009)
14. Zhang, Z., Cheng, R., Papadias, D., Tung, A.K.H.: Minimizing the communication cost for continuous skyline maintenance. In: SIGMOD, Providence, Rhode Island, June 29-July 2, pp. 495–508 (2009)
15. Lin, X., Yuan, Y., Wang, W.: Stabbing the sky: Efficient skyline computation over sliding windows. In: ICDE, Tokyo, Japan, April 5-8, pp. 502–513 (2005)
16. Tao, Y., Papadias, D.: Maintaining sliding window skylines on data streams. IEEE TKDE 18(3), 377–391 (2006)
17. Sarkas, N., Das, G., Koudas, N., Tung, A.K.H.: Categorical skylines for streaming data. In: SIGMOD, Vancouver, BC, Canada, June 10-12, pp. 239–250 (2008)

Malleability-Aware Skyline Computation
on Linked Open Data

Christoph Lofi[1], Ulrich Güntzer[2], and Wolf-Tilo Balke[1]

[1] Institut für Informationssysteme
Technische Universität Braunschweig, 38106 Braunschweig, Germany
{lofi,balke}@ifis.cs.tu-bs.de
[2] Institut für Informatik
Universität Tübingen, 72076 Tübingen, Germany
ulrich.guentzer@informatik.uni-tuebingen.de

Abstract. In recent years, the skyline query paradigm has been established as a reliable and efficient method for database query personalization. While early efficiency problems have been approached, new challenges in its effectiveness continuously arise. Especially, the rise of the Semantic Web and linked open data leads to personalization issues where skyline queries cannot be applied easily. In fact, the special challenges presented by linked open data establish the need for a new definition of object dominance that is able to cope with the lack of strict schema definitions. However, this new view on dominance in turn has serious implications on the efficiency of the actual skyline computation, since transitivity of the dominance relationships is no longer granted. Therefore, our contributions in this paper can be summarized as a) we design a novel, yet intuitive skyline query paradigm to deal with linked open data b) we provide an effective dominance definition and establish its theoretical properties c) we develop innovative skyline algorithms to deal with the resulting challenges and extensively evaluate the our new algorithms with respect to performance and the enriched skyline semantics.

Keywords: Query Processing; Personalization; Skyline Queries; Linked Open Data.

1   Introduction

The continuous efforts to put the Semantic Web vision into practice have led to two important insights: implementing a full-fledged machine understandable Web has largely failed, but focusing only on the 'reasonable' part already reveals a vast variety of valuable data [1]. This area of so-called linked open data (LOD) [2] has immediately spawned interesting efforts like for instance the DBPedia knowledge base[1] that currently describes more than 3.64 million things, out of which 1.83 million are classified in a consistent ontology. Moreover, the potential applications also promoted the

---

[1] http://www.dbpedia.org/About

S.-g. Lee et al. (Eds.): DASFAA 2012, Part II, LNCS 7239, pp. 33–47, 2012.
© Springer-Verlag Berlin Heidelberg 2012

development of innovative methods to make such data available to users in a structured way. IR-style or rule-based extraction frameworks like ALICE [3], Xlog [4], or SOFIE [5] can already crawl the Web and extract structured relationships from unstructured data with largely sufficient accuracy.

However, when it comes to retrieval of the now structured information also the typical query paradigms have to be adapted. This is not only because extracted knowledge is usually represented in some form of knowledge representation language (with RDF triples as most prominent example), but also due to the semantic loss of focus that results from ambiguities in the extraction process. For instance when querying for a person's *place of birth*, the information where somebody *grew up* is generally heavily related, but definitely less focused regarding the original query intention. Still, whenever the exact place of birth is unknown, the information where a person grew up is still much more helpful than an empty result set. Thus, it should be retrieved as relevant, but of course should always get a penalty in the ranking. This desirable facet of retrieval is known as *schema malleability* [6, 7].

While current retrieval paradigms for example in SOFIE's retrieval engine NAGA [8] or Xlog's DBLife [9], only focus on SQL-style retrieval (usually SPARQL over RDF) and keyword search with top-k ranking, the problem of preference-based retrieval paradigms like skyline queries over linked open data has not yet been solved. In this paper we tackle the problem of *malleability-aware skyline queries* over linked open data. The problem is twofold: first a viable semantics has to be defined trading a user's value preferences against the extracted relationships' loss of focus with respect to the original query, then efficient algorithm(s) have to be designed to solve the retrieval task in practical runtimes.

In a nutshell the problem is the intuitive interleaving of each individual user's attribute value preferences with the generally applicable preferences on attribute semantics as specified in the query. Whereas skyline queries up to now only dealt with relaxing value preferences, the new additional relaxation in attribute semantics is owed to the linked open data. Let's extend our example from above:

Example: A user might be interested in famous Nobel laureates in physics that were *born* in Munich, Germany. Querying the DBPedia knowledge base retrieves only two entries: Rudolf Mössbauer and Arno Allan Penzias. However, a similar query for Nobel laureates in physics *growing up* in Munich also retrieves Werner Heisenberg (who went to school in Munich) and a further relaxation to Nobel laureates in physics *living in* Munich finally retrieves Wilhelm Conrad Röntgen. With a different degree of relevance (with respect to famousness and having a relationship with Munich) all these are possible answers that, however, with respect to the original query are getting less focused and thus should be displayed accordingly. That means the final result including schema malleability may be a trade-off between the famousness of the physicists and their relationship to Munich, which is best represented by a skyline query result.

To model this paradigm in databases (and schema malleability as such), each query attribute can be considered as a database column holding not only tuples based on the strict relationships given by the query, but also tuples from semantic similar relationships. However, to prepare for later retrieval each such malleable attribute has to be associated with a second attribute measuring the semantic loss of focus for each tuple. This can be done by either automatically measuring semantic loss of focus by instance-based precision/recall tests like shown in [10], testing the relationships' semantic relatedness with externally available ontologies like in [11], or simply denoting possible relationships and allowing users to define a (partial) order over these relationships with respect to their queries.

In any case, the new associated attribute columns have to be considered by retrieval algorithms, but in contrast to the attribute value columns have a slightly different quality. This is because relaxations on preferred *values* for cooperative query processing might change a tuple's desirability, but larger relaxations in attribute *senses* might render tuples utterly useless. Consider the example above where a Nobel laureate's *place of birth* is relaxed in terms of the preferred *value*, e.g. from 'Munich' to 'Bavaria' or 'Germany', or in terms of the *relationship with Munich* e.g., from 'born in' to 'lived in'. Whether a broader relaxation of the sense like 'visited' is still of any use is doubtful. Thus, classical skyline query processing following Pareto-optimality cannot readily be applied. Moreover, by basically doubling the problem dimensionality also the well-known efficiency problems of skyline processing in terms of runtimes and result set manageability, see e.g., [12], are bound to be encountered.

The contribution of this paper is threefold: we design an intuitive notion of skyline dominance with respect to malleability in the form of semantically typed links in linked open data and discuss its characteristics. We develop innovative algorithms to efficiently process skyline queries even over large data repositories. And we extensively evaluate these algorithms with respect to runtime behavior and skyline manageability. In fact, our experiments show that in the general case, our algorithm can achieve significant performance improvements over the baseline. However, when slightly restricting general malleability, we can even show that performance indeed can increased by several orders of magnitude, even rivaling the runtime behavior of classical skyline algorithms over strictly transitive preferences.

This paper is structured as follows: after briefly surveying related work in section 2, we discuss the necessary foundations and theoretical characteristics of skylines over linked open data in section 3. Section 4 then presents and evaluates skyline algorithms over several malleability attributes, whereas section 5 deals with the special case of a single aggregated malleability attribute. We close with a short summary and outlook.

2   Related Work

Due to its potential usefulness linked open data has received a lot of attention and even inspired a taskforce[2] of the World Wide Web Consortium (W3C). Current

---

[2] http://www.w3.org/wiki/SweoIG/TaskForces/CommunityProjects/LinkingOpenData

research is often focused on the area of business intelligence, but also for the collection of common knowledge. The basic idea is using the Web to create typed links between data items from different sources. Once extracted, these links represent semantic relationships which can in turn be exploited for querying. However when querying (or reasoning over) such relationships the exact nature of the relationship and its semantic correspondence to the query is often difficult. Therefore apart from typical exact match queries (usually performed in SPARQL[3] over RDF triples) many approaches for ranking the best matching information have been designed.

The first notable approach to rank queries on extracted entity properties was EntitySearch [13] proposing an elaborate ranking model combining keywords and structured attributes. When it comes to exploiting also semantic relationships NAGA [8] used a scoring model based on the principles of generative language models, from which measures such as confidence, informativeness, and compactness are derived, which are subsequently used to rank query results. Finally [14] develop a general model for supporting approximate queries on graph-modeled data, with respect to both attribute values and semantic relationships and derive a first top-k algorithm to implement the ranking efficiently. However, like in all top-k frameworks even farfetched semantic relationships can be compensated for by good matching attribute values. Moreover, all these approaches directly work on graph-structured data relying on the path-based semantic relatedness e.g., defined by [15], whereas our approach works on relationship malleability quantifying the respective loss of focus.

To our knowledge the only algorithm similar to skyline queries on linked open data is given by [16]. However, the developed algorithm has been designed for optimizing skyline queries over RDF data stored using a vertically partitioned schema model and thus presents an efficient scheme to interleave the skyline operator with joins over multiple relational tables. Unfortunately it does not offer any techniques with respect to personalization and the problem of semantic linkage and thus is not really related to our work here. In brief, to our knowledge our approach features the first skyline algorithm respecting semantic malleability.

3   Theoretical Foundations of Malleability-Aware Skylines

In the following we will briefly revisit the notion of Pareto skylines as given by [17]. Assume a database relation $R \subseteq D_1 \times ... \times D_n$ on n attributes.

- A preference P_i on some attribute A_i with domain D_i is a strict partial order over D_i. If some attribute value $a \in D_i$ is *preferred over* another value $b \in D_i$, then $(a, b) \in P_i$, written as $a >_i b$ (read "a dominates b wrt. to P_i"). The set of all preferences is denoted as P.
- Analogously, also an equivalence relation on D_i compatible with P_i can be defined. Then, two attribute values attribute $a, b \in D_i$ can be defined as being equivalent with respect to the domain: $a \approx_i b$. Moreover, if some attribute value $a \in D_i$ is either *preferred over or equivalent to* some ue $b \in D_i$, we write $a \gtrsim_i b$.

---

[3] http://www.w3.org/TR/rdf-sparql-query/

Assuming preferences $P_1, \ldots, P_n$ for each attribute in R, the concept of Pareto dominance between two tuples $\vec{x}, \vec{y} \in R$ with $\vec{x} = (x_1, \ldots, x_n)$ can be defined as:

Definition 1. Pareto Dominance

$$(\vec{x} >_P \vec{y}) := \forall_{i \in D}(x_i \gtrsim_i y_i) \wedge \exists_{i \in D}(x_i >_i y_i)$$

The classical *skyline set* [18] can now be defined as all those tuples in each database instance, which are not dominated by any other tuple:

Definition 2. Pareto Skyline for some relation **R** and preferences **P**

$$sky(R, P) := \{\vec{x} \in R \mid \neg \exists \, \vec{y} \in R : \vec{y} >_P \vec{x}\}$$

Now we are ready to extend the semantics by introducing the concept of *malleability-aware dominance*, which specifically respects the semantic challenges introduced by linked open data entities. Like motivated above the intuition is that regarding each queried attribute personalized skyline queries consist of a user-specific value preference and a certain meaning of the attribute that may more or less correspond to some number of extracted attribute types in the database instance. Thus, for getting acceptable results, not only the entities' attribute values, but also the loss of focus with respect to each attribute's semantics has to be taken into account. The baseline approach for this would be to simply compute skylines with the double dimensionality (one attribute value and a malleability score for each attribute).

However, apart from the obvious scalability problems, also the semantics are unclear. Whereas attribute values like dates, prices, or ratings are usually crisp and follow a certain preference order (users want the cheapest price for some product or the highest quality rating), labels for semantic relationships are usually fuzzy and to some degree ambiguous depending on their labels. Often *grew_up_in* may be used synonymously with *born_in*, but *lived_in* definitely is not. Thus, the *relative loss of focus* (or δ-distance) between semantic labels needs to be considered: if two labels are at most differing by δ they should be considered semantically equivalent, but once two labels are too far apart a different class of semantic relationship has to be assumed.

Definition 3. δ-preferences for modeling malleability over linked open data

A δ-preference δP_i on some attribute A_i with metric domain D_i and metric $dist_i(.,.)$ is a reflexive and transitive binary relation $>_i$ over D_i, together with an *intransitive* form of equivalence with the notion of indifference: for all $a, b \in D_i$: $a \approx_i b \Leftrightarrow dist_i(a, b) \leq \delta$ see e.g [19]. If some attribute value $a \in D_i$ is *preferred over* another value $b \in D_i$ and $dist_i(a, b) > \delta$ we write $a >_i b$ (read "a strictly δ-dominates b wrt. to δP_i"). The combination of several δP_i can easily be achieved using the normal Pareto product and will be denoted as $>_{\delta P}$. Likewise, we write $a \succsim_i b$ if either $a \approx_i b$ or $a >_i b$.

It is easy to mix δ-preferences and normal strict partial order preferences to create a product preference over some relation (which for ease of use we will again simply denote by '>') and we will define the respective domination relationships for malleability-aware skylines in section 4 and 5. But up to now such δ-preferences with relative distances have not been considered in skyline queries, because their use directly contradicts the generally assumed *transitivity* of domination relationships between tuples. Actually, since long results in psychology show that in contrast to common belief intransitivity often occurs in a person's system of values or preferences, potentially leading to unresolvable conflicts, see e.g., [20] or [21]. Analogously, in economics intransitivity may occur in a consumer's preferences. While this may lead to consumer behavior that does not conform to perfect economic rationality, in recent years economists have questioned whether violations of transitivity must necessarily lead to 'irrational' behavior, see for instance [22].

Indeed, from an order-theoretical point of view it is easy to show that whenever δ-distances are used in at least one preference and $(\vec{x} >_P \vec{y})$ and $(\vec{y} >_P \vec{z})$ are given, $(\vec{x} >_P \vec{z})$ does not necessarily follow:

Lemma 1. Dominance relationships are not transitive using δ-distances.
Proof: Transitivity for dominance regarding any product preference P is violated, if three tuples $\vec{x}$, $\vec{y}$, and $\vec{z}$ can be constructed, for which holds: $(\vec{x} >_P \vec{y} \wedge \vec{y} >_P \vec{z})$, but $\vec{x} \not>_P \vec{z}$.
Assume a product preference P over some relation R and assume there is one attribute m for which a δ-preference δP_m is declared stating the equivalence of values within the relative distance of some fixed δ. Now define preference $P^\wedge$ by removing δP_m from P and construct three tuples $\vec{x}$, $\vec{y}$, and $\vec{z}$ such that $(\vec{x} >_{P^\wedge} \vec{y} >_{P^\wedge} \vec{z})$. Now assign values of $\vec{x}$, $\vec{y}$, and $\vec{z}$ for attribute m as follows $y_m := (x_m + \delta)$ and $z_m := (y_m + \delta)$. Then with respect to P holds $\vec{x} >_P \vec{y}$ because of $(\vec{x} >_{P^\wedge} \vec{y})$ and x_m and y_m are equivalent with respect to the chosen δ. Analogously holds $\vec{y} >_P \vec{z}$. However, $\vec{x} \not>_P \vec{z}$ because of $\vec{x} >_{P^\wedge} \vec{z}$, but $(z_m = (x_m + 2\delta) >_{\delta P m} x_m)$. Hence $\vec{x}$ and $\vec{z}$ are incomparable with respect to P and the domination relationship is not transitive.   □

While the resulting preference orders are not transitive, at the same time domination relationships within the intransitive product order are sensible, since there can never exist any cyclic base preferences. However, this is only the case when strict partial-order preferences and δ-preferences are used conjointly to build the product order, product orders built only from δ-preferences will inevitably lead to cycles. In order to guarantee acyclic product orders, some observations can be made: a) no cycles can ever emerge between tuples showing dominance with respect to any attributes, over which a strict partial-order preference is defined (due to their guaranteed transitivity), b) cycles can only occur, if tuples are equivalent with respect to all partial-order preferences. In this case, *strict* δ-dominance ($\succ$) must be enforced, and none of the tuples are allowed to dominate by simple δ-dominance alone ($\succeq$). This leads to our formal definition of malleability-aware dominance (Definition 4) in Section 4.1.

3.1   Implications for Algorithm Design

The danger on intransitivity of dominance relationships is that it may lead to *non-deterministic behavior* when computing skylines using standard skyline algorithms. According to Definition 2, the skyline contains all tuples of a given relation which are not dominated by any other tuple, assuming that preferences are partial orders. Naïvely, this would need an algorithm pairwise comparing all tuples with respect to the chosen dominance criterion. In practice however, most skyline algorithms increase efficiency by pruning large numbers of tuple comparisons (e.g. basic block-nested-loop (BNL) algorithms [18], branch-and-bound algorithms [23], distributed algorithms [24], or online algorithms [25]). These optimizations usually all rely on the transitivity of dominance.

Example. When using non transitive dominance with for instance a BNL algorithm the result will vary *non-deterministically* depending on the order of the tuples in the database instance (and therefore, also the order of the tests for dominance). For example, when assuming $(\vec{x} >_P \vec{y})$, $(\vec{y} >_P \vec{z})$, but $(\vec{x} \not>_P \vec{z})$, then a skyline computed by some BNL algorithm just contains $\{\vec{x}\}$, if the test for $(\vec{y} >_P \vec{z})$ is performed first and thus $\vec{z}$ is immediately pruned from the database. Otherwise, if $(\vec{x} >_P \vec{y})$ is tested first, the resulting skyline contains $\{\vec{x}, \vec{z}\}$, because $\vec{y}$ is removed prematurely before also $\vec{z}$ could be removed by testing $(\vec{y} >_P \vec{z})$; and due to $(\vec{x} \not>_P \vec{z})$, $\vec{z}$ incorrectly remains in the skyline set.

However, the idea of skylines is still sensible since as we will prove in lemma 2, cyclic preferences cannot occur and thus a skyline based on the notion of containing all non-dominated objects can be computed. Since pruning may cause difficulties, the obvious way is by simply comparing all tuples in the database instance pairwise (with quadratic runtime). But as we will see in the next section, far more efficient algorithms can be designed and thus skylines over linked open data are indeed practical.

4   Malleability-Aware Skylines

Before delving into designing skyline algorithms capable of dealing with intransitivity as described above, we have to formalize our concept of product orders also built from δ-preferences in form of a dominance criterion usable in skyline algorithms.

4.1   Malleability-Aware Skylines with Individual Attribute Malleability

Assuming preferences P that can be decomposed into strict partial-order preferences $P^\wedge$, and δ-preferences δP, the concept of malleability-aware dominance between two tuples $\vec{x}, \vec{y}$ can be defined as

Definition 4. Malleability-aware dominance over individual attributes

$$(\vec{x} >_P \vec{y}) :\Leftrightarrow \big((\vec{x} >_{P^\wedge} \vec{y}) \wedge (\vec{x} \succcurlyeq_{\delta P} \vec{y})\big) \vee \big((\vec{x} \approx_{P^\wedge} \vec{y}) \wedge (\vec{x} >_{\delta P} \vec{y})\big)$$

In this definition, there is a malleability-aware dominance *a)* if all non-malleable attribute values of $\vec{x}$ show Pareto dominance over $\vec{y}$ and all malleable attributes of $\vec{x}$ are at least equivalent to those of $\vec{y}$ with respect to the δ-preferences (i.e. all malleable attributes encoding the tuple's loss-of-focus are tested for "soft" dominance here, allowing a certain δ of flexibility) or *b)* if all data attributes are equivalent with respect to the Pareto preferences, but show strict dominance with respect to the malleable attributes for the δ-preferences (this means: all malleable attributes encoding loss-of-focus have to show real Pareto dominance, i.e. no δ-distances are considered. This important property is required to prevent cycles to form in P:

Lemma 2. Product orders of strict partial order preferences and δ-preferences following Definition 4 cannot contain cyclic preferences.

Proof: We have to show that the dominance relation of the product order does not induce cycles, more precisely, if $\overrightarrow{x_1} >_P \ldots >_P \overrightarrow{x_k}$ with $k > 1$, then neither $\overrightarrow{x_k} >_P \overrightarrow{x_1}$ nor $\overrightarrow{x_k} \approx_P \overrightarrow{x_1}$ is possible. Please note that $\vec{x} \approx_P \vec{y}$ means $\overrightarrow{x_i} \approx_{Pi} \overrightarrow{y_i}$ for all non-malleable attributes and $\overrightarrow{x_j} = \overrightarrow{y_j}$ for all malleable attributes, i.e. no malleability is allowed for equivalence.

For $1 \le t \le k$ let $\overrightarrow{x_t} := (x_{t,1}, \ldots, x_{t,n}, x_{t,n+1}, \ldots, x_{t,n+m})$ where the first n attributes are non-malleable and the following m attributes are malleable. We distinguish two cases:

a) There is a strict preference in the non-malleable part between two objects in an assumed cycle, i.e. there are $1 \le t < k$ and $1 \le i \le n$ such that $x_{t,i} >_{Pi} x_{t+1,i}$. Then within the cycle we have $x_{1,i} \gtrsim_{Pi} \ldots \gtrsim_{Pi} x_{t,i} >_{Pi} x_{t+1,i} \gtrsim_{Pi} \ldots \gtrsim_{Pi} x_{k,i}$ and therefore $x_{1,i} >_{Pi} x_{k,i}$, rendering both $\overrightarrow{x_k} >_P x_1$ and $\overrightarrow{x_k} \approx_P x_1$ impossible.

b) If there is no strict preference in the non-malleable part, for all $1 \le t < k$ and $1 \le i \le n$ we have $x_{t,i} \approx_{Pi} x_{t+1,i}$. Thus following Definition 4 for the malleable attributes for all $1 \le t < k$ holds: $(x_{t,n+1}, \ldots, x_{t,m}) >_{\delta P} (x_{t+1,n+1}, \ldots, x_{t+1,m})$ which means $(x_{1,n+1}, \ldots, x_{1,m}) >_{\delta P} (x_{k,n+1}, \ldots, x_{k,m})$ due to the strictness of $>_{\delta P}$.   Hence $\overrightarrow{x_k} >_P x_1$ is   impossible. In the same way it is easy to see that also $\overrightarrow{x_k} \approx_P x_1$ is impossible.   □

Now, the respective malleability-aware skyline can be computed analogously to definition 2 by

$$sky(R, P) := \{\vec{x} \in R \mid \neg \exists \, \vec{y} \in R : \vec{y} >_P \vec{x}\}$$

Unfortunately, actually implementing such malleability-aware skyline computations algorithmically poses several challenges. Therefore, in the following we demonstrate how such algorithms can be designed. For the sake of cleaner notions and without loss of generality, we will assume that all our preferences are encoded in the database tuples by normalized scores in $[0,1]$, where 1 represents the most preferable attribute values, and 0 the least preferable ones. Any database tuple $\vec{x}$ is given by $\vec{x} = (x_1, \ldots, x_m)$, and the individual attributes can be separated into non-malleable data attributes x_i with $i \in D$ (corresponding to $P^\wedge$), and malleable attributes x_i with $i \in M$ (corresponding to δP). Then, the dominance criterion of definition 4 can be reformulated as:

$$(\vec{x} >_P \vec{y}) :\Leftrightarrow \left[(\forall_{i \in D} \, x_i \ge y_i \wedge \exists_{i \in D} x_i > y_i) \wedge (\forall_{i \in M} x_i \ge (y_i - \delta_i)) \right]$$
$$\vee \left[(\forall_{i \in D} x_i = y_i) \wedge (\forall_{i \in M} x_i \ge y_i \wedge \exists_{i \in D} x_i > y_i) \right]$$

It is easy to see that this definition is equivalent to the Pareto dominance as given by definition 1 for the cases of $M = \emptyset$ or $\delta = 0$. If $M \neq \emptyset$ and $\delta > 0$, then malleability-aware dominance allows for additional tuples being dominated compared to Pareto dominance, hence the resulting Skyline is a subset of the Pareto skyline.

4.2   Computing Non-transitive Skylines

As already indicated in section 3.1, modern Skyline algorithms have come to rely on the transitivity of dominance criteria: For sake of improved performance, many tuple comparisons are avoided by pruning objects early, relying on transitivity for the computational correctness, i.e. a tuple shown to be dominated can be fully excluded from the further execution of the algorithm. However, without guaranteed transitivity, even basic algorithms like the well-known Block-Nested Loop Algorithm [18] fail. Therefore, the need arises to develop new algorithms being able to cope with these new requirements. In this section, we will therefore present a general purpose algorithm designed for use with any non-transitive dominance criteria, including dominance for malleability-aware skylines.

The naïve solution to the given problem is relying on exhaustive pair-wise comparison, i.e. each possible tuple pair has to be tested for dominance. However, this algorithm shows prohibitive practical performance requiring $\frac{1}{2}(n\,(n-1))$ expensive tests for dominance, with n being the size of the database (and assuming that each test for dominance is bi-directional, i.e. by testing $a >_P b$, we can test $b >_P a$ at the same time).

Hence, we propose a novel algorithm which is capable of dealing with any transitive or non-transitive preferences P. Our algorithm is derived from this naïve implementation by carefully avoiding any tuples comparisons which are guaranteed to show no effect. This can be formalized as follows:

Algorithm 1. Non-Transitive Skyline Algorithm

$T := R;\ \ L := \emptyset;\ \ S := \emptyset;$
while $(T \setminus L \neq \emptyset)$ **do**
    choose $t \in (T \setminus L)$;
    $C := T \setminus \{t\}$;
    $failed := false$;
    while $(C \neq \emptyset)$ **do**
        choose $c \in C$;
        if $t >_P c$ **then** $L := L \cup \{c\}$
        elseif $c >_P t$ **then**
            $L := L \cup \{t\};\ failed = true;$
        $C := C \setminus \{c\}$;
        if $failed$ **then** $C := C \setminus L$
    If $not(failed)$ **then** $S := S \cup \{t\}$;
    $T := T \setminus \{t\}$;

Given is a database relation R with n tuples and preferences P. Furthermore, we need the set T of all tuples which need further testing for *a)* if any $t \in T$ is dominated by any other tuple and *b)* if any $t \in T$ dominates any tuples itself; T is initialized with $T = R$. Furthermore, we use the set S of all tuples which are the final skyline, and the set L (i.e. losers) of those tuples which have already been shown to be dominated by any other tuple. In contrast Skyline algorithms with transitive dominance, we cannot exclude tuples in L from further computation without additional guarantees. This results in following algorithm:

The algorithm contains two loops, the outer one iterating t over all objects to be tested which have not already been shown to be dominated. For finding new dominance relationships, the second loop iterates c over the set C (C is initialized in each run with $T \setminus \{t\}$.) By testing t and each c for dominance, objects can be marked to be dominated by adding them the set L of all losers. As soon as t is dominated, any subsequent comparisons of t with any other tuple which has been shown to be dominated can be avoided as those yield no new information. If t was not dominated within the inner loop, it can safely be added to the skyline. Compared to the naïve approach, this algorithm saves a significant number of superfluous tuple comparisons (see evaluation in the next section).

Furthermore, this algorithm can be implemented efficiently by representing the membership of a tuple in the different sets by simple flags attached to the tuples in R, thus minimizing the overhead of additional book-keeping.

5   Evaluations

5.1   Evaluating General Malleability-Aware Skylines

In this section, we evaluate the effects of malleability-aware dominance respecting any number of malleable attributes on the properties of skylines. Furthermore, we will also measure the performance of respective skyline algorithms.

Skyline Size: For the first set of experiments, we examined the impact malleability-aware dominance (represented by varying values δ) on the Skyline size. For this purpose, we relied on syntetic data, and in each experimental run generated new database tuples with 12 independently-distributed numeric attributes. Six of these attributes represent non-malleable (data) attributes, while the other six attributes are malleable ones representing loss-of-focus. Using the operationalized dominance criterion of Section 4.1, skylines are computed for δ-values ranging from $\delta = 0$ (the baseline; equivalent to Pareto skylines as in definition 2) to $\delta = 0.3$. For each value of δ, the experiment is repeated 50 times with newly generated tuples (to ensure comparability, the same random seed is used for each δ, resulting in the same sequence of generated tuples). The averaged results are shown in Fig. 1. It is clearly obvious that the skyline resulting from the baseline ($\delta = 0$, identical to Pareto skyline of the same data) is not practically manageable: from the 50,000 database tuples, 26,981 are contained in the skyline (53%). This can be attributed to high dimensionally of $d = 12$. But with growing δ, the skyline sizes dramatically decrease: already with $\delta = 0.15$, the skyline is reduced to 11,959 tuples in average - a clearly more manageable result. Similar behavior can also be observed for smaller database sizes. Therefore, we can conclude that malleability-aware skyline indeed efficiently address the issue of overly large skylines when considering on malleable loss-of-focus attribute per data attribute.

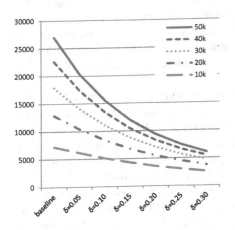

Fig. 1. Skyline Size wrt. to δ using 6 malleable and 6 non-malleable attributes and varying database sizes; y-axis shows skyline size

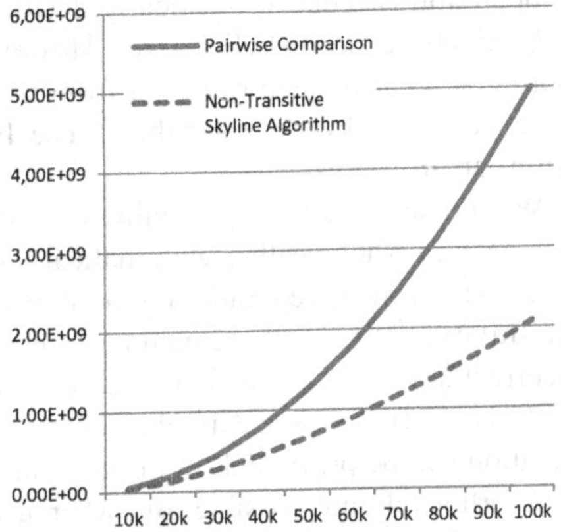

Fig. 2. Performance using 6 malleable and 6 non-malleable attributes; x-axis shows #tuples in database; y-axis shows number of required tests for dominance; $\delta = 0.15$

Performance of Algorithms: In the second set of experiments, we examined the performance of the naïve baseline and our non-transitive skyline algorithm (measured in the required number of tests for dominance). Similar to the last experiment, we again relied on synthetic data with 12 independent-attributes (6 malleable, 6 non-malleable), and incrementally increased the size n of the database from 10,000 tuples up to 100,000 tuples. The results are shown in Fig. 2: Clearly, our non-transitive algorithm shows significantly better performance than the baseline using pairwise comparisons. Furthermore, this performance advantage increases with growing database sizes. But still, the total time required by both algorithms is quite high (272 seconds with $n = 100k$ using our non-transitive skyline algorithm vs. 637 seconds for pairwise comparisons; tests performed on a 1.86 GHz Dual-Core CPU, using Java 6 and just a single core.) Therefore, additional optimizations must be found for application domains with tighter time constraints.

5.2   Malleability-Aware Skylines with a Single Malleable Attribute

As demonstrated in the last section, the runtime of general non-transitive skyline algorithms with one malleable loss-of-focus attribute for each non-malleable data attribute can be quite high. Thus, for time-critical applications, we suggest reducing the number of malleable attributes to using just a single attribute. This single attribute then represents the overall loss-of-focus of a given database tuple with respect to the query in an aggregated form. This reduction can be implemented by different methods: a) by combining multiple malleable attributes by some combining function or b) by directly eliciting just a single attribute representing loss-of-focus using one of the established frameworks for this task (e.g.[10] or [11]).

As an immediate effect, the number of dimensions to be respected during skyline computation is reduced drastically, leading to direct performance advantages due to respectively reduced skyline sizes. However, there is a less obvious and significantly more crucial advantage resulting from this reduction which allows us to build vastly more efficient skyline algorithms. The basic considerations leading to these algorithms are as follows:

When using established skyline algorithms like BNL, the only problem which is encountered when dealing with malleability-aware dominance is that tuples are eliminated early which are required to dominate another non-skyline tuple, and due to non-transitivity, none of the remaining tuples can lead to the same dominance; thus an incorrect skyline is computed (e.g. see example in Section 3.1). Therefore, we could use a more efficient standard algorithm like BNL if it could be made "safe", i.e. if this situation can be prevented. In the general case with multiple malleable attributes, this is unfortunately not possible. But when using just one malleable attribute, the correctness of BNL depends only on the order in which the tuples are inserted into the window: for example, consider three tuples with preferences already encoded in scores $\vec{x} = (0.8, 0.8, 0.4)$, $\vec{y} = (0.7, 0.7, 0.6)$, and $\vec{z} = (0.6, 0.6, 0.8)$; the grey score represents the single malleable attribute. When computing a malleability-aware skyline with $\delta = 0.1$, then $\vec{x} >_P \vec{y} >_P \vec{z}$, and the resulting skyline is just $\{\vec{x}\}$. But due to $\vec{x} \not>_P \vec{z}$, the BNL algorithm could first test $\vec{x} >_P \vec{y}$, removing $\vec{y}$, and resulting in the skyline $\{\vec{x}, \vec{z}\}$ (because $\vec{z}$ cannot be dominated anymore). Obviously, the skyline result would be correct if the tested order was $(\vec{x} >_P (\vec{y} >_P \vec{z}))$. It is easy to see that this observation can be generalized, i.e. problems in BNL can only occur if tuples with a lower malleability score are removed before they have been tested for dominance against all tuples with a higher malleability score. Therefore, for the case that there is only one malleable attribute, we can use established algorithms like BNL if all tuples are processed in descending order with respect to the malleability attribute (preventing the situation leading to incorrect skylines described above), i.e. the skyline algorithm is therefore *stratified* with respect to the malleability attribute. This can be implemented by pre-sorting the data before executing e.g. a BNL algorithm. The effectiveness of this approach is tested later in this section.

Skyline Size: Before dealing with performance issues, similar to the last section, we also measured the skyline sizes for varying δ and database sizes n. Again, we generate tuples using 6 non-malleable independently distributed attributes, but just one single malleable attribute. As now the number of overall dimensions is reduced from $d = 12$ down to $d = 7$, the respective skyline sizes are also reduced dramatically to only 4,017 tuples (8% of database) for the baseline $\delta = 0$ with $n = 50,000$ (see Fig. 4). But still, by slightly increasing δ, the skyline can be furthermore decreased to more manageable levels (e.g. 2,809 for $\delta = 0.15$ and $n = 50,000$).

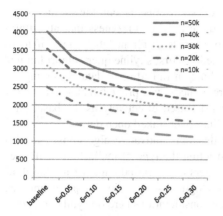
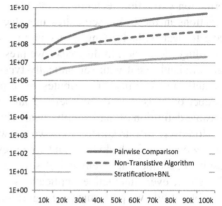

Fig. 3. Skyline Size wrt. to δ using one malleable and 6 non-malleable attributes and varying database sizes; y-axis shows skyline size

Fig. 4. Performance using one malleable and 6 non-malleable attributes; x-axis shows #tuples in database; y-axis shows number of required tests for dominance on a logarithmic scale. $\delta = 0.15$

Performance of Algorithms: In this last set of experiments, we examined the performance of the naïve baseline, our non-transitive skyline algorithm, and the stratified BNL algorithm as described above. Again, performance is measured by the required number of tests for dominance. We also relied on synthetic data with 7 independent-attributes (one malleable, 6 non-malleable), and incrementally increased the size n of the database from 10,000 tuples up to 100,000 tuples. The results are shown in Fig. 4 (using a logarithmic y-axis). Here, we can see that the stratified BNL-algorithm needs roughly two orders of magnitudes fewer dominance tests than the naïve baseline, and is also one order of magnitude more efficient than our general non-transitive skyline algorithm. In terms of absolute runtime, the general non-transitive algorithm needed 218 seconds for $n = 100$ and $\delta = 0.15$, which is still quite long. In contrast, the stratified BNL algorithm could be executed in less than 1.4 seconds using the same hardware (the time needed for sorting the 50,000 tuples before executing the algorithm is negligible). This significant result clearly shows that malleability-aware skylines can even be used in interactive environments having tight constraints with respect to response time like for example web applications.

6   Summary and Outlook

In this paper we discussed the case of query processing over linked open data. Whereas traditional query processing algorithms are usually graph-based and use exact matches on typed links between data items in SQL-like languages like SPARQL, the fuzzy nature of semantic links calls for approximate query processing algorithms. In particular, the exact labels of links cannot always be taken at face value, because information extraction techniques, the use of different concept ontologies, and slight variations in the links' semantics introduce quite a bit of fuzzyness that algorithms have to deal with. Relying on techniques to estimate different labels' loss of focus regarding each other, in this paper we presented the first skyline query algorithm that can efficiently deal with semantically typed links in linked open data. Modeling the semantic malleability of attributes by δ-preferences, we proved that the resulting product order is indeed well-defined and can be used effectively as the basis for a sensible definition of malleability-aware skylines over linked open data.

Moreover, in our experiments we show that our innovative algorithms can efficiently evaluate such skylines and when restricting the type of malleability will even result in runtime improvements of several orders of magnitude against the baseline. Therefore, even interactive applications with tight response time requirements are possible. While we performed the algorithmic considerations here on synthetic data to test our algorithms in an unbiased environment, our future work will focus on the integration of our algorithmic framework into practical linked open data sets. Our aim is to use potential bias in the data for a tighter integration of the attribute malleability respective to each individual query. It seems that different query intensions might need different degrees of admissible malleability to stay semantically meaningful.

References

1. Hitzler, P., van Harmelen, F.: A reasonable Semantic Web. Semantic Web 1, 39–44 (2010)
2. Heath, T., Hepp, M., Bizer, C. (eds.): Special Issue on Linked Data. International Journal on Semantic Web and Information Systems (IJSWIS) 5 (2009)
3. Banko, M., Etzioni, O.: Strategies for lifelong knowledge extraction from the web. In: Int. Conf. on Knowledge Capture (K-CAP). ACM Press, Whistler (2007)
4. Shen, W., Doan, A.H., Naughton, J.F., Ramakrishnan, R.: Declarative information extraction using datalog with embedded extraction predicates. In: Int. Conf. on Very Large Data Bases (VLDB), Vienna, Austria (2007)
5. Suchanek, F.M., Sozio, M., Weikum, G.: SOFIE: a self-organizing framework for information extraction. In: Int. World Wide Web Conf. (WWW), Madrid, Spain (2009)
6. Dong, X., Halevy, A.Y.: Malleable schemas: A preliminary report. In: Int. Workshop on Web Databases (WebDB), Baltimore, Maryland, USA (2005)
7. Dong, X., Halevy, A.Y.: A platform for personal information management and integration. In: Conf. on Innovative Data Systems Research (CIDR), Asilomar, California, USA (2005)
8. Kasneci, G., Suchanek, F.M., Ifrim, G., Ramanath, M., Weikum, G.: Naga: Searching and Ranking Knowledge. In: Int. Conf. on Data Engineering (ICDE), Cancún, México (2008)
9. DeRose, P., Shen, W., Chen, F., Lee, Y., Burdick, D., Doan, A.H., Ramakrishnan, R.: DBLife: A community information management platform for the database research community. In: Conf. on Innovative Data Systems Research (CIDR). Citeseer, Asilomar (2007)
10. Mena, E., Kashyap, V., Illarramendi, A., Sheth, A.: Imprecise Answers in Distributed Environments: Estimation of Information Loss for Multi-Ontology Based Query Processing. International Journal on Cooperative Information Systems 9 (2000)
11. Gracia, J., Mena, E.: Web-Based Measure of Semantic Relatedness. In: Bailey, J., Maier, D., Schewe, K.-D., Thalheim, B., Wang, X.S. (eds.) WISE 2008. LNCS, vol. 5175, pp. 136–150. Springer, Heidelberg (2008)

12. Godfrey, P., Shipley, R., Gryz, J.: Algorithms and analyses for maximal vector computation. The VLDB Journal 16, 5–28 (2007)
13. Cheng, T., Chang, K.C.-C.: Entity search engine: Towards agile best effort information integration over the web. In: Conf. on Innovative Data Systems Research (CIDR), Asilomar, California, USA (2007)
14. Mandreoli, F., Martoglia, R., Villani, G., Penzo, W.: Flexible Query Answering on Graph-modeled Data. In: Int. Conf. on Extending Database Technology (EDBT), St. Petersburg, Russia (2009)
15. Cohen, S., Mamou, J., Kanza, Y., Sagiv, Y.: A Semantic Search Engine for XML. In: Int. Conf. on Very Large Data Bases (VLDB), Berlin, Germany (2003)
16. Chen, L., Gao, S., Anyanwu, K.: Efficiently Evaluating Skyline Queries on RDF Databases. In: Antoniou, G., Grobelnik, M., Simperl, E., Parsia, B., Plexousakis, D., De Leenheer, P., Pan, J. (eds.) ESWC 201. LNCS, vol. 6644, pp. 123–138. Springer, Heidelberg (2011)
17. Balke, W.-T., Güntzer, U., Lofi, C.: Eliciting Matters – Controlling Skyline Sizes by Incremental Integration of User Preferences. In: Kotagiri, R., Radha Krishna, P., Mohania, M., Nantajeewarawat, E. (eds.) DASFAA 2007. LNCS, vol. 4443, pp. 551–562. Springer, Heidelberg (2007)
18. Börzsönyi, S., Kossmann, D., Stocker, K.: The Skyline Operator. In: Int. Conf. on Data Engineering (ICDE), Heidelberg, Germany (2001)
19. Fishburn, P.C.: Intransitive indifference in preference theory: A survey. Operations Research 18 (1970)
20. Tversky, A.: Intransitivity of preferences. Psychological Review 76 (1969)
21. Fishburn, P.C.: The irrationality of transitivity in social choice. Behavioral Science 15 (1970)
22. Anand, P.: Foundations of Rational Choice Under Risk. Oxford University Press (1995)
23. Papadias, D., Tao, Y., G.F., Seeger, B.: Progressive skyline computation in database systems. ACM Transactions on Database Systems 30, 41–82 (2005)
24. Balke, W.-T., Güntzer, U., Zheng, J.X.: Efficient Distributed Skylining for Web Information Systems. In: Bertino, E., Christodoulakis, S., Plexousakis, D., Christophides, V., Koubarakis, M., Böhm, K. (eds.) EDBT 2004. LNCS, vol. 2992, pp. 256–273. Springer, Heidelberg (2004)
25. Kossmann, D., Ramsak, F., Rost, S.: Shooting stars in the sky: an online algorithm for skyline queries. In: Int. Conf. on Very Large Data Bases (VLDB), Hongkong, China (2002)

Effective Next-Items Recommendation
via Personalized Sequential Pattern Mining

Ghim-Eng Yap[1], Xiao-Li Li[1], and Philip S. Yu[2]

[1] Institute for Infocomm Research, 1 Fusionopolis Way #21-01 Connexis Singapore 138632
{geyap,xlli}@i2r.a-star.edu.sg
[2] Department of Computer Science, University of Illinois at Chicago, IL 60607-7053
psyu@cs.uic.edu

Abstract. Based on the intuition that frequent patterns can be used to predict the next few items that users would want to access, sequential pattern mining-based next-items recommendation algorithms have performed well in empirical studies including online product recommendation. However, most current methods do not perform *personalized* sequential pattern mining, and this seriously limits their capability to recommend the best next-items to each specific target user. In this paper, we introduce a personalized sequential pattern mining-based recommendation framework. Using a novel Competence Score measure, the proposed framework effectively learns user-specific sequence importance knowledge, and exploits this additional knowledge for accurate personalized recommendation. Experimental results on real-world datasets demonstrate that the proposed framework effectively improves the efficiency for mining sequential patterns, increases the user-relevance of the identified frequent patterns, and most importantly, generates significantly more accurate next-items recommendation for the target users.

1 Introduction

With the rapid growth of online information sources and e-commerce businesses, users increasingly need reliable recommender systems [25,4] to highlight relevant next-items, i.e., the next few items that the users most probably would like. Fortunately, the user consumption histories offer crucial clues that help us to tackle this important problem. Whenever we visit web pages or purchase things from online stores, we leave a time-ordered sequence of items that we have seen or bought. When these historical sequences are consolidated into a *sequence database* (SDB), we can employ a powerful data mining process - *sequential pattern mining* (SPM) [1] - to discover temporal patterns that are frequently repeated among different users. Sequential pattern mining-based next-items recommendation works on the intuition that if many users have accessed an item j *after* an item i, it makes sense to recommend item j to someone who just seen item i.

Most sequential pattern mining algorithms focus on efficiently finding *all* the sequential patterns with frequencies above a given threshold in a sequence database. For example, algorithms exist for exact sequential pattern mining [1,3,11,16,20,30,33], approximate mining [14], constraint-based mining [5,7,21], as well as sequential pattern mining from incremental [8] and progressive databases [13]. A significant shortcoming is that all current methods do not perform *user-specific* sequential pattern mining and, as a result, they cannot give accurate *personalized* recommendation to users.

S.-g. Lee et al. (Eds.): DASFAA 2012, Part II, LNCS 7239, pp. 48–64, 2012.

In this paper, we address the important problem of *personalized* sequential pattern mining-based next-items recommendation, where the system discovers the *user-specific* frequent sequential patterns from past users' sequences, and it uses the mined patterns to predict the next few items which a target user would access. We establish the theoretical relation between (1) *support* of a sequential pattern p which recommends an item i, and (2) *predictive power* of p in terms of whether i is visited. Our analysis confirms that the predictive power of high-support (*frequent*) patterns is bounded by greater information gain [9], which justifies the *inductive assumption* [18] in sequential pattern-based recommendation - that the higher-support sequential patterns predict next-items better.

Current sequential pattern-based recommendation methods are ineffective because the simple frequency counts of patterns are used to compute the *support* they get from the sequence database (SDB). These methods treat all sequences in the SDB as equally important. However, in real applications, sequences in SDB often carry varying significance (or *weights*) with respect to each target user. To mine the *personalized* sequential patterns for a target user, we have to effectively model this varying relevance among the historical sequences for that specific user. Since each sequence in the SDB belongs to a different user, we can weight the sequences based on available knowledge about, for example, the target user's social affiliation to other users, or in most cases, how relevant is each of the sequences compared to the known sequence of the target user. For instance, in a book recommender system, the transaction records (i.e., sequences) of users who have borrowed similar books as the target user should be assigned greater weights because they are more likely to lead to accurate recommendations. Likewise, web browsing sequences with pages related to those the target user has visited should carry greater weights when we mine for sequential patterns to make recommendation.

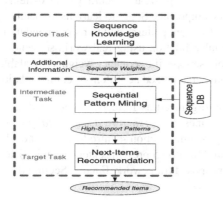

Fig. 1. Proposed personalized learning framework - effective learning is achieved by providing sequential pattern mining with an additional information source (the personalized sequence weights from a related source task) apart from the standard training data (the sequence database (DB))

We shall show that significantly more accurate next-items recommendations can be achieved through using our novel framework (Figure 1) to enable *personalized sequence weight learning and sequential pattern mining-based next-items recommendation.*

The proposed framework personalizes the sequence database (SDB) by assigning a user-specific weight to each sequence in the SDB (*source task*). This effectively enables the sequential pattern mining algorithms to perform a personalized search in sequence space for user-specific frequent patterns (*intermediate task*), which leads to significantly more accurate personalized next-items recommendation (*target task*). This approach is modeled after the inherent ability of human to recognize and leverage additional knowledge from related *source tasks* to improve the performance of a *target task* [31]. The next-items recommendation performance improves due to the sequence weight learning that makes the resulting sequential patterns much more relevant to every target user.

Our framework is novel because it *inverts the current paradigm by advocating the deep mining of user-specific patterns*, unlike traditional methods which generate all frequent patterns and prune them according to user relevance. The traditional post-filtering methods will not work well for *personalized* recommendation because the meaningful patterns for one user may not be frequent among the sequences of all users, so the traditional methods miss important patterns that really match the target user's preferences. By exploiting sequence weight knowledge to adjust the support of patterns, we effectively personalize the hypothesis space in which frequent patterns are searched, thereby enabling the discovery of more user-relevant frequent sequential patterns which cannot be found by traditional sequential pattern mining algorithms. The resulting user-specific sequential patterns are then more useful for personalized next-items recommendation.

We summarize our major research contributions as follows:

1. We propose a novel personalized sequential pattern mining-based next-items recommendation framework that learns and exploits additional user-specific sequence importance knowledge to improve the accuracy of next-items recommendation.
2. We propose a novel Competence Score to learn user-specific sequence weights for mining personalized frequent patterns and to recommend user-relevant next-items.
3. Through experimental validation using real-world sequence datasets, we demonstrate that learning of user-specific sequence weights is scalable, and it yields significantly more accurate personalized next-items recommendation for the target users.

The rest of this paper is organized as follows. Section 2 discusses the related work and Section 3 formally defines the *personalized sequential pattern mining-based next-items recommendation* problem. Section 4 introduces our novel personalized recommendation framework, and Section 5 presents extensive empirical results on real-world datasets to validate the proposed framework. Finally, Section 6 concludes this paper.

2   Related Works

The sequential pattern mining-based next-items recommendation is an inductive learning task where the objective is to induce a predictive model (i.e., the next-items recommender system) from a set of examples (i.e., the sequences in database). The usefulness of sequential pattern mining-based recommendation has, to a certain extent, been

Table 1. An example sequence database

No.	Sequence
1	<(1,4),(3),(2,8),(1,5)>
2	<(5,6),(1,2),(4,9),(3),(8)>
3	<(5),(7),(1,6),(3),(2),(8)>

demonstrated empirically by past studies on various domains such as e-commerce [12], web browsing [34] and IPTV programs scheduling [23], but there remains a lack of theoretical analysis on why high-support sequential patterns are useful for recommending next-items to users. We provide the necessary theoretical justification in this paper.

The proposed framework leverages on the learned user-specific sequence weights to improve the next-items recommendation. This is similar in concept to Bayesian transfer, where inductive transfer improves Bayesian learning [31]. In Bayesian learning, the *prior* distribution is combined with training data to get the predictive model. Bayesian transfer provides informative priors as additional knowledge from source task, so that the resulting model is more accurate. The learning of sequence weights for sequential pattern mining-based recommendation is similar to the learning of priors for Bayesian prediction in that the additional sequence weight knowledge helps the sequential pattern mining algorithms to better interpret the data and thus yield better recommendations.

Weighted sequential pattern mining [17,32] recognizes the difference in importance between sequences, but assumes that the importance of a sequence depends only on its items. The importance of items are derived from their domain value (e.g. price and popularity). The same weights are assigned to sequences and patterns with the same item characteristics regardless of the target user, which obviously cannot generate accurate *personalized predictions* since user-specific information are not used. Capelle et al. [5] proposed a method where they require the discovered patterns to be similar to a reference pattern. Specifically, they mandate that each user has to supply a reference pattern as well as specify a minimum similarity threshold. The similarity constraint introduces restrictions that only their proposed sequential pattern mining algorithm can satisfy, making their solution non-generalizable to other sequential pattern mining algorithms. In contrast, our framework extends to any source task that is related to sequence weight learning (not restricted to reference sequence matching and no need to ask users to state the similarity threshold etc), and can be readily integrated with all the existing sequential pattern mining algorithms. Additionally, our novel Competence Score method not only emphasizes other users' sequences that are relevant to the target user by computing their compatibility scores, but also takes the sequences' recommendation ability (or extensibility) into consideration, thus giving much better next-items recommendations.

3 Problem Definition

We now formally introduce the personalized sequential pattern mining-based next-items recommendation problem. A sequence database (SDB) is a collection of sequences, i.e. SDB=$\{s_1, s_2, \ldots, s_n\}$ where $|SDB| = n$ denotes the number of sequences (also the number of users as each user has a corresponding sequence in SDB). A sequence $s_i \in$

$SDB(i = 1, 2, \ldots, n)$, can be represented as $s_i =<s_{i1}, s_{i2}, \ldots, s_{i|s_i|}>$, which is an ordered list of itemsets (each s_{ij} in s_i ($j = 1, 2, \ldots, |s_i|$) is an itemset) associated with an user U_i. Each itemset s_{ij} comprises transacted items with the same timestamp $t(s_{ij})$, and the different itemsets s_{ij1} and s_{ij2} in the same sequence s_i cannot have the same timestamp, i.e., $t(s_{ij1}) \neq t(s_{ij2}), (j_1 \neq j_2)$. A sequence $s_i =<s_{i1}, s_{i2}, \ldots, s_{i|s_i|}>$ contains a sequence $s_j =<s_{j1}, s_{j2}, \ldots, s_{j|s_j|}>$, or s_j is a *subsequence* of s_i, if there exist integers $d_1 < d_2 < \ldots < d_{|s_j|}$ such that $s_{j1} \subseteq s_{id1}, s_{j2} \subseteq s_{id2}, \ldots, s_{j|s_j|} \subseteq s_{id|s_j|}$ [1]. Table 1 shows an example sequence database comprising three sequences. Each row is a sequence representing a user's consumption histories; items in each ()-bracket form an itemset with same timestamp, e.g. (1,4). Sequence 1 describes a user who accessed the items 1 and 4 at timestamp t_0, item 3 at timestamp t_1, items 2 and 8 at timestamp t_2, and items 1 and 5 at timestamp t_3, etc. Using *patterns* to refer to subsequences generated during mining process, the *count* of a pattern p is the number of sequences in SDB containing p. The *support* for pattern p is then defined as $support(p) = count(p)/|SDB|$.

Given a target user's past sequence s_q and the database of all other users' past sequences, the next-items recommendation task is to predict items that the target user is most likely to access *in the near future*, i.e., those items in the next few itemsets that he or she will access. We do not restrict the prediction to just the immediate next item that a target user will access, nor the immediate itemset that the user will access in the next visit [24]. The *personalized* next-items recommendation task can be formalized by identifying and ranking the candidate next-items through a novel framework of personalized sequential pattern mining that takes into account the relative importance (i.e., weights) of other users' sequences in the database according to each target user, so that the top-m items (may be from multiple itemsets) are recommended to the target user.

4   Personalized Sequential Pattern Mining-Based Recommendation

Given a sequence database SDB and a minimum support δ, the traditional sequential pattern mining task is to discover all the patterns having a support not less than δ, under the assumption that all the sequences in the SDB should be treated equally. The required most-probable next-items are then recommended from the patterns having the highest-support. In contrast, *the problem of next-items recommendation via personalized sequence weight learning takes into account the fact that sequences in the real-world have different importance to each specific user; sequence weights should be effectively learned and exploited to generate more user-relevant sequential patterns, which can in turn significantly improve the next-items recommendation accuracy for the end-users.*

While numerous empirical studies (e.g. [12,34,23]) have demonstrated that sequential pattern mining-based next-items recommendation is effective in many different domains, we take the analysis further by formally explaining how the predictive power and the support of sequential patterns are related, similar to how the discriminative power of features in pattern-based classification is related to the pattern/feature frequency [9]. Specifically in our case, we need to establish that reliably computing the support values of patterns can result in better next-items recommendations, in order to explain why our proposed framework, which enables the resulting pattern supports to reliably reflect the user-relevance, is so effective in practice for personalized next-items recommendations.

4.1 The Predictive Power and Support of Patterns

Let $V \in \{0,1\}$ and $X \in \{0,1\}$ be the events of *visiting an item i* and *predicting item i*, respectively. The information gain (a standard measure of predictive power [9]) for V (i.e., whether i is visited) from knowing X (i.e., whether i is predicted) is computed by:

$$IG(V|X) = H(V) - H(V|X) \tag{1}$$

where $H(V)$ and $H(V|X)$ are the entropy and conditional entropy for V. Specifically,

$$H(V|X) = - \sum_X P(X) \sum_V P(V|X) log(P(V|X)) \tag{2}$$

Given a sequence database, $H(V)$ is fixed as the prior probability of i, so the upper bound of the information gain (i.e., $IG_{UB}(V|X)$) depends only on the lower bound of the conditional entropy for V given X (i.e., on $H_{LB}(V|X)$). $P(X = 1)$ is the probability of predicting item i. Item i is predicted if a pattern that predicts i is considered *frequent*, i.e., if $P(X = 1) = \theta \geq minsup$, where $minsup$ is a minimum support threshold and θ is the support of such a pattern. $P(V = 1)$ is the probability that i is visited. Given a sequence database, $P(V = 1) = \alpha$ is a fixed value equal to i's support in the database ($\alpha \geq 0$). By Apriori lemma, i appears in any pattern that predicts it, so $\theta \leq \alpha$.

Letting $P(V = 1|X = 1) = \beta$, the conditional entropy $H(V|X)$ is at its minimum if knowing $X = 1$ completely predicts V, i.e., if $\beta = 1$ or $\beta = 0$. For the case $\beta = 1$,

$$H_{LB}(V|X) = (\alpha - 1)log(\frac{1 - \alpha}{1 - \theta}) - (\alpha - \theta)log(\frac{\alpha - \theta}{1 - \theta}) \tag{3}$$

The partial derivative of $H_{LB}(V|X)_{\beta=1}$ with respect to the support θ is:

$$\frac{\partial H_{LB}(V|X)_{\beta=1}}{\partial \theta} = log(\frac{\alpha - \theta}{1 - \theta}) \leq log 1 \leq 0 \tag{4}$$

So, $H_{LB}(V|X)$ reduces as the support θ increases. For the case $\beta = 0$,

$$H_{LB}(V|X) = (\theta - (1 - \alpha))log(\frac{1 - \alpha - \theta}{1 - \theta}) - \alpha log(\frac{\alpha}{1 - \theta}) \tag{5}$$

The partial derivative of $H_{LB}(V|X)_{\beta=0}$ with respect to the support θ is:

$$\frac{\partial H_{LB}(V|X)_{\beta=0}}{\partial \theta} = log(\frac{1 - \alpha - \theta}{1 - \theta}) \leq log 1 \leq 0 \tag{6}$$

The above analysis reveals that lower bound conditional entropy $H_{LB}(V|X)$ monotonically decreases as θ increases, so theoretical upper bound information gain increases with θ. Hence, the predictive power of a higher-support pattern is upper-bounded by a higher information gain, justifying the strategy of predicting items from more frequent sequential patterns rather than less frequent patterns. This explains why our proposed framework can be so effectively used for personalized next-items recommendation.

4.2   Sequence Weight Learning

We shall now discuss the source task which performs the sequence weight learning (Figure 1). Given a target user's sequence s_q, for each s_i in sequence database SDB ($s_i = <s_{i1}, s_{i2}, \ldots, s_{i|s_i|}>$ is user U_i's sequence), we assign a sequence weight $w(s_i, s_q)$ to s_i that represents its significance with respect to the target user sequence s_q. We first discuss two existing methods and then present our novel Competence Score method.

Method 1: User-independent Weight Learning. Weighted sequential pattern mining [6] weighs s_i based on the characteristics of items in s_i, e.g., price [32], popularity [23] (application oriented). These methods give the same weight for s_i regardless of the target user's sequence s_q. Weight $w(s_i, s_q)$ can be computed using the mean popularity of s_i's itemsets — popularity of an itemset s_{i_j}, $s_{i_j} \in s_i$, is computed as $count(s_{i_j})/|SDB|$.

Method 2: User-dependent Weight Learning. This method accounts for the differences between target users to learn more user-specific sequence weights. In this learning task, each $s_i \in SDB$ is compared to s_q to determine their similarity, and similar sequences in SDB are deemed more relevant for making recommendation [15,27] to the target user. The weight of s_i w.r.t. s_q is computed using the following pattern similarity functions:

1. Longest Common Subsequence (LCS) [29,5]: Let the longest common subsequence of s_i and s_q be denoted $LCS_{i,q} = <lcs_{i,q_1}, lcs_{i,q_2}, \ldots, lcs_{i,q_{|LCS_{i,q}|}}>$ (each lcs_{i,q_j} is an itemset that is common to sequences s_i and s_q, such that timestamp $t(lcs_{i,q_{j+1}}) > t(lcs_{i,q_j})$, $j = 1, 2, \ldots, |LCS_{i,q}| - 1$). The weight $w(s_i, s_q)$ is defined as:

$$w(s_i, s_q) = \frac{|LCS_{i,q}|}{|s_q|} \tag{7}$$

    where $|\cdot|$ denotes the number of itemsets in a sequence.
2. Cosine Similarity [23]: The weight $w(s_i, s_q)$ for s_i w.r.t. s_q can be measured as the cosine similarity [26] between their respective itemset vectors (denoted v_i and v_q):

$$w(s_i, s_q) = \frac{v_i \cdot v_q}{|v_i||v_q|} \tag{8}$$

    where v_i and v_q are the vector representations of s_i and s_q, respectively, such that each dimension in v_i and v_q denotes an unique itemset.

Method 3: Our Proposed Weight Learning Method. Our novel *Competence Score* method computes a personalized score for each sequence s_i in the sequence database (SDB) to reflect its competency in recommending next-items for the target user. Every s_i is a temporally-ordered list of itemsets with timestamps from oldest to most recent. Instinctively, s_i is highly-competent if it is not only compatible to the user sequence s_q, but also is capable of readily extending beyond s_q to offer more next-items. To satisfy both these requirements, our proposed Competence Score is computed in three steps:

1. **Step 1: Compute backward-compatibility score (BCS) of each sequence $s_i \in SDB$ w.r.t. the user sequence s_q.** We iterate over every sequence s_i in sequence database (SDB) to evaluate its *string-compatibility* and *temporal-compatibility* to s_q. String compatibility of s_i and s_q is derived by Longest Common Subsequence ($LCS_{i,q}$, Eq. 7), so that the longer the $LCS_{i,q}$, the more compatible s_i is to s_q. However, string-compatibility alone does not take into account that the most recently transacted itemset by the target user is the most relevant for predicting next-items to that user [34,17]. Our BCS score also considers the *temporal*-compatibility of the itemsets in s_i and s_q by weighting the itemsets in s_q such that its more recent itemsets are weighted heavier. This is analogous to a Markov Chain where the future items of the target user depend mainly on the most recent items the user saw.

   Letting the longest common subsequence of s_i and s_q be $LCS_{i,q} = < lcs_{i,q_1}, lcs_{i,q_2}, \ldots, lcs_{i,q_{|LCS_{i,q}|}} >$, the weight of each itemset $lcs_{i,q_j} \in LCS_{i,q}$ is $w(t_j, t_l)$, where t_j is the timestamp of lcs_{i,q_j} in s_q and t_l is the timestamp of the last itemset in s_q. For a given s_q, t_l is a constant so we can simply refer to the weight as $w(t_j)$. This itemset weight decay is modeled using the normalized Gaussian distribution:

$$w(t_j) = ae^{-\frac{(t_j-t_l)^2}{2\sigma^2}}, 0 \leq w(t_j) \leq 1 \qquad (9)$$

   where $a = \frac{1}{\sigma\sqrt{2\pi}}$ and $\sigma = (t_f - t_l)/(2\sqrt{2\ln 2})$; the constants t_f and t_l denote the timestamps of the first and the last itemsets in the user sequence s_q, respectively.

   The time-weighted BCS of s_i w.r.t. target sequence s_q is computed as follows:

$$b_{i,q}(t) = \frac{\sum_{j=1}^{|LCS_{i,q}|} w(t_j), lcs_{i,q_j} \in LCS_{i,q}}{\sum w(t_k), s_{q_k} \in s_q}, 0 \leq b_{i,q}(t) \leq 1 \qquad (10)$$

   where t_k is the timestamp of itemset s_{q_k} in s_q. This BCS score favors s_i sequences whose $LCS_{i,q}$ contains itemsets closer in time to the most recent itemset in s_q.

2. **Step 2: Compute the forward-extensibility score (FES) of each sequence $s_i \in SDB$ w.r.t. the user sequence s_q.** We define forward-extensibility score (FES) of s_i to answer the following important question: having seen some of the itemsets in s_q, does s_i have sufficient suitable next-items to recommend to the target user?

   Similar to our BCS score, our FES score computation involves the identification of the longest common subsequence ($LCS_{i,q}$) between each $s_i \in SDB$ and s_q. Let $LCS_{i,q} =< lcs_{i,q_1}, lcs_{i,q_2}, \ldots, lcs_{i,q_{|LCS_{i,q}|}} >$. We identify the set of unique candidate next-items that s_i is able to recommend to the target user, by finding the unique items from those itemsets in s_i that are later than the last common itemset $lcs_{i,q_{|LCS_{i,q}|}}$. The basic formula for computing the FES score is therefore:

$$f_{i,q} = \frac{|CNI_{i,q}|}{|CNI_{i,q}|_{max}}, 0 \leq f_{i,q} \leq 1 \qquad (11)$$

   where $CNI_{i,q} = \{cni_{i,q_1}, cni_{i,q_2}, \ldots, cni_{i,q_{|CNI_{i,q}|}}\}$ is the set of unique candidate next-items in s_i with respect to s_q. To keep $f_{i,q}$ between 0 to 1, we only consider up to at most the first $|CNI_{i,q}|_{max}$ unique candidate next-items after $lcs_{i,q_{|LCS_{i,q}|}}$, where $|CNI_{i,q}|_{max}$ is the maximum desired number of new items, a standard parameter in recommenders. In this way, $|CNI_{i,q}|$ (i.e., the number of candidate next-items in $CNI_{i,q}$) cannot be more than $|CNI_{i,q}|_{max}$. We further default parameter $|CNI_{i,q}|_{max}$ to the number of items in s_q, $||s_q||$, since it is reasonable for heavier users to demand more recommendations (e.g., books). The FES formula becomes:

$$f_{i,q} = \frac{|CNI_{i,q}|}{||s_q||}, 0 \leq f_{i,q} \leq 1 \tag{12}$$

where $||s_q||$ denotes the number of items in sequence s_q.

Furthermore, we need to take into account the different suitability of each candidate next-item cni_{i,q_j} ($cni_{i,q_j} \in CNI_{i,q}$) in terms of how much time difference exists between the timestamp t_k when the last common itemset $lcs_{i,q|LCS_{i,q}|}$ (i.e., the last itemset in $LCS_{i,q}$) was seen in s_i and the timestamp t_j when candidate next-item cni_{i,q_j} was seen in s_i. The intuition is that items transacted nearer in time from the most recent common itemset (i.e., $lcs_{i,q|LCS_{i,q}|}$) are more relevant to the target user and should get more weight, whilst candidate next-items that are further in time from $lcs_{i,q|LCS_{i,q}|}$ are not quite as relevant to the target user and should be weighted less. This decay (over time) in item weight $w(t_j, t_k)$, which we simply refer to as $w(t_j)$ since t_k (timestamp of $lcs_{i,q|LCS_{i,q}|}$ in s_i) is a constant given the sequence pair s_i and s_q, is modeled using the normalized Gaussian distribution:

$$w(t_j) = ae^{-\frac{(t_j - t_k)^2}{2\sigma^2}}, 0 \leq w(t_j) \leq 1 \tag{13}$$

where $a = \frac{1}{\sigma\sqrt{2\pi}}$ and $\sigma = (t_l - t_k)/(2\sqrt{2\ln 2})$; the constants t_l and t_k denote the timestamps of the last itemset in s_i and the last itemset in $LCS_{i,q}$, respectively.

The time-weighted FES of s_i w.r.t. target sequence s_q is computed as follows:

$$f_{i,q}(t) = \frac{\sum w(t_j), cni_{i,q_j} \in CNI_{i,q}}{||s_q||}, 0 \leq f_{i,q}(t) \leq 1 \tag{14}$$

where $||s_q||$ denotes number of items in sequence s_q.

3. **Step 3: Compute Competence Score (CS) of sequence s_i with respect to user sequence s_q by merging its BCS and FES**. The sequence weight $w(s_i, s_q)$ is given by the *Competence Score* ($c_{i,q}(t)$) of s_i w.r.t. s_q. We define s_i's Competence Score $c_{i,q}(t)$ as the harmonic mean between its time-sensitive BCS and FES scores:

$$c_{i,q}(t) = \frac{b_{i,q}(t) \times f_{i,q}(t)}{\frac{1}{2} \times (b_{i,q}(t) + f_{i,q}(t))}, 0 \leq c_{i,q}(t) \leq 1 \tag{15}$$

The BCS $b_{i,q}$ is high only if the given sequence s_i is highly-compatible to the target user's sequence, while the FES $f_{i,q}$ is high only if s_i has sufficient next-items to recommend. Being their harmonic mean, CS $c_{i,q}$ is high only if both $b_{i,q}$ and $f_{i,q}$ are high, i.e., both compatibility and extensibility requirements are satisfied. In specific domains where there exists a certain preferred trade-off between BCS and FES, the more general weighted harmonic mean (where different weights can be allocated to

BCS and FES) can also be used. Besides the harmonic mean, alternative methods to merge BCS and FES have been investigated but are found to be not as suitable. For instance, using multi-objective optimization (i.e., skyline queries [22]) may not maintain the balance between the two important requirements, e.g. a high skyline score may be given to a sequence s_i with a high BCS but a very low FES [19].

4.3 Exploiting the Sequence Weight Knowledge for Next-Items Recommendation

In the proposed framework, personalized weights are first assigned to all sequences in the sequence database (SDB) as an off-line operation using the aforementioned learning methods. We now define the personalized *count* and *support* to take into account the learned sequence weights so that the user-specific sequence knowledge is effectively exploited by the proposed framework to personalize the sequential pattern mining:

[*Pattern Count*]. Given that each sequence $s_i \in SDB$ has a learned weight $w(s_i, s_q)$ with respect to user sequence s_q, the count of pattern x in sequence database (SDB) is:

$$count(x, s_q) = \sum_{s_i:(x \subseteq s_i)} w(s_i, s_q) \tag{16}$$

Equation 16 sums the weights of all sequences which contain the pattern x, such that a higher $count(x, s_q)$ means a higher support for pattern x in the sequence database.

[*Pattern Support*]. The support of x in SDB is defined as:

$$support(x, s_q) = \frac{count(x, s_q)}{\sum_{s_i \in SDB} w(s_i, s_q)} \tag{17}$$

where the denominator is the total sequence weight in the sequence database.

The above support definition satisfies the monotonically-decreasing property: given patterns A and B, if $B \subseteq A$, then $support(B) \geq support(A)$, as B must be part of all sequences containing A. Hence, it is readily applicable to any of the conventional Apriori-inspired sequential pattern mining algorithms to mine the high-support (*frequent*) patterns (the intermediate task in Figure 1). An example of such sequential pattern mining algorithms is PrefixSpan [20], the popular pattern-growth approach in which a sequence database is recursively projected into a set of smaller databases, and sequential patterns are grown in each projected database by exploring only locally frequent fragments [20].

During sequential pattern mining, patterns with personalized *support* not less than the specified minimum support δ are output as the user-specific frequent patterns F_q:

$$F_q = \{x | support(x, s_q) \geq \delta, \delta > 0\} \tag{18}$$

In Eq. 18, we used sequence weight knowledge to eliminate all the user-irrelevant records in SDB which are not related to the target user (since we will not include x in F_q if $support(x, s_q) = 0$). As such, we can significantly improve the mining efficiency.

The frequent pattern set F_q is then used in our target task (Figure 1) to recommend next-items to the target user [34,17], where the most recent items of the target user are considered more valuable for predicting the next-items. Let the target user's known sequence be $s_q = <i_1, i_2, \ldots, i_n>$ where i_j ($i_j \in s_q$) is an itemset. We find candidate pattern set $CP = \{x | i_n \in x, x \in F_q\}$ which is the frequent patterns containing i_n. We eliminate redundant patterns from CP by keeping only closed patterns starting with i_n.

Let a candidate pattern be $cp = <i_n, i_{n+1}, \ldots, i_{n+k}>$, $cp \in CP$. Items in the itemsets after i_n are candidate items for recommendation. The support for each candidate item i_c ($i_c \in i_w, w > n, i_w \in cp$) is the sum of *support* of all cp that promote i_c, i.e.,

$$support_{i_c} = \sum_{i_c \in i_w, w>n, i_w \in cp, cp \in CP} support(cp, s_q) \qquad (19)$$

The overall algorithm is presented in Algorithm 1, where our proposed Competence Score method is used to learn the personalized sequence weights for each target user. The personalized sequence weights are then effectively exploited to discover the user-specific frequent sequential patterns for personalized next-items recommendation.

Algorithm 1.   Personalized Sequential Pattern Mining-based Recommendation

---

INPUT: The target user u_q's sequence, s_q, and the database SDB of other users' sequences
BEGIN
`// sequence weight learning step (source task)`
for all sequence $s_i \in SDB$ **do**
    Compute $b_{i,q}(t)$ // backward-compatibility (Eq. 10)
               $f_{i,q}(t)$ // forward-extensibility (Eq. 14)
               $c_{i,q}(t)$ // Competence Score (Eq. 15)
end for // now we have the SDB with sequence weights personalized to user u_q
`// sequential pattern mining step (intermediate task)`
    Apply a state-of-the-art sequential pattern mining algorithm to get personalized frequent patterns, F_q, with Eqs. 16 and 17 to compute the personalized support
`// next-items recommendation step (target task)`
    Use Eq. 19 to compute the personalized support of each candidate next-item in frequent patterns F_q, and recommend the items with highest support to user u_q
END

---

5   Experimental Evaluation

We present a two-part evaluation of our proposed personalized sequential pattern mining-based next-items recommendation framework (Figure 1). In the first part (Section 5.2), we evaluate the effectiveness of learning sequence weights with respect to the sequential pattern mining. In the second part (Section 5.3), we evaluate the framework's effectiveness in terms of prediction accuracy of the next-items recommendation.

5.1   Experimental Setup

- **Algorithms:** We compare the sequence weight learning methods with the traditional sequential pattern mining-based recommendation [34,17] with no weight learning. The state-of-art PrefixSpan [20] algorithm is used for the pattern mining.

- **Datasets:** We use two real-world datasets for the performance evaluation. The first dataset is the msnbc.com dataset from UCI [2]. It captures the time-ordered sequence of webpages visited by msnbc users on a day. There are a total of 989,818 sequences, each one corresponding to a different user. The second dataset is a book-loan dataset comprising the borrowing records for 126,714 users on 144,966 books (126,714 sequences), all during a six-month period.
- **Evaluation Metrics:** The measures to evaluate if sequence weight learning helps the next-items recommendation include [31]: (1) reduction in time for mining sequential patterns, and (2) improvement in accuracy for next-items recommendation.

5.2 Evaluating the Framework Efficacy

We choose msnbc.com dataset to evaluate the efficacy of our framework as it has much more sequences (close to 1 million) than the book-loan dataset. Note that the learning of user-specific (personalized) sequence weights should decrease the time taken for pattern mining, because by eliminating all the user-irrelevant sequences (i.e. their weights are equal to zero) in SDB, we only need to handle a much smaller personalized hypothesis space consisting of sequences that are more relevant to the target users. On the contrary, without exploiting personalized sequence weights, sequential pattern mining algorithm will be performed inefficiently since it has to go through all the transactions in SDB.

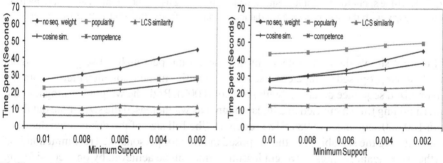

(a) Shorter time taken to mine patterns (lower is better) due to the sequence weight learning.

(b) Total time taken (lower is better) for the sequence weight learning and the pattern mining.

Fig. 2. Improvement in time taken as a result of our proposed personalized learning framework

Figure 2(a) shows the time taken to mine sequential patterns over different minimum support values when there is no personalized sequence weight learning, versus with sequence weight learning using the different methods presented in Section 4.2. Indeed, our framework significantly reduced the time for the sequential pattern mining, especially with our Competence Score weighting method. Figure 2(b) compares the total time taken for both learning sequence weights and running the sequential pattern mining algorithm. Taking into account the sequence weight learning time, the user-independent learning method, which completely ignores the target users and computes sequence weights based on their items' popularity, added significantly to the time taken, making it less attractive compared to sequential pattern mining without sequence weight learning. However, with reference to Figures 2(a) and 2(b), the extra time spent on the more user-specific weight learning methods (LCS, cosine similarity, Competence Score) is worthwhile since it greatly reduced the time used in the subsequent pattern mining step. More specifically, our proposed Competence Score method added only an average of 6.37 seconds for the weight learning but led to an average reduction of more than 20 seconds in pattern mining time. Considering that the learning of Competence Score and other methods to compute personalized sequence weights can be completed off-line, the time saving in sequential pattern mining due to our framework (Figure 2(a)) is even more significant.

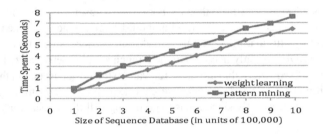

Fig. 3. The time spent for weight learning and pattern mining using the Competence Score method grows linearly with the size of sequence database (varying from 100,000 to 989,818 sequences).

Figure 3 shows the total time spent in learning the personalized sequence weights based on our proposed Competence Score, and in mining the frequent patterns from the personalized sequence database (with *minsup*=0.002). Both the weight learning and the pattern mining time increased in a linear manner as we increased the size of the database from 100,000 sequences to 989,818 sequences (the full size of the msnbc.com dataset), demonstrating the scalability of the proposed method for personalized recommendation. Further significant savings in weight learning time can be achieved by only considering those sequences in the database which have at least one common itemset with the target sequence (e.g., using the standard inverted index approach in search engines [35] so that sequences without any of the itemsets in the target sequence are pruned immediately).

The sequence weight learning methods in Section 4.2 use different relevance criteria (i.e., popularity, LCS, cosine similarity, and Competence Score) to compute personalized sequence weights. For each method, we evaluate the *quality* of the high-support sequential patterns in terms of how well they satisfy the corresponding criterion used to learn the sequence weights (similar results are observed for the top 5, 10, 15, 20 and 25 high-support patterns, and for *minsup* 0.002-0.010). We compare the results to the corresponding cases where personalized sequence weights are not used. The proposed Competence Score method improved pattern quality by five times, the LCS and cosine similarity methods improved pattern quality by three times, and the user-independent weight learning method based on item popularity also managed to improve the pattern quality, albeit marginally. The results clearly show that, for each of the sequence weight learning methods, the relevance criterion used to compute the sequence weights has been effectively exploited to improve the quality of the resulting high-support patterns, such that the discovered patterns inherit the corresponding desirable relevance criterion.

The improvement in mining time and pattern quality from the item popularity-based weight learning method, which ignores the target user, is much less than our personalized methods (LCS, cosine similarity, and Competence Score) that learn user-specific sequence weights . This shows that even when using the same sequential pattern mining algorithm, the more user-specific sequence knowledge helps to generate high-support sequential patterns that are more relevant to the target users. This could explain why simply picking patterns to match the users *after* sequential pattern mining is inadequate [34,23]; since patterns that are meaningful for a user may not be frequent among the sequences for all users, directly using standard sequential pattern mining algorithm (without learning and exploiting sequence weights) would miss those *personalized* patterns which really match the users' preferences, indicating that it is crucial to integrate user relevant sequence knowledge into the pattern mining process.

5.3   Accuracy of the Next-Items Recommendation

We can now compare the different methods for next-items recommendation using both the msnbc.com and book-loan datasets. Using a standard *backlog* evaluation method, we partition each target user's test sequence into two portions: release only the earlier portion to the recommendation methods, and evaluate the accuracy by comparing the recommended items to the held-out portion. We need to experiment with sequences having sufficient items (at least ten items in our experiments). For each sequence, we release the first five items (known portion) and hold-out the remaining items as the ground-truth for evaluation. We allow each method to recommend up to ten items with the highest support values and evaluate results in terms of *recall, precision (precision@1)* and *F1-measure*, all of which are standard evaluation metrics for recommendation systems [12,23,34]. As there are test sequences (or cases) for which certain methods do not give any recommendation, we also measure the *applicability* of each method in terms of the percentage of test cases for which recommendations are given. We set the *minsup* for msnbc.com and book-loan as high as possible to 0.5 and 0.01, respectively, so that most of the less frequent patterns are effectively filtered off by the sequential pattern mining, while maintaining a minimum applicability of around 40% for the recommendation.

The results for the msnbc.com and book-loan datasets are presented in Table 2. The experimental results clearly demonstrate that the methods for learning of sequence weights which are more related to the target users can indeed yield significantly more accurate next-items recommendations. In particular, sequence knowledge learning using our proposed recommendation Competence Score in Section 4.2 produced the greatest improvement in performance among the competing methods; it provided recommendations for almost 100% of the test cases (a vast improvement in *applicability*), and significantly increased the recall, precision, F1-measure, and precison@1 (i.e., proportion of test cases where the top-1 recommended item is correct). Specifically, for the msnbc.com dataset, 70.6% of the test cases contained the first recommended item when Competence Score was used, compared to just around 40% for the competing

Table 2. Recommendation performance. Indicated in brackets are the % improvement over and above sequential pattern mining with no sequence weight learning (denoted as "no seq. weight").

(a) Performance on msnbc.com dataset.

	Applicability	Recall	Precision	F1-measure	Precision@1
no seq. weight	45.4%	0.178	0.313	0.227	0.375
popularity	45.4% (0%)	0.178 (0%)	0.313 (0%)	0.227 (0%)	0.375 (0%)
cosine sim.	49.1% (8%)	0.223 (25%)	0.342 (9%)	0.270 (19%)	0.404 (8%)
LCS similarity	52.3% (15%)	0.229 (29%)	0.366 (17%)	0.282 (24%)	0.435 (16%)
competence	100% (120%)	0.547 (207%)	0.502 (60%)	0.524 (131%)	0.706 (88%)

(b) Performance on book-loan dataset.

	Applicability	Recall	Precision	F1-measure	Precision@1
no seq. weight	39.5%	0.017	0.073	0.028	0.075
popularity	43% (9%)	0.024 (41%)	0.063 (-14%)	0.035 (25%)	0.07 (-7%)
LCS similarity	93% (135%)	0.069 (306%)	0.083 (14%)	0.075 (168%)	0.135 (80%)
cosine sim.	95% (141%)	0.074 (335%)	0.092 (26%)	0.082 (193%)	0.135 (80%)
competence	99% (151%)	0.077 (353%)	0.094 (29%)	0.085 (204%)	0.155 (107%)

methods. Likewise for the book-loan dataset, our proposed Competence Score method was able to outperform all the competing methods in terms of the various evaluation metrics. In particular, while the baseline method without any sequence weight learning managed a mere 4% accuracy in terms of predicting one or more of the next-books that were subsequently borrowed by the target users in the six-months period, our proposed personalized recommendation framework using the Competence Score method for sequence weight learning was able to achieve a significantly higher accuracy of 57%. The experimental results thus demonstrate that our personalized framework predicts future items reliably and can be used to automatically recommend next-items to target users.

6 Conclusions

Learning about the different importance of sequences can improve the sequential pattern mining-based next-items recommendation in real-world applications. We present a novel *personalized sequence weight learning and sequential pattern mining-based next-items recommendation framework* that learns every sequence's importance as weights, and then effectively exploits this additional sequence knowledge to improve the efficiency and the quality of learning in sequential pattern mining algorithms. Experimental results using real-world sequence datasets demonstrate that the proposed framework is highly effective in learning and exploiting the sequence knowledge for sequential pattern mining, and that this significantly improves the performance of the personalized next-items recommendation, especially with our novel Competence Score method.

The proposed framework can be readily applied for next-items recommendation in domains such as web mining, financial mining, and product/service consumption analysis, etc. An interesting future work is to explore sequence weighting methods which exploits user knowledge beyond what is present in the sequences, e.g., social influence measures [10] and similarity in user vocabulary [28]. It is also interesting to investigate the combination of weight knowledge from multiple methods (i.e., early fusion) in a single recommender, and the efficacy of a recommender ensemble that inherits weight knowledge from many methods (i.e., late fusion). Other measures of recommendation performance like coverage, diversity and serendipity (e.g. [19]) can also be investigated. More generally, the benefits of personalized mining can be realized in data mining problems beyond sequential pattern mining (e.g. in classification, clustering); the question is how we can effectively personalize inputs to these algorithms for user-specific results.

References

1. Agrawal, R., Srikant, R.: Mining sequential patterns. In: Proceedings of IEEE International Conference on Data Engineering (ICDE 1995), pp. 3–14 (1995)
2. Asuncion, A., Newman, D.: UCI ML Repository (2007),
   http://www.ics.uci.edu/~mlearn
3. Ayres, J., Gehrke, J., Yiu, T., Flannick, J.: Sequential pattern mining using a bitmap representation. In: Proceedings of ACM SIGKDD International Conference on Knowledge Discovery and Data Mining (KDD 2002), pp. 429–435 (2002)
4. Burke, R.: Hybrid Recommender Systems: Survey and Experiments. User Modeling and User-Adapted Interaction 12, 331–370 (2002)
5. Capelle, M., Masson, C., Boulicaut, J.-F.: Mining Frequent Sequential Patterns under a Similarity Constraint. In: Yin, H., Allinson, N.M., Freeman, R., Keane, J.A., Hubbard, S. (eds.) IDEAL 2002. LNCS, vol. 2412, pp. 1–6. Springer, Heidelberg (2002)
6. Chang, J.H.: Mining weighted sequential patterns in a sequence database with a time-interval weight. Knowledge-Based Systems (2010) Available Online
7. Chen, E., Cao, H., Li, Q., Qian, T.: Efficient strategies for tough aggregate constraint-based sequential pattern mining. Information Sciences 178(6), 1498–1518 (2008)
8. Cheng, H., Yan, X., Han, J.: IncSpan: Incremental mining of sequential patterns in large databases. In: Proceedings of ACM SIGKDD International Conference on Knowledge Discovery and Data Mining (KDD 2004), pp. 527–532 (2004)
9. Cheng, H., Yan, X., Han, J., Hsu, C.-W.: Discriminative Frequent Pattern Analysis for Effective Classification. In: Proceedings of IEEE International Conference on Data Engineering (ICDE 2007), pp. 716–725 (April 2007)
10. Goyal, A., Bonchi, F., Lakshmanan, L.V.S.: Learning influence probabilities in social networks. In: Proceedings of ACM International Conference on Web Search and Data Mining (WSDM 2010), pp. 241–250 (2010)
11. Han, J., Pei, J., Mortazavi-Asl, B., Chen, Q., Dayal, U., et al.: FreeSpan: Frequent pattern-projected sequential pattern mining. In: Proceedings of ACM SIGKDD International Conference on Knowledge Discovery and Data Mining (KDD 2000), pp. 355–359 (2000)
12. Huang, C.-L., Huang, W.-L.: Handling sequential pattern decay: Developing a two-stage collaborative recommender system. ECRA 8(3), 117–129 (2009)
13. Huang, J.-W., Tseng, C.-Y., Ou, J.-C., Chen, M.-S.: A general model for sequential pattern mining with a progressive database. IEEE TKDE 20(9), 1153–1167 (2008)
14. Kum, H.-C., Pei, J., Wang, W., Duncan, D.: ApproxMAP: Approximate mining of consensus sequential patterns. In: Proceedings of SIAM Intl. Conf. on Data Mining, pp. 311–315 (2003)
15. Li, C., Lu, Y.: Similarity measurement of web sessions by sequence alignment. In: Procs. of IFIP Intl. Conf. on Network and Parallel Comp. Workshops (NPC 2007), pp. 716–720 (2007)
16. Lin, M.-Y., Hsueh, S.-C., Chang, C.-W.: Fast discovery of sequential patterns in large databases using effective time-indexing. Information Sciences 178(22), 4228–4245 (2008)

17. Lo, S.: Binary prediction based on weighted sequential mining method. In: Proceedings of International Conference on Web Intelligence (WI 2005), pp. 755–761 (2005)

18. Mitchell, T.: Machine Learning. McGraw Hill (1997)

19. Onuma, K., Tong, H., Faloutsos, C.: TANGENT: A novel, 'surprise me', recommendation algorithm. In: Proceedings of ACM SIGKDD International Conference on Knowledge Discovery and Data Mining (KDD 2009), pp. 657–665 (2009)

20. Pei, J., Han, J., Mortazavi-Asl, B., et al.: Mining sequential patterns by pattern-growth: The PrefixSpan approach. IEEE TKDE 16(11), 1424–1440 (2004)

21. Pei, J., Han, J., Wang, W.: Mining sequential patterns with constraints in large databases. In: Procs. of ACM Intl. Conf. on Info. and Knowl. Management (CIKM 2002), pp. 18–25 (2002)

22. Pei, J., Fu, A.W.-C., Lin, X., Wang, H.: Computing compressed multidimensional skyline cubes efficiently. In: Procs. of IEEE Intl. Conf. on Data Eng. (ICDE 2007), pp. 96–105 (2007)

23. Pyo, S., Kim, E., Kim, M.: Automatic recommendation of (IP)TV program schedules using sequential pattern mining. In: Adjunct Proceedings of EuroITV 2009, pp. 50–53 (2009)

24. Rendle, S., Freudenthaler, C., Schmidt-Thieme, L.: Factorizing personalized Markov chains for next-basket recommendation. In: Proceedings of International Conference on World Wide Web (WWW 2010), pp. 811–820 (2010)

25. Resnick, P., Varian, H.R.: Recommender Systems. Comms. of the ACM 40(3), 56–58 (1997)

26. Salton, G.: Automatic text processing: The transformation, analysis, and retrieval of information by computer. Addison-Wesley Longman Publishing (1989)

27. Saneifar, H., Bringay, S., Laurent, A., Teisseire, M.: S^2MP: Similarity measure for sequential patterns. In: Procs. of Australasian Data Mining Conf. (AusDM 2008), pp. 95–104 (2008)

28. Schifanella, R., Barrat, A., Cattuto, C., et al.: Folks in folksonomies: Social link prediction from shared metadata. In: Proceedings of ACM International Conference on Web Search and Data Mining (WSDM 2010), pp. 271–280 (2010)

29. Sequeira, K., Zaki, M.: Admit: Anomaly-based data mining for intrusions. In: Proceedings of ACM Intl. Conf. on Knowledge Discovery and Data Mining (KDD 2002), pp. 386–395 (2002)

30. Srikant, R., Agrawal, R.: Mining Sequential Patterns: Generalizations and Performance Improvements. In: Apers, P.M.G., Bouzeghoub, M., Gardarin, G. (eds.) EDBT 1996. LNCS, vol. 1057, pp. 3–17. Springer, Heidelberg (1996)

31. Torrey, L., Shavlik, J.: Transfer Learning. In: Handbook of Research on Machine Learning Applications. IGI Global (2009)

32. Yun, U.: A new framework for detecting weighted sequential patterns in large seq. databases. Knowledge-Based Systems 21, 110–122 (2008)

33. Zaki, M.J.: SPADE: An efficient algorithm for mining frequent sequences. Machine Learning 42(1/2), 31–60 (2001)

34. Zhou, B., Hui, S.C., Chang, K.: An intelligent recommender system using sequential web access patterns. In: Procs. of IEEE Conf. on Cybernetics and Intell. Sys., pp. 393–398 (2004)

35. Zobel, J., Moffat, A.: Inverted files for text search engine. ACM Comp. Surveys 38(2) (2006)

Scalable Top-k Keyword Search in Relational Databases

Yanwei Xu[1], Jihong Guan[1], and Yoshiharu Ishikawa[2,3,4]

[1] Department of Computer Science and Technology, Tongji University, Shanghai 201804, China
[2] Information Technology Center, Nagoya University, Nagoya-shi, 464-8601 Japan
[3] Graduate School of Information Science, Nagoya University, Nagoya-shi, 464-8601 Japan
[4] National Institute of Informatics, Chiyoda-ku, Tokyo, 101-8430 Japan

Abstract. *Keyword search in relational databases* has been widely studied in recent years because it does not require users neither to master a certain structured query language nor to know the complex underlying database schemas. There would be a huge number of valid results for a keyword query in a large database. However, only the top 10 or 20 most relevant matches for the keyword query –according to some definition of "Relevance"– are generally of interest. In this paper, we propose an efficient method for answering top-k keyword queries over relational databases. The proposed method is built on an existing scheme of keyword search on relational data streams, but incorporates the ranking mechanisms into the query processing methods and makes two improvements to support top-k keyword search in relational databases. Experimental results validate the effectiveness and efficiency of the proposed method.

Keywords: Relational databases, keyword search, top-k query.

1 Introduction

With the proliferation of text data available in relational databases, simple ways to exploring such information effectively are of increasing importance. *Keyword search in relational databases*, with which a user specifies his/her information need by a set of keywords, is a popular information retrieval method because the user needs to know neither a complex query language nor the underlying database schemas. Given l-keyword query $Q = \{w_1, w_2, \cdots, w_l\}$, the task of keyword search in a relational database is to find structural information constructed from tuples in the database [21].

Example 1. Consider a sample publication database shown in Fig. 1. Figure 1(a) shows the three relations *Papers*, *Authors*, and *Writes*. In the following, we use the initial of each relation name (P, A, and W) as its shorthand. There are two foreign key references: $W \rightarrow A$ and $W \rightarrow P$. Figure 1(b) illustrates the tuple connections based on the foreign key references. For the keyword query "James P2P" consisting of two keywords "James" and "P2P", there are six matched tuples in the database (underlined in Fig. 1(a)), which can be regarded as the query results. However, they can be joined with other tuples according to the foreign key references to form more meaningful results that are shown in Fig. 1(c). The arrows represent the foreign key references between tuples. Finding such results which are formed by the tuples containing the keywords is the task of keyword search in relational databases. As described later, results are often ranked by relevance scores evaluated by a certain ranking strategy. □

S.-g. Lee et al. (Eds.): DASFAA 2012, Part II, LNCS 7239, pp. 65–80, 2012.
© Springer-Verlag Berlin Heidelberg 2012

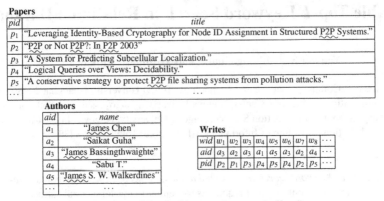

Papers

pid	title
p_1	"Leveraging Identity-Based Cryptography for Node ID Assignment in Structured P2P Systems."
p_2	"P2P or Not P2P?: In P2P 2003"
p_3	"A System for Predicting Subcellular Localization."
p_4	"Logical Queries over Views: Decidability."
p_5	"A conservative strategy to protect P2P file sharing systems from pollution attacks."
...	...

Authors

aid	name
a_1	"James Chen"
a_2	"Saikat Guha"
a_3	"James Bassingthwaighte"
a_4	"Sabu T."
a_5	"James S. W. Walkerdines"
...	...

Writes

wid	w_1	w_2	w_3	w_4	w_5	w_6	w_7	w_8	...
aid	a_3	a_2	a_3	a_1	a_5	a_3	a_2	a_4	...
pid	p_2	p_1	p_3	p_4	p_5	p_4	p_2	p_5	...

(a) Database (Matched keywords are underlined)

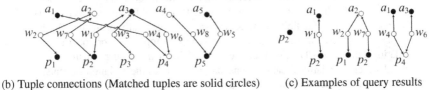

(b) Tuple connections (Matched tuples are solid circles)   (c) Examples of query results

Fig. 1. A sample database with a keyword query "James P2P"

There would be a huge number of valid results for a keyword query in a large database. However, only the top 10 or 20 most relevant matches for the keyword query –according to some definition of "Relevance"– are generally of interest [4]. [4] and [11] have tried to avoid exhaustive processing by introducing a top-k processing algorithm such as a pipelined algorithm. Although their algorithms can stop early before all the results are generated, they still suffer from a huge number of join checking which cannot produce results. In this paper, we describe an efficient method which can efficiently find the top-k results. Major contributions of this paper are:

- We incorporate the ranking mechanisms into the query processing methods of the existing query processing scheme of [14], and then the keyword query can be evaluated in pipelined way as in [4] and [11].
- For each tuple that contains keywords, a tree consisting of the tuples that can join it through the foreign key references is constructed through the query processing, which can highly accelerate the following processing involving that tuple.
- Extensive experiments are conducted to evaluate the proposed approach.

The rest of this paper is organized as follows. In Section 2 some basic concepts are introduced. Section 3 presents the details of the proposed method. Section 4 gives the experimental results. Section 5 discusses related work. Finally, in Section 6 we conclude this paper.

2 Preliminaries

We consider a relational database schema as a directed graph $G_S(V, E)$, called a schema graph, where V represents the set of relation schemas $\{R_1, R_2, \ldots\}$ and E represents the foreign key references between pairs of relation schemas. Given two relation schemas, R_i and R_j, there exists an edge in the schema graph, from R_j to R_i, denoted $R_i \leftarrow R_j$, if the primary key of R_i is referenced by the foreign key defined on R_j. For example, the schema graph of the publication database in Fig. 1 is *Papers* $\leftarrow$ *Writes* $\rightarrow$ *Authors*. A relation on relation schema R_i is an instance of R_i (a set of tuples) conforming to the schema, denoted $r(R_i)$. Below, we use R_i to denote $r(R_i)$ if the context is obvious.

The results of keyword queries in relational databases are a set of connected trees of tuples, each of which is called a *joint-tuple-tree* (*JTT* for short). A JTT represents how the *matched tuples*, which contain the specified keywords in their text attributes, are interconnected through foreign key references. Two adjacent tuples of a JTT, $t_i \in R_i$ and $t_j \in R_j$, are interconnected if they can be joined based on a foreign key reference defined on relational schema R_i and R_j. For example, the second JTT in Fig. 1(c) can be denoted as $p_2 \leftarrow w_1 \rightarrow a_3$ or $p_2 \bowtie w_1 \bowtie a_3$. To be a valid result of a keyword query Q, each leaf of a JTT is required to be a matched tuple. The number of tuples in a JTT T is called the *size* of T, denoted by $size(T)$.

Each JTT belongs to the results of a relational algebra expression, which is called a *candidate network* (CN) [4,11,17]. For example, the CN that corresponds to JTT $p_2 \leftarrow w_1 \rightarrow a_3$ is $P^Q \bowtie W \bowtie A^Q$, where $\bowtie$ represents a equi-join between relations. In the following, we also denote $P^Q \bowtie W \bowtie A^Q$ as $P^Q \leftarrow W \rightarrow A^Q$, the arrows represent the foreign key references. Relations in CNs are also called as *tuple sets*. Given a keyword query Q, the *query tuple set* R_i^Q of a relation R_i is defined as the set of matched tuples in R_i, and the *free tuple set* R_i^F is defined as the set of un-matched tuples in R_i. In Example 1, $P^Q = \{p_1, p_2, p_5\}$, $A^Q = \{a_1, a_3, a_5\}$, $P^F = \{p_3, p_4, \ldots\}$ and $A^F = \{a_2, a_4, \ldots\}$. If a relation R_i has no text attributes (e.g., relation W in Fig. 1), R_i is used to denote R_i^F. Hence, a CN corresponds to a join expression on tuple sets that produces JTTs as results, where each join clause $R_i^{QorF} \bowtie R_j^{QorF}$ corresponds to an edge $\langle R_i, R_j \rangle$ in the database schema graph. As the leaf nodes of JTTs must be matched tuples, the leaf nodes of CNs must be query tuple sets. The *size* of CN C, denoted as $size(C)$, is defined as the size of the JTTs it produces. A CN can be easily transformed into an equivalent SQL statement and executed by an RDBMS.[1]

When a keyword query Q is specified, all the non-empty query tuple sets are firstly computed using the full-text indices. Then they and the database schema are used to generate the set of valid CNs. There is always a constraint CN_{max} to restrict the maximum size of CNs. In Example 1, there are two non-empty query tuple sets P^Q and A^Q. If $CN_{max} = 5$, seven CNs are generated: $CN_1 = P^Q$, $CN_2 = A^Q$, $CN_3 = P^Q \leftarrow W \rightarrow A^Q$, $CN_4 = P^Q \leftarrow W \rightarrow A^Q \leftarrow W \rightarrow P^Q$, $CN_5 = P^Q \leftarrow W \rightarrow A^F \leftarrow W \rightarrow P^Q$, $CN_6 = A^Q \leftarrow W^1 \rightarrow P^Q \leftarrow W \rightarrow A^Q$ and $CN_7 = A^Q \leftarrow W \rightarrow P^F \leftarrow W \rightarrow A^Q$. Then, these CNs are evaluated by sending the corresponding SQL statements to the RDBMS for finding the query results. An efficient evaluation process is necessary because the number of candidate networks is huge. For example, [5] shows how generating a CN can reduce the cost of execution by exploiting reused common sub-expressions [6].

There would be a huge number of valid results for all the CNs. However, only the top 10 or 20 most relevant matches for the keyword query –according to some definition of

---

[1] For example, we can transform CN $P^Q \leftarrow W \rightarrow A^Q$ as: SELECT * FROM W w, P p, A a WHERE w.pid = p.pid AND w.aid = a.aid AND p.pid in (p_1, p_2, p_5) and a.aid in (a_1, a_3, a_5).

"Relevance"– are generally of interest [4]. Thus, many efforts have been dedicated to the top-k keyword search [1,4,11,17]. The problem of *top-k keyword search* is to find the top-k JTTs based on a certain scoring function. In this paper, we adopt the scoring method employed in [4], which is an ordinary ranking strategy in the information retrieval area. The following function $score(T, Q)$ is used to score JTT T for query Q, which is based on the TF-IDF weighting scheme:

$$score(T, Q) = \frac{\sum_{t \in T} tscore(t, Q)}{size(T)} \tag{1}$$

where $t \in T$ is a tuple (a node) contained in T, $tscore(t, Q)$ is the *tuple score* of t with regard to Q defined as follows:

$$tscore(t, Q) = \sum_{w \in t \cap Q} \frac{1 + \ln(1 + \ln(tf_{t,w}))}{(1 - s) + s \cdot \frac{dl_t}{avdl}} \cdot \ln\left(\frac{N}{df_w + 1}\right) \tag{2}$$

where $tf_{t,w}$ is the *term frequency* of keyword W in tuple t, df_w is the number of tuples in relation $r(t)$ (the relation corresponds to tuple t) that contain W. df_w is interpreted as the *document frequency* of W. dl_t represents the size of tuple t, i.e., the number of letters in t, and is interpreted as the *document length* of t. N is the *total number* of tuples in $r(t)$, $avdl$ is the *average tuple size* (*average document length*) in $r(t)$, and s ($0 \le s \le 1$) is a constant which usually be set to 0.2. The function in Eq. (1) has the property of *tuple monotonicity*, which is defined as: for any two JTTs $T = t_1 \bowtie t_2 \bowtie \ldots \bowtie t_l$ and $T' = t'_1 \bowtie t'_2 \bowtie \ldots \bowtie t'_l$ generated from the same CN C, if for any $1 \le i \le l$, $tscore(t, Q) \le tscore(t', Q)$, then we have $score(T, Q) \le score(T', Q)$.

Table 1 shows the tuple scores of the six matched tuples in Example 1. We suppose all the matched tuples are shown in Fig. 1, and the numbers of tuples of relations A and P are 150 and 170, respectively. Therefore, the top-3 results are $T_1 = p_2$ ($score = 7.04$), $T_2 = a_1$ ($score = 4.00$) and $T_3 = p_2 \leftarrow w_1 \rightarrow a_3$ ($score = 3.48$).

Table 1. Tuple scores of tuples of P^Q and A^Q

Tuple	p_1	p_2	p_5	a_1	a_3	a_5
$tscore$	3.28	7.04	3.33	4.00	3.40	3.36

3   Top-k Keyword Search

3.1   Evaluating CNs Using Lattice

KDynamic [14,16] formalizes each CN as a rooted tree, whose root is defined to be the node r such that the maximum path from r to all leaf nodes is minimized[2]. Figure 2(a) shows the rooted tree of CN_6. Each node V_i in the rooted trees is associated with an output buffer, denoted by $V_i.output$, which contains the tuples of V_i that can join at least one tuple in the output buffer of its each child. Tuples in the output buffer are called the

---

[2] Note that the CN defined in KDynamic has some differences with ours.

output tuples of the node. Thus, each output tuple of the root can form JTTs with the output tuples of its descendants. Tuples of CNs are processed in a two-phase approach in the rooted tree. In the filter phase, as illustrated in Fig. 2(a), when a tuple t is processed at the node W^1, KDynamic uses selections to check if (1) t can join at least an output tuple of each child of W^1; and (2) t can join at least an output tuple of the ancestors of W^1. The tuples that can not pass the checks are pruned; otherwise, in the join phase (shown in Fig. 2(b)), a joining process is initiated from each output tuple of the root node that can join t, in a top-down manner, to find the JTTs involving t. KDynamic achieves full tuple reduction by pruning the tuples that cannot form JTTs, and thus the join operations can always produce results. In order to share the computation cost among CNs, all the rooted trees are compressed into a $\mathcal{L}$-Lattice by collapsing their common subtrees. Thus, the output tuples of a node are shared by more than one nodes, among different CNs. Figure 2(c) shows the lattice of the seven CNs. We use V_i^Q to denote a node of query tuple set particularly. The dual edges between two nodes, for instance, V_1^Q and V_5, indicate that V_5 is a dual child of V_1^Q.

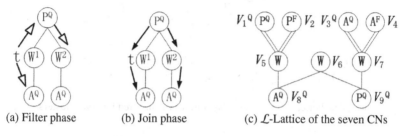

(a) Filter phase   (b) Join phase   (c) $\mathcal{L}$-Lattice of the seven CNs

Fig. 2. Query processing in KDynamic

3.2   Candidate Network Clustering

According to Eq. (1), relevance scores of JTTs of different CNs have great differences. For example, relevance scores of JTTs of CN_5 and CN_7 are smaller than that of JTTs of CN_3 due to their large sizes. And then the same tuple set can have different numbers of processed tuples in different CNs if they are evaluated separately. If the seven CNs are evaluated separately, A^Q of CN_7 would have no processed tuples. However, in the lattice, a node V_i^Q can be shared by multiple CNs. For instance, the node V_8^Q in Fig. 2(c) is shared by CN_2, CN_3, CN_6, and CN_7. We use $V_i.CN$ to denote the set of CNs that node V_i belongs to. Then, when processing a tuple t at node V_8^Q, t is processed in all the CNs in $V_8^Q.CN$; hence some results of CN_7 can be computed, which would have very small relevance scores and cannot contribute to the top-k results.

The essence of the above problem is that CNs have different potentials in producing the top-k results. Thus, the CNs that have very different such potentials can not share tuple sets. The optimal method is merely to share the tuple sets which have the same set of processed tuples if CNs are evaluated separately. However, we can not get these sets without evaluating them. As an alternative, we attempt to estimate these sets according to two heuristic rules:

- If $Max(C) = \dfrac{\sum_{1 \le i \le m} C.R_i^Q .t_1.tscore^u}{size(C)}$, which indicates the maximum $score^u$ of JTTs
  that C can produce, is high, tuple sets of C have more processed tuples.
- If two CNs have the same $Max(C)$ values, tuple sets of the CN with larger size have
  more processed tuples.

Therefore, we can use $Max(C) \cdot \ln(size(C))$ to measure the potential of a CN in producing
top-k results, where $\ln(size(C))$ is used to normalize the effect of CN sizes. Then, we can
cluster the CNs using their $Max(C) \cdot \ln(size(C))$ values, and only the subtrees of CNs in
the same cluster can be collapsed when constructing the lattice. For instance, $Max(C) \cdot$
$\ln(size(C))$ of the seven CNs are: 5.15, 2.93, 5.39, 6.84, 5.32, 5.70 and 3.03; hence
they can be clustered into two clusters: $\{CN_2, CN_7\}$ and $\{CN_1, CN_3, CN_4, CN_5, CN_6\}$.
Figure 3 shows the lattice after they are clustered.

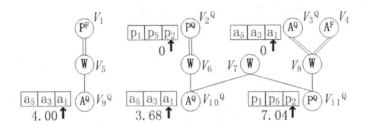

Fig. 3. The lattice of the seven CNs if they are clustered into two clusters

The CNs are clustered using the K-means clustering algorithm [10], which needs an
input parameter to indicate the number of expected clusters. We use $Kmean$ to indicate
this parameter. The value of $Kmean$ represents the trade-off between sharing the computation cost among CNs and considering their different potentials in producing top-k
results. The CNs is not clustered when $Kmean = 1$, then the computation cost is shared
at the maximum extent. When $Kmean = MAX$, all the CNs are evaluated separately. As
shown in the experimental section, clustering the CNs can highly improve the efficiency
in computing the initial top-k results, and the optimal $Kmean$ depends on CN_{max}.

3.3   Pipelined Evaluation of the Lattice

In order to find the top-k results in a pipelined way, we first sort tuples in each query
tuple set $V_i^Q.R^Q$ in non-increasing order of $tscore$. We use $V_i^Q.cur$ to denote the current
tuple such that the tuples before its position are all processed, and we use $V_i^Q.cur \leftarrow$
$V_i^Q.cur + x$ to move $V_i^Q.cur$ to the next x position. Initially, for each node V_i^Q in $\mathcal{L}$,
$V_i^Q.cur$ is set as the top tuple in $V_i^Q.R^Q$. In Fig. 3, $V_i^Q.cur$ of the four nodes are denoted
by arrows. Note that, for a node V_i that is of a free tuple set R_i^F, we regard all its tuples
as processed tuples for all the times.

The key to evaluate queries in a pipelined way in [4] and [11] is to compute an upper
bound for the relevance score of the un-found results. For a keyword query Q, given a

CN C, let the set of query tuple sets of C be $\{R_1^Q, R_2^Q, \ldots, R_m^Q\}$. For each tuple $R_i^Q.t_j$, [4] computes the upper bound score for all the JTTs of C that contain $R_i^Q.t_j$ as:

$$\overline{score}(C.R_i^Q.t_j, Q) = \frac{t_j.tscore + \sum_{1 \le i' \le m \wedge i' \ne i} C.R_{i'}^Q.t_1.tscore}{size(C)} \tag{3}$$

where $C.R_{i'}^Q.t_1$ indicates the top-most tuple of query tuple set $C.R_{i'}^Q$. Using this equation, for each node V_i^Q, this paper computes the maximum $score$ of the found JTTs by processing the un-processed tuples at V_i^Q as:

$$\overline{score}\left(V_i^Q, Q\right) = \begin{cases} 0, & \text{a child node of } V_i^Q \text{ has an empty output buffer,} \\ \max_{C \in V_i^Q.CN}\left(\overline{score}\left(C.V_i^Q.cur, Q\right)\right), & \text{otherwise} \end{cases} \tag{4}$$

If a child of V_i^Q has an empty output buffer, processing any tuple at V_i^Q can not produce JTTs; hence $\overline{score}\left(V_i^Q, t_j, Q\right) = 0$ in such cases, which chokes the tuple processing at V_i^Q until all its child nodes have non-empty output buffers. This property of $\overline{score}\left(V_i^Q, Q\right)$ can be seen as our version of the event-driven evaluation in KDynamic, which is firstly proposed in S-KWS [13] and can noticeably reduce the query processing cost. In Fig. 3, $\overline{score}\left(V_i^Q, Q\right)$ values of the four V_i^Q nodes are shown next to the arrows. For example, $\overline{score}\left(V_9^Q, Q\right)$ is computed as $\max_{C \in \{CN_2, CN_7\}}\left(\overline{score}\left(C.A^Q.a_1, Q\right)\right) = 4.00$.

Algorithm 1 outlines our pipelined algorithm of evaluating lattice $\mathcal{L}$ to find the top-k results. Lines 1-1 are the initialization steps, which sort tuples in each query tuple set and initialize each $V_i^Q.cur$. Then in each while iteration (lines 1-1), $step$ un-processed tuples in the node V^Q which maximizes $\overline{score}\left(V_i^Q, Q\right)$ are processed. Processing tuples at a node is done by calling the procedure $Insert$. Algorithm 1 stops when $\max_{V_i^Q \in \mathcal{L}} \overline{score}(V_i^Q, Q)$ is not larger than the relevance score of the top-k-th found result because no results with larger relevance scores can be found in the further evaluation. The procedure $Insert(V_i, \mathbb{S})$ is firstly provided in KDynamic, which updates the output buffers for V_i (line 1) and all its ancestors (lines 1-1), and finds all the JTTs containing tuples of $\mathbb{S}'$ by calling the procedure $EvalPath$ (line 1), which is firstly provided by KDynamic too. In KDynamic, the second parameter of $Insert$ and $EvalPath$ is one tuple. As shown by the BP algorithm of [11], processing tuples in batch can achieve high efficiency due to the reduced numbers of database accesses. Hence, tuples are processed in batch in Algorithm 1: $step$ tuples are processed when $Insert$ is called in line 1, and $EvalPath$ also handles a set of tuples. However, it is not the larger $step$ is, the higher efficiency of Algorithm 1 has. Because a larger $step$ can result in un-necessary tuple processing in some lattice nodes. We will experimentally to study how to select a proper $step$ value. The recursive procedure $EvalPath(V_i, \mathbb{S}, path)$ constructs JTTs using the output tuples of V_i's descendants that can join tuples in $\mathbb{S}$. The stack $path$ is used to record the join sequence for reducing the join cost (see line 1).

Example 2. Figure 4 shows the lattice after finding the top-3 results. Suppose $step = 1$, then in the first round, tuple $V_{11}^Q.p_2$ is processed by calling $Insert(V_{11}^Q, \{p_2\})$. Since V_{11}^Q is the root of CN_1, $EvalPath(V_{11}^Q, \{p_2\})$ is called and JTT $T_1 = p_2$ is found. Then, $Insert(V_7, \{w_1, w_7\})$ and $Insert(V_8, \{w_1, w_7\})$ are called. $V_7.output$ is not updated because $V_{10}^Q.output = \emptyset$; $V_8.output$ is updated to $\{w_1, w_7\}$. And then, for the two father

---

Algorithm 1. *EvalStatic-Pipelined* (lattice $\mathcal{L}$, the top-k value k, integer *step*)

---

1 $topk \leftarrow \emptyset$: the priority queue for storing found JTTs ordered by *score*;
2 Sort tuples of each $V_i^Q.R^Q$ in non-increasing order of *tscore*";
3 **foreach** node V_i^Q in $\mathcal{L}$ **do** let $V_i^Q.cur \leftarrow V_i^Q.R^Q.t_1$;
4 **while** $\max_{V_i^Q \in \mathcal{L}} \overline{score}(V_i^Q, Q) > topk[k].score$ **do**
5 $\quad$ Suppose $\overline{score}(V_0^Q, Q) = \max_{V_i^Q \in \mathcal{L}} \overline{score}(V_i^Q, Q)$;
6 $\quad$ $path \leftarrow \emptyset$;$\qquad\qquad\qquad$ // A stack which records the join sequence
7 $\quad$ $Insert(V_0^Q, \{V_0^Q.cur, \cdots, V_0^Q.cur + step - 1\})$; // Processing *step* tuples at V_0^Q
8 $\quad$ $V_0^Q.cur \leftarrow V_0^Q.cur + step$;
9 Output the first k results in *topk*;
10 $\mathcal{L}.\theta \leftarrow topk[k].score$;

11 Procedure *Insert*(lattice node V_i, set of tuples $\mathbb{S}$)
12 Let $\mathbb{S}' \leftarrow \{t | t \in \mathbb{S}, t$ can join at least one outputted tuple of every child of V_i }；
13 $V_i.output \leftarrow V_i.output \bigcup \mathbb{S}'$;
14 **if** $\mathbb{S}' \neq \emptyset$ **then**
15 $\quad$ Push $(V_i, \mathbb{S}')$ to *path*;
16 $\quad$ **if** V_i is a root node **then** $topk \leftarrow topk \bigcup EvalPath(V, \mathbb{S}', path)$;
17 $\quad$ **foreach** father node of V_i, $V_{i'}$ in $\mathcal{L}$ **do**
18 $\quad\quad$ Let $\mathbb{S}''$ be the set of processed tuples of $V_{i'}$ that can join tuples in $\mathbb{S}'$;
19 $\quad\quad$ $Insert(V_{i'}, \mathbb{S}'')$;
20 $\quad$ Pop $(V, \mathbb{S}')$ from *path*;
21

22 Procedure *EvalPath*(lattice node V_i, set of tuples $\mathbb{S}$, stack *path*)
23 $\mathcal{T} \leftarrow \mathbb{S}$;$\qquad\qquad\qquad\qquad$ // The set of found JTTs
24 **foreach** child node of V_i, $V_{i'}$ in $\mathcal{L}$ **do**
25 $\quad$ **if** $V_{i'} \in path$ **then** $\mathbb{S}' \leftarrow$ the set of output tuples of $V_{i'}$ that are stored in *path*;
26 $\quad$ **else** $\mathbb{S}' \leftarrow$ the set of output tuples of $V_{i'}$ that can join tuples in $\mathbb{S}$;
27 $\quad$ $\mathcal{T}' \leftarrow EvalPath(V_{i'}, \mathbb{S}', path)$;
28 $\quad$ $\mathcal{T} \leftarrow \mathcal{T} \bowtie \mathcal{T}'$;$\qquad\qquad$ // Join the JTTs in the two sets
29 **return** $\mathcal{T}$;

---

nodes of V_8, V_3^Q and V_4, $V_3^Q.output$ is not updated since V_3^Q has no processed tuples, and $V_4.output$ is set as $\{a_2\}$ because there is only one tuple a_2 in A^F that can join w_1

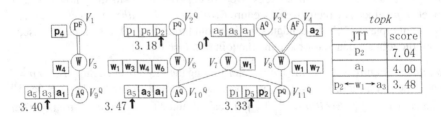

Fig. 4. The lattice after finding the top-3 results (output tuples are shown in bold)

and w_7. Since V_4 is the root node of CN_5, $EvalPath(V_4, \{a_2\}, path)$ is called but the found JTT $p_2 \leftarrow w_7 \rightarrow a_2 \leftarrow w_7 \rightarrow p_2$ is not a valid result. After processing $V_{11}^Q.p_2$, $\overline{score}\left(V_3^Q, Q\right) = 3.61$ and $\overline{score}\left(V_{11}^Q, Q\right) = \overline{score}(CN_1.P^Q.p_5, Q) = 3.33$. In the second round, $Insert(V_9^Q, a_1)$ is called, $\cdots$. Lastly, Algorithm 1 stops because $topk[3].score$ is larger than all the $\overline{score}\left(V_i^Q, Q\right)$ values.

3.4 Caching Joined Tuples

In Algorithm 1, procedure $Insert$ may be called multiple times upon multiple nodes for the the same tuple. And the procedure $EvalPath$ may also be called multiple times for the same tuple in procedure $Insert$. The core of these two procedures are the $select$ $operations$. For example, line 1 selects the tuples that can join tuples of $\mathbb{S}$ from the output buffer of each child node of V_i. Although such select operations can be done efficiently by the RDBMS using indexes, the cost is high due to the large number of database accesses. For example, in our experiments, for a tuple t, the maximal number of database accesses can be up to several hundred.

In this paper, the selections in $Insert$ and $EvalPath$ are done efficiently by caching the joined tuples for each tuple. Algorithm 2 shows our procedure to find the tuples in $\mathbb{S}$ that can join at least one output tuple of node V_i, which is called in line 1 of procedure $Insert$. For each tuple t in $\mathbb{S}$, if the joining tuples of relation R_i are not cached, they are queried from the database and are stored into t in line 2. The procedures of doing the selections in line 1 of $Insert$, and line 1 of $EvalPath$ are also designed in this pattern, which are omitted due to the space limitation. Since the two procedures are called recursively, for each tuple t, a tree rooted at t and consist of all the tuples that can join t is created temporarily, which can be seen as the cached localization information of t and is denoted as $\mathcal{T}$. Since $\mathcal{T}$ of different tuples can share the same tuples, fractions of the database graph are created.

Algorithm 2. $CanJoinOneOutputTuple$(lattice node V_i, set of tuples $\mathbb{S}$)

---

1 Let R_i be the relation corresponding to the tuple set of V_i;
2 Let $\mathbb{S}' \leftarrow \{t | t \in \mathbb{S}, t$ does not store the joining tuples of relation $R_i\}$;
3 **if** $\mathbb{S}' \neq \emptyset$ **then** Query the joining tuples of relation R_i for tuples in $\mathbb{S}'$;
4 **foreach** $Tuple\ t\ in\ \mathbb{S}$ **do**
5 | **if** the stored joining tuples of relation R_i in t has empty intersection with $V_i.output$
   | **then** Remove t from $\mathbb{S}$;
6 **return** $\mathbb{S}$;

Assume procedure $Insert$ is called three times at V_3^Q, V_9^Q and V_{10}^Q for a tuple a_0, which would incur at most seven selections denoted by arrows in the left part of Fig. 5. For instance, the arrow form V_3^Q to V_8 selects the output tuples of V_8 that can join a_0. There are three selections denoted by dashed arrows because they would not be done if the results of the three selections: form V_9^Q to V_5, form V_{10}^Q to V_6 and form V_{10}^Q to V_7, are empty. If both the two selections, from V_9^Q to V_5 and from V_5 to V_1, have non-empty results, *EvalPath* is called and would incur the two selections denoted by dotted arrows in Fig. 5. The right part of Fig. 5 shows the created $\mathcal{T}$ for the tuple w_0, where tuples in the dashed rectangle are queried in the dashed arrows and tuples in the dotted rectangle are queried in *EvalPath*.

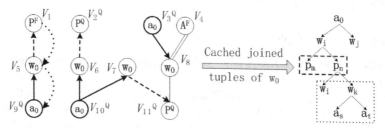

Fig. 5. Selections done in *Insert* and *EvalPath* and the cached joined tuples for a tuple a_0 of A^Q

$\mathcal{T}$ is created on-the-fly, i.e., along the execution of procedures *Insert* and *EvalPath*, and its depth is determined by the recursion depths of them. Therefore, $\mathcal{T}$ is not complete in Fig. 5. The maximum recursion depth of procedure *Insert* is $\lceil \frac{CN_{max}}{2} \rceil$ [14]. And the recursion depth of procedure *EvalPath* is $\lfloor \frac{CN_{max}}{2} \rfloor$. Hence, the height of $\mathcal{T}$ is bounded by CN_{max}. If we use M_1 and M_2 to indicate the maximum number of adjacent relations that each relation R_i can have and the maximum number of tuples that a tuple of R_i can join in its each adjacent relation, respectively. Note that M_1 and M_2 are often rather small compared to the number of CNs. Then in the worst case, there are tuples of M_1^l relations in level l (the level of the root is 0) of $\mathcal{T}$; and for each relation, there are M_2^l tuples. Hence, the total number of tuples in $\mathcal{T}$ is:

$$O\left(\sum_{l=1}^{CN_{max}-1} M_1^l \cdot M_2^l \right) = O\left((M_1 \cdot M_2)^{CN_{max}} \right) \qquad (5)$$

In practice, the height of $\mathcal{T}$ is smaller than $\lceil \frac{CN_{max}}{2} \rceil$ for most of the tuples. We use n to denote the number of lattice nodes that the procedure *Insert* or *EvalPath* is called for a tuple t. In the best case, we only need to query tuples from the database in the first time of calling *Insert*, and then the cached tuple in T can be reused. The cost of doing the select operations is the majority of the cost of executing *Insert* and *EvalPath*; hence the total time cost to process t can be reduced to $O(1/n)$, compared to KDynamic.

4   Experimental Study

We conducted extensive experiments to test the efficiency of our methods. We use the DBLP dataset[3]. The downloaded XML file is decomposed into relations. We retrieve the data in the XML file sequentially until number of tuples in the relations reach the numbers shown in Table 2. The DBMS used is MySQL (v5.1.44) with the default "Dedicated MySQL Server Machine" configuration. All the relations use the MyISAM storage engine. Indexes are built for all primary key and foreign key attributes, and full-text

---

[3] http://dblp.mpi-inf.mpg.de/dblp-mirror/index.php/

indexes are built for all text attributes. All the algorithms are implemented in C++. We conducted all the experiments on a 2.53 GHz CPU and 4 GB memory PC running Windows 7.

Table 2. Tuple numbers of relations

Papers	PaperCite	Write	Authors	Proceedings	ProcEditors	ProcEditor
354,804	19,438	899,565	320,579	3,904	5,208	8,439

We use the following five parameters in the experiments: (1) k: the top-k value; (2) l: the number of keywords in a query; (3) IDF: the $\frac{df_w}{N}$ value of Eq. (3); (4) CN_{max}: the maximum size of the generated CNs; (5) $Kmean$: the number of clusters of CNs; and (6) $step$: the number of tuples being processed one time in Algorithm 1. The parameters with their default values (bold) are shown in Table 3. The keywords selected are listed in Table 4 with their IDF values, where the keywords in bold fonts are keywords popular in author names. For each l value, ten queries are constructed by selecting l keywords from the set of keywords of $IDF = 0.013$. We run the Algorithm 1 on different values of each parameter while keeping the other five parameters in their default values. Ten top-k queries are selected for each combinations of parameters, and report the average time cost (*Time*) and number of computed JTTs (*#R*).

Table 3. Parameters

Name	Values
k	100, **200**, 250, 300
l	2, 3, 4, 5
IDF	0.003, 0.007, **0.013**, 0.025
CN_{max}	4, 5, **6**, 7
$Kmean$	1, 4, 8, **14**, 20, MAX
$step$	1, 50, 100, **200**, 400

Table 4. Keywords and their *IDF* values

Keywords	IDF
ATM, collaboration, cluster, Java, navigation, ontology, privacy, QoS, scal-able, Spatial **Charles**, **Eric**	0.004
embedded, fuzzy, genetic, Intenet, machine, mining, semantic, sensor, video, XML, **James**, **Zhang**	0.007
adaptive, architecture, database, evaluation, mobile, oriented, optimization, process, security, simulation, wireless, **John**, **Wang**	0.013
algorithm, design, distributed, information, learning, networks, performance, software, time, web, **David**, **Michael**	0.025

Figure 6 shows how the two measures change while varying the five parameters. The values of *#R* are all plotted on the right Y-axis. Figure 6(a), (b) and (c) show that the two measures all increase while k, idf and CN_{max} grow. However, they do not show rapid increase in Fig. 6(a), (b) and (c), which imply the good scalability of our method. Figure 6(d) shows that the effect of l seems more complicated: all the two measures may decrease when l increases, and they even both achieve the minimum values when $l = 5$. This is because the probability that the keywords to co-appear in a tuple and the matched tuples can join is high when the number of keywords is large. Therefore, there are more JTTs that have high relevance scores, which results in larger θ and small values of the two measures. Figure 6(e) presents the importance of processing tuples in batch (or block), due to the highly reduced number of database accesses. Because $\overline{score}\left(V_i^Q, Q\right)$ is computed using the first un-processed tuple, larger values of $step$ can result in more un-necessary tuple processing at node V_i^Q. Hence, as can be seen from

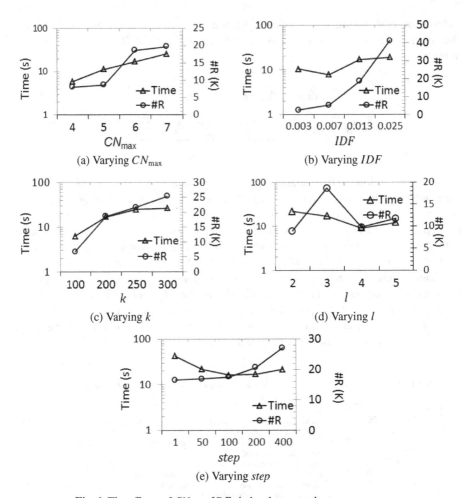

Fig. 6. The effects of CN_{max}, IDF, k, l and $step$ on the two measures

Fig. 6(e), #R increases while *step* grows, and the time cost increases while *step* grows from 200 to 400.

In Fig. 7a, we draw the changes of the two measures while varying *Kmean* when $CN_{max} = 6$ (indicated by "Time (6)" and "#R (6)"), and the changes of the time cost while varying *Kmean* when CN_{max} is 4, 5 and 7 (indicated by "Time (4)", "Time (5)" and "Time (7)", respectively). Since the results of the *K*-means clustering may be affected by the starting condition [10], for each *Kmean* value, we run Algorithm 1 five times on different starting condition for each keyword query and report the average result. Note that, KDynamic corresponding to *Kmean* = 1 since there is no CN clustering in its method. From Fig. 7a, we can find that clustering the CNs can considerably improve the efficiency of computing the top-*k* results. The two meaures decreases quickly while *Kmean* growing from 1 to 14. However, increasing *Kmean* from 14 to *MAX* can not distinctly reduce the time cost, because the CNs that have big differences in their $Max(C) \cdot \ln(size(C))$ values have been clustered into different clusters when

Kmean > 14. When *Kmean* grows from 14 to *MAX* (i.e., the CNs are evaluated separately), the time cost changes differently on different CN_{max} values. When CN_{max} is 6 or 7, the time cost is increased. When CN_{max} = 5, the time cost unchange. When CN_{max} = 4, the time cost is decreased to the minimum value. This is because the number of CNs is small when CN_{max} is 4 and 5, but is large when CN_{max} is 6 and 7; hence sharing the time cost among CNs can achieve improvement when CN_{max} is 6 and 7. Therefore, for a small CN_{max} value, the CNs can be evaluated separately, each as a rooted tree; and for a large CN_{max} value, the lattice is constructed while clustering the CNs on *Kmean* = 14. It worth noting that #R continually decreases as *Kmean* grows, which implies the effectiveness of clustering CNs using $Max(C) \cdot \ln(size(C))$ values.

Figure 7b compares the time cost of computing the initial top-k results of Algorithm 1, denoted by "LP", with that of the *Block pipeline (BP)* algorithm of SPARK (which is the state-of-art top-k keyword search algorithm [20]) and KDynamic, respectively, while varying CN_{max}. Figure 7b shows that, compared to SPARK, Algorithm 1 and KDynamic are more efficient in finding the top-k results, because evaluating the CNs using the lattice can achieve complete reduction since all the output tuples of the root nodes can form JTTs [14]. The time costs of KDynamic in Fig. 7b are all obtained when *Kmean* = 14. Hence, the difference between our approach and KDynamic reflects the effect of caching the joined tuples. We can see that caching the joined tuples highly improves the efficiency of computing the top-k results. More importantly, the improvement increases as CN_{max} grows. The reason is that when CN_{max} grows, the number of lattice nodes at which the procedure *Insert* is called for each tuple increases exponentially; hence the saved cost for storing the joined tuples increases as CN_{max} grows.

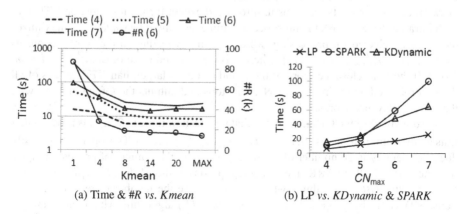

(a) Time & #R vs. *Kmean*   (b) LP *vs. KDynamic & SPARK*

Fig. 7. Comparison in calculating top-k results

5 Related Work

Keyword search in relational databases has attracted substantial research effort in recent years, which can be categorized into two approaches. The *graph-based methods* [1,3,7,8,9,15] model and materialize the entire database as a directed graph where the nodes are relational tuples and the directed edges are foreign key references between tuples. Fig. 1(b) shows such a database graph of the example database. Then for each keyword query, they find a set of structures (either Steiner trees [1], distinct rooted trees [7], r-radius Steiner graphs [8], or multi-center subgraphs [15]) from the database graph. For the details, please refer to the survey papers [6,21]. The *schema-based approaches* [2,4,5,11,12,14,17,18,19] in this area utilize the database schema to generate SQL queries. After receiving a keyword query, they first utilize the database schema to generate a set of *candidate networks* (*CN*s), which can be interpreted as select-project-join views and all have explicit meanings. Then, these CNs are evaluated by sending the corresponding SQL statements to the RDBMS to find JTTs. A data graph cannot exploit the semantics of the underlying database schema directly. Another drawback of the data graph model is that a graph of the tuples must be materialized and maintained; therefore, it may not be scalable when maintaining a large size database [6]. This paper adopts the schema-based framework for query processing, but materializes small fractions of the entire database graph in the process of query processing.

There would be a huge number of valid results for a keyword query in a large database. However, only the top 10 or 20 most relevant matches for the keyword query –according to some definition of "Relevance"– are generally of interest [4]. DISCOVERII [4], SPARK [11] and SPARKII [12] efficiently execute top-k queries by avoiding the creation of all the query results. DISCOVERII proposed the *Global-Pipelined* (*GP*) algorithm. For a keyword query Q, given a CN C, let the set of query tuple sets of C be $\{R_1^Q, R_2^Q, \ldots, R_m^Q\}$. Tuples in each R_i^Q are sorted in non-increasing order of their scores computed by Eq. (2). For each tuple $R_i^Q.t_j$, the upper bound score for all the JTTs of C that contain $R_i^Q.t_j$ is computed using Eq. (3). Algorithm GP initially mark all tuples in each tuple set as un-processed except for the top-most one. Then in each iteration (one round), the un-processed tuple, assume it be $C_0.R_s^Q.t_{pro}$, maximizes the $\overline{score}$ is selected for processing, which is done by testing all the combinations as $(t_1, t_2, \ldots, t_{s-1}, R_s^Q.t_{pro}, t_{s+1} \ldots, t_m)$, where t_i is a processed tuple of $C_0.R_i^Q$ ($1 \le i \le m$, $i \ne s$). If the k-th relevance score of the found results is larger than $\overline{score}$ values of all the un-processed tuples in all the CNs, GP stops and outputs the k found results with the largest relevance scores.

One drawback of the GP algorithm is that when a new tuple $C.R_s^Q.t_{pro}$ is processed, it tests all the combinations as $(t_1, t_2, \ldots, t_{s-1}, R_s^Q.t_{pro}, t_{s+1} \ldots, t_m)$. This operation is costly due to extremely large number of combinations when the number of processed tuples becomes large [20]. SPARK and SPARKII proposed the *Skyline-Sweeping* (*SS*) and *Block-pipeline* (*BP*) algorithms, which highly reduce the number of tested combinations. SPARKII also proposed the *Tree-pipeline* (*TP*) algorithm, which can share the computational cost among CNs in some extent, using the *binary decompositions* of them. However, SPARK and SPARKII still can not avid testing a huge number of combinations which cannot produce results.

This paper incorporates the ranking mechanisms into the query processing schema of KDynamic, which was introduced in Section 3.1. We adopt the basic idea of the pipelined query evaluation method of DISCOVERII, SPARK and SPARKII. The novelties of our method can be summarized as:

- The CNs are clustered according to their potentials in producing the top-*k* results;
- A pipelined evaluation method of the lattice is provided; and
- Fractions of the database graph are materialized to accelerate the query processing.

6 Conclusion

In this paper, we have studied the problem of finding the top-*k* results in relational databases. We adopt an existing scheme of finding all the results in a relational database stream, but incorporate the ranking mechanisms in the query processing methods and make several improvements that can facilitate efficient top-*k* keyword search. Hence, the keyword query can be evaluated in a pipelined way and the evaluation process can stop early after finding the top-*k* result. The experimental results validate the effectiveness and efficiency of the proposed method. Therefore, it can be used to solve the problem of answering top-*k* keyword search in relational databases.

Acknowledgements. The authors would like to thank Prof. Shuigeng Zhou and Ph.D. Fengrong Li for their revision and comments on the manuscript. This research was partly supported by the National Natural Science Foundation of China (NSFC) under grant No. 60873040. Jihong Guan was also supported by the "Shu Guang" Program of Shanghai Municipal Education Commission and Shanghai Education Development Foundation. Yoshiharu Ishikawa was supported by KAKENHI (22300034).

References

1. Aditya, B., Bhalotia, G., Chakrabarti, S., Hulgeri, A., Nakhe, C., Parag: BANKS: Browsing and keyword searching in relational databases. In: VLDB, pp. 1083–1086 (2002)
2. Agrawal, S., Chaudhuri, S., Das, G.: DBXplorer: A system for keyword-based search over relational databases. In: ICDE, pp. 5–16 (2002)
3. He, H., Wang, H., Yang, J., Yu, P.S.: Blinks: ranked keyword searches on graphs. In: ACM SIGMOD, pp. 305–316. ACM, New York (2007)
4. Hristidis, V., Gravano, L., Papakonstantinou, Y.: Efficient IR-style keyword search over relational databases. In: VLDB, pp. 850–861 (2003)
5. Hristidis, V., Papakonstantinou, Y.: DISCOVER: Keyword search in relational databases. In: VLDB, pp. 670–681 (2002)
6. Jaehui, P., Sang-goo, L.: Keyword search in relational databases. Knowledge and Information Systems 26(2), 175–193 (2011)
7. Kacholia, V., Pandit, S., Chakrabarti, S., Sudarshan, S., Desai, R., Karambelkar, H.: Bidirectional expansion for keyword search on graph databases. In: VLDB, pp. 505–516 (2005)
8. Li, G., Ooi, B.C., Feng, J., Wang, J., Zhou, L.: EASE: An effective 3-in-1 keyword search method for unstructured, semi-structured and structured data. In: ACM SIGMOD, pp. 903–914 (2008)
9. Li, G., Zhou, X., Feng, J., Wang, J.: Progressive keyword search in relational databases. In: ICDE, pp. 1183–1186 (2009)
10. Lloyd, S.P.: Least squares quantization in PCM. IEEE Transactions on Information Theory 28, 129–136 (1982)
11. Luo, Y., Lin, X., Wang, W., Zhou, X.: SPARK: Top-k keyword query in relational databases. In: ACM SIGMOD, pp. 115–126 (2007)

12. Luo, Y., Wang, W., Lin, X., Zhou, X., Wang, J., Li, K.: SPARK2: Top-k keyword query in relational databases. IEEE Trans. Knowl. Data Eng. 23(12), 1763–1780 (2011)
13. Markowetz, A., Yang, Y., Papadias, D.: Keyword search on relational data streams. In: ACM SIGMOD, pp. 605–616 (2007)
14. Qin, L., Yu, J.X., Chang, L.: Scalable keyword search on large data streams. VLDB J. 20(1), 35–57 (2011)
15. Qin, L., Yu, J.X., Chang, L., Tao, Y.: Querying communities in relational databases. In: ICDE, pp. 724–735 (2009)
16. Qin, L., Yu, J.X., Chang, L., Tao, Y.: Scalable keyword search on large data streams. In: ICDE, pp. 1199–1202 (2009)
17. Xu, Y., Ishikawa, Y., Guan, J.: Effective Top-k Keyword Search in Relational Databases Considering Query Semantics. In: Chen, L., Liu, C., Zhang, X., Wang, S., Strasunskas, D., Tomassen, S.L., Rao, J., Li, W.-S., Candan, K.S., Chiu, D.K.W., Zhuang, Y., Ellis, C.A., Kim, K.-H. (eds.) WCMT 2009. LNCS, vol. 5731, pp. 172–184. Springer, Heidelberg (2009)
18. Xu, Y., Ishikawa, Y., Guan, J.: Efficient Continuous Top-k Keyword Search in Relational Databases. In: Chen, L., Tang, C., Yang, J., Gao, Y. (eds.) WAIM 2010. LNCS, vol. 6184, pp. 755–767. Springer, Heidelberg (2010)
19. Xu, Y., Ishikawa, Y., Guan, J.: Efficient continual top-k keyword search in relational databases. Journal of Information Processing 20(1), 1–14 (2012)
20. Yu, J.X., Qin, L., Chang, L.: Keyword Search in Databases. Synthesis Lectures on Data Management. Morgan & Claypool Publishers (2010)
21. Yu, J.X., Qin, L., Chang, L.: Keyword search in relational databases: A survey. Bulletin of the IEEE Technical Committee on Data Engineering 33(10) (2010)

Composition and Efficient Evaluation of Context-Aware Preference Queries

Patrick Roocks, Markus Endres, Stefan Mandl, and Werner Kießling

Institut für Informatik, Universität Augsburg,
D-86159 Augsburg, Germany
{roocks,endres,mandl,kiessling}@informatik.uni-augsburg.de

Abstract. This paper presents a modular approach to context-aware preference query composition based on a novel kind of preference generator. We introduce a constructive model to generate preference terms within the Preference SQL framework. Given several sources for preference related knowledge like explicit user input, information extracted from a preference repository, domain-specific application knowledge, location-based sensor data, or web service feeds for weather data our preference generator can compile a user search request into one rather complex context-aware Preference SQL query. Choosing as use case a commercial e-business platform for outdoor activities, we demonstrate how such queries despite the power and complexity of this approach can be evaluated efficiently on a practical data set.

1 Introduction and Related Work

Preferences in Databases [Kie05] – as shown by a recent survey [SKP11] – are a well established framework to create personalized information systems. By using well designed preference models, users can be provided with just the information they need, thereby overcoming the dreaded empty result set and flooding effect [Kie02]. These improvements are starting to show up in real world applications. For instance, today in the area of tourism – which is used in this paper as an example application domain – there are many portals which provide information about flights, hotels or outdoor activities by parametrized queries which either result in an abundant number of items or no answer at all. Clearly, this state of affairs is non-satisfactory for both the users and the owners of the portals. In new tourist portals (like presented in [KSH+11]), preferences allow for richer queries which define a strict partial order on the items available for the purpose of selecting just the best items available, resulting strongly improved user experience.

In domains like tourism, the notion of preference varies between users and strongly depends on users' situations. Hence, recently models of context-aware preferences emerged [LKM11, PSV11, BFA07, SPV07], which are taking into account factors like users' personalities, situation parameters (e.g. location, time, season, weather), or even options of users' acquaintances. [LKM11] introduces CareDB, that provides scalable personalized location-based services to users

S.-g. Lee et al. (Eds.): DASFAA 2012, Part II, LNCS 7239, pp. 81–95, 2012.

based on their preferences and current surrounding context. In particular it deals with the problem that some contextual values may be very expensive to compute. In [Cho07] relation-algebraic aspects of database queries under changing preferences are discussed. [BFA07] suggests a context-model for preference queries, where context is represented by a variant of description logic.

There is the open challenge to create a comprehensive framework to derive preferences from context. In this paper, we meet this challenge by presenting an approach for context-aware preference composition, which 1) produces inductively structured preference queries that are intuitive such that they can be verified by domain experts, 2) allows for easy adaption of the composition process for the specific application domain and 3) covers the whole path from the application model to the preference query language.

[SPV07] suggests a discrete context model and introduces profile trees as a data structure for the context resolution problem, hence retrieving the most appropriate preferences depending on context. Similarly [PSV11] suggests a model for a contextual preference selection based on the idea of a situation hierarchy. These models can be seen as top-down approaches in contrast to our constructive bottom-up model. In our approach all preferences are generated dynamically instead of performing a look-up in a set of predefined preferences.

Throughout the paper we refer to the following example from the domain of hiking tour recommendation which informally demonstrates the lines of reasoning about context-aware preferences of domain experts (like touristic companies) we want to model: social network (recommendations from friends), history (already visited regions by the user), and external knowledge sources (weather services, like the snow line and the altitude-dependent temperature).

Example 1 (Running Use Case). Consider John, who is planning a hiking tour in the touristic region of the Bavarian Alps. He was already last year there in the sub-region "Tannheim Mountains", and a friend of him made a tour in the "Walser Valley" which he was very excited about. Because now John wants to see something new and he trusts his friend he wants to avoid regions already visited by himself and prefers regions recommended by friends.

As he is not very experienced, he only specifies a duration for the tour, while for the ascent and length he relies on the recommender. John prefers not to hike in the snow, hence only mountain peaks below the current snow line are preferable for him. Because it is a sunny day, the temperatures in low altitudes are too high, so the tour should mainly be in a convenient altitude range.

John looks out for an online portal which picks out a hiking tour from a large database which perfectly matches the preferences formulated above. In case there are no "perfect tours" he is able to accept compromises, e.g. tours whose altitude exceeds the snow line a little bit.

The remainder of the paper is structured as follows: Section 2 presents informally the preference model used in this paper. Thereafter Section 3 describes our context-aware preference generation process. Section 4 shows how the generated preference queries can be evaluated using *Preference SQL*. Finally, Section 5 summarizes our claimed contributions and outlines further research directions.

2   Preference Modeling

Preference modeling has been in focus for some time, leading to diverse approaches, e.g. [Cho03, Kie02, Kie05]. We follow the preference model from [Kie05] which is a direct mapping to relational algebra and declarative query languages. It is semantically rich, easy to handle and very flexible to represent user preferences which are ubiquitous in our life.

A *preference* $P = (A, <_P)$ is a *strict partial order* on the domain values of the attributes A of a database relation. The term $x <_P y$ is interpreted as "y is better than x according to P" where x, y are domain values of the attribute set A. The result of a preference is computed by the *preference selection*.

Definition 1 (Preference Selection, BMO-set). *The Best-Matching-Objects (BMO-set) of a preference $P = (A, <_P)$ are all tuples from a database relation R that are not dominated w.r.t. the preference. It is computed by the preference selection operator $\sigma[P](R)$ (called winnow in [Cho03]) and finds all best matching tuples t for P, where $t[A]$ is the projection to the attribute set A:*

$$\sigma[P](R) := \{t \in R \mid \neg\exists t' \in R : t[A] <_P t'[A]\}$$

Preference selection offers a cooperative query answering behavior by automatic matchmaking: The BMO query result adapts to the quality of the data in the database, defeating the empty result effect and reducing the flooding effect by filtering off worse results.

To specify a preference a choice of intuitive base preference constructors together with some complex preference constructors has been defined. Subsequently, we present some selected preference constructors used in this paper. More preference constructors as well as their formal definition can be found in [Kie02, Kie05].

Base Preference Constructors. Preferences on single attributes are called *base preferences*. There are base preference constructors for discrete and for continuous domains.

Definition 2 (POS and NEG Preference). *The discrete Positive-preference POS$(A, POS\text{-}set)$ states that the user has a set of preferred values, the POS-set, in the domain of A. The Negative-preference constructor is the counterpart to the POS-preference, formally NEG$(A, NEG\text{-}set) :=$ POS$(A, \mathrm{dom}(A)\backslash NEG\text{-}set)$.*

We now present some preference constructors for continuous domains.

Definition 3 (AROUND$_d$ Preference). *In the AROUND$_d(A, z)$ preference the desired value should be $z \in \mathrm{dom}(A)$. If this is infeasible, values within a distance of d are acceptable.*

Definition 4 (LESS_THAN$_d$ and MORE_THAN$_d$ Preference).

a) *In the MORE_THAN$_d(A, z)$ preference the desired values are greater or equal to z. If this is infeasible, values within a distance of d are acceptable.*

b) *LESS_THAN$_d(A, z)$ is the dual preference to MORE_THAN$_d(A, z)$. The desired values are less or equal to z w.r.t. to the d-Parameter.*

Complex Preference Constructors. If one wants to combine several preferences into more *complex preferences*, one has to decide the relative importance of these given preferences. Intuitively, people speak of "this preference is more important to me than that one" or "these preferences are all equally important to me". Equal importance is modeled by the so-called *Pareto preference*.

Definition 5 (Pareto Preference). *In a Pareto preference*

$$P := P_1 \otimes \ldots \otimes P_m = (A_1 \times \cdots \times A_m, <_P)$$

all preferences $P_i = (A_i, <_{P_i})$, $i = 1, \ldots, m$ *on the attributes* A_i *are of equal importance.*

The *Prioritization preference* allows the modeling of combinations of preferences that have different importance.

Definition 6 (Prioritization Preference). *Let* $P_i = (A_i, <_{P_i})$, $i = 1, \ldots, m$ *be preferences. In a Prioritization preference*

$$P := P_1 \& \ldots \& P_m = (A_1 \times \cdots \times A_m, <_P)$$

preferences expressed earlier are more important than preferences expressed later in the preference term.

Example 2 (Running Use Case). John prefers a hiking tour with a *duration* of 5 hours which is *more important* than all other preferences. Thereby he specifies a d-paramter of 10%, i.e. 0.5 hours. John prefers tours recommended by his friends and *equally important* is that the *altitude* should be *less than* 2400 meters because of the snow line. Using the preference constructors this complex preference can be formulated as:

$$P_{\text{John}} := \text{AROUND}_{0.5}(duration, 5) \; \&$$
$$(\text{POS}(recommended, \text{'yes'}) \otimes \text{LESS\_THAN}_{d_a}(altitude, 2400))$$

The variable d_a is the d-parameter which will be used for all preferences regarding altitude. A typical value could be $d_a = 200$, implying a tolerance of 10% at a typical altitude of 2000 meters.

For the rest of this paper we also need the notion of a *preference term*.

Definition 7 (Preference Terms). *We define the set of preference terms* $\mathbb{P}$ *inductively:*

- *A base preference* $P = (A, <_P)$ *is a preference term.*
- *If* P_1 *and* P_2 *are preference terms, then the complex preferences* $P_1 \otimes P_2$ *and* $P_1 \& P_2$ *are preference terms.*

The semantically well-founded approach of this preference model is essential for personalized preference term composition. As a unique feature this preference model allows multiple preferences on the same attribute without violating the strict partial order. Furthermore, the *inductive* preference construction preserves strict partial order, too.

3   Context-Aware Preference Generation

Based on the preference framework presented in Section 2 we present our model for creating preference terms dependent on the context. This model depends on context-based triggers which are described by a discrete situation model with few states. Compared to continuous models such models are particularly useful when communicating with domain experts as the small number of states makes it possible to systematically consider all combinations of context state and therefore to guarantee that the system behaves in the desired way for all possible context values. For example, concerning the weather, it is intuitive to distinguish between "good" and "bad". For our domain of a hiking tour recommender we introduced the states: "good", "bad" and "warning". Whereas "bad" implies less convenient outdoor activities, the state "warning" discourages from activities like hiking because of safety reasons. Dependent on the context (time of year, region) alternatives like city tours or visiting the spa could be suggested.

3.1   A Constructive Approach for Preference Generation

According to our running use case of a hiking tour recommender we have three different kinds of influences for the preference generation: The user input from the search mask, the user profile from the database and the contextual information from external knowledge sources. In the following we describe how they interact in our model. As described in Example 1, the weather conditions lead to certain preferences. But weather warnings do not mean that the user input is overridden. The rough concept of the preference composition always follows this prioritization-schema, where each $\langle ... \rangle$-part is called *preference component*.

$$\langle \textbf{pref\_term} \rangle := \langle \text{user\_input} \rangle \, \& \, \langle \text{context} \rangle \, \& \, \langle \text{profile} \rangle \qquad (1)$$

This schema is supported by the conclusion of [BD03] that users might feel a "lack of control" in context-aware systems, therefore the user input is prioritized. In the same study was shown that users prefer context-awareness to personalization, thus the profile is less prioritized.

The gravity of this external influences depends on the user: A well specified user input, i.e. all fields of a search mask all filled out, implies that the context can just do slight modifications and the profile even less. If a "lazy" user leaves all fields blank, a "default request" based on the context and the profile will be generated.

In the following the components of Eq. 1 are stepwise filled with preferences which underlines our *constructive approach*. It is one of the big advantages of our model that preferences are closed under Prioritization and Pareto-composition, even "seeming contradictions" keep the strict partial order property. In our running use case of the hiking tour recommender, the user input consists of the tour parameters length, duration, and ascent, unless they are not specified. The components $\langle \text{context} \rangle$ and $\langle \text{profile} \rangle$ are sub-divided:

$$\langle \text{context} \rangle := \langle \text{weather\_safety} \rangle \, \& \, \langle \text{children} \rangle \, \& \, \langle \text{weather\_conv} \rangle \qquad (2)$$

$$\langle \text{profile} \rangle := (\langle \text{social\_net} \rangle \otimes \langle \text{history} \rangle) \, \& \, \langle \text{default\_input} \rangle \qquad (3)$$

In the ⟨context⟩ component in Eq. 2 the influence of the weather is divided into ⟨weather_safety⟩, the safety-relevant preferences (e.g. snow line), and ⟨weather_conv⟩, the convenience-relevant (abbr. conv) preferences (e.g. convenient temperature). The ⟨children⟩ component is relevant for families with young children and will set preferences for tours with playgrounds, etc.

In the ⟨profile⟩ component in Eq. 3 the recommendations from friends are contained in ⟨social_net⟩ and the already visited tours in ⟨history⟩.

The components ⟨user_input⟩ in Eq. 1 and ⟨default_input⟩ in Eq. 3 only depend on what the user filled out in the search mask, which is shown in the following example.

Example 3 (Running Use Case). Consider the tourist John who has only specified a duration for his tour, let us say 5 hours. In our tour recommender this preference is modeled with the AROUND-Constructor. Thereby we assume a tolerance of 10% which leads to a d-value of $0.5h$. Thus we have the following user input preference:

$$\langle\text{user\_input}\rangle := \text{AROUND}_{0.5}(duration, 5)$$

Depending on empirical values like average speed and condition of his user type (e.g. "tourist" or "athlete"), the system selects defaults for length and ascent according to the duration, let us say 12 kilometers for the length and 600 meters for the ascent, both with a d-parameter of 10 %.

$$\langle\text{default\_input}\rangle := \text{AROUND}_{1.2}(length, 12) \otimes \text{AROUND}_{60}(ascent, 600)$$

The Pareto-composition of these preferences is based on the assumption that the attributes of length and ascent are of equal importance for the user. This ordering may depend on the user type, e.g. an athletic user is focused on the ascent of the tour whereas an unexperienced tourist is more used to the duration, i.e. the preferences in our use-case also depend on the "tourist role", which is investigated in [GY02].

The social network and the history in Eq. 3 are Pareto-composed to allow compromises. In the following example we will show, how conflicting preferences in this components can be handled.

Example 4 (Running Use Case). We assume the attributes *recommended* and *visited*, which are not attributes of the tour, but represent the recommendations from the social network and the visited tours for the current user. For *recommended* we have the values "yes", "no" and "unknown", representing that the region is recommended, disadvised or that there are no or contradicting informations in the social network. For the *visited* attribute we just have the values "yes" and "no" to mark the regions already visited by John. Recommended regions are preferred whereas already visited regions are not preferred, because the tourist wants to see something new.

$$\langle\text{social\_net}\rangle := \text{POS}(recommended, \text{'yes'})$$
$$\langle\text{history}\rangle := \text{NEG}(visited, \text{'yes'})$$

According to the Pareto composition $\langle\text{social\_net}\rangle \otimes \langle\text{history}\rangle$ a region which was both visited and recommended is considered equal to a region which was neither visited nor recommended.

The preference model from Section 2 gives us the freedom for this component-based model, where the preferences in the components may contradict each other or may be defined on the same attribute.

To fill the $\langle\text{context}\rangle$ component with context-dependent preferences, a formal model based on *context-aware generators* is introduced in the following.

3.2 Context-Aware Generators

The generation of preferences in our approach is context-aware in two regards:

1. Preferences in the components are triggered by a discretized context.
2. Parameters of this preferences can change with the (continuous) context.

By this breakdown the model is kept clear. Whereas in the "discretized world" the rough structure of the preference term is decided, the "continuous world" influences the fine structure of the preference term. Thus it is easy to see how the configuration of the recommender changes the final output.

Definition 8 (Context Variables, (Current) Situation). *Let Ω the set of world states and $\omega \in \Omega$ a world state, containing user input and contextual knowledge. We define context variables by mappings*

$$v_i : \Omega \to V_i \ \text{for} \ i \in \{1, ..., n\} =: I$$

where V_i are finite sets modeling the discretization of the context. The set $\mathbb{S} := \times_{i \in I} V_i$ represents possible situations and $s \in \mathbb{S}$ is a single situation. The aggregation of all context variables leads to the mapping

$$\text{sit} : \Omega \to \mathbb{S}, \omega \mapsto (v_1(\omega), ..., v_n(\omega))$$

If ω is the current world state, then $\text{sit}(\omega)$ is called current situation. For $a \in V_i$ we define

$$\{v_i = a\} := V_1 \times ... \times V_{i-1} \times \{a\} \times V_{i+1} \times ... \times V_n.$$

We illustrate the concept of context variables in the following example.

Example 5 (Running Use Case). To describe the current weather state of the hiking region selected by a hard constraint, we introduce a context variable *weather* : $\omega \mapsto \{\text{good}, \text{bad}, \text{warning}\}$. To this end, we assume ω contains informations from an online weather service, which is retrieved automatically after the user submitted the search request.

By *children* : $\omega \mapsto \{\text{yes}, \text{no}\}$ we describe if young children are participating in the hiking tour. We assume that the user ticks this in the search mask and this information is stored in ω.

With the context variables we will trigger the rough structure of the ⟨context⟩ component. To form the fine structure, we introduce the concept of ω-*dependent preference terms.*

Definition 9 (ω-Dependent Preference Terms). *Functions $f : \Omega \to M$, where M is either a finite set or the real numbers, are called ω-dependent functions. They extract a single part of the world context.*

$\mathbb{P}[\omega]$ is the set of ω-dependent preference terms. If $p \in \mathbb{P}[\omega]$ is a base preference, then for all parameters of base preferences, ω-dependent functions may occur. Analogous to Def. 7 complex preferences which are inductively constructed from ω-dependent base preferences, are also in $\mathbb{P}[\omega]$.

For $p \in \mathbb{P}[\omega]$ the evaluation of all ω-dependent functions in p is denoted by $p(\omega)$, where $p(\omega) \in \mathbb{P}$ holds.

Example 6 (ω-Dependent Preference Terms). An ω-dependent function representing the altitude of the snow line is $snowline : \Omega \to \mathbb{R}$. The term LESS_THAN$_{d_a}(alt\_max, snowline(\omega))$ is an ω-dependent base preference, where alt_max is the attribute for the maximal altitude of the tour.

The interplay of the rough and the fine structure of preference generation is done by the *context-aware generators* introduced next.

Definition 10 (Context-Aware Generator). *We define context-aware generators as tuples*

$$g = (S, p) \in \mathbb{G}, \quad \mathbb{G} := \mathcal{P}(\mathbb{S}) \times \mathbb{P}[\omega]$$

where $S \in \mathcal{P}(\mathbb{S})$ ($\mathcal{P}$ denotes the power set) is the set of associated situations and $p \in \mathbb{P}[\omega]$ is the associated ω-dependent preference term. A generator $g = (S, p)$ is called active, if $\mathrm{sit}(\omega) \in S$.

Example 7 (Running Use Case). To realize the components ⟨children⟩ and ⟨weather_safety⟩ we define the following generators:

$$g_{\text{ch}} := (\{children = \text{yes}\}, \text{POS}(difficulty, \text{'easy'}))$$
$$g_{\text{ws},1} := (\mathbb{S}, \qquad\qquad \text{LESS\_THAN}(alt\_max, snowline(\omega)))$$
$$g_{\text{ws},2} := (\{weather = \text{bad}\}, \text{LESS\_THAN}_{d_a}(alt\_max, \min(snowline(\omega), 1500)))$$
$$g_{\text{ws},3} := (\{weather = \text{wrn}\}, \text{POS}(activity, \text{'CityTrail'}) \,\&$$
$$\text{LESS\_THAN}_{d_a}(alt\_max, \min(snowline(\omega), 1500)))$$

Thereby wrn is short for warning. Whereas $g_{\text{ws},1}$ is always active (the situation set is $\mathbb{S}$), the generators g_{ch}, $g_{\text{ws},2}$, and $g_{\text{ws},3}$ are restricted to situations depending on the weather and the presence of children. The sets of generators $\{g_{\text{ch}}\}$ and $\{g_{\text{ws},i} \mid i = 1, 2, 3\}$ are associated with ⟨children⟩ and ⟨weather_safety⟩ respectively.

3.3 Constructing Preference Terms from Generators

Now we have components and associated sets of generators. The next step is to select the active generators and to create a preference term from it. To this end we have to specify how the preferences of the generators shall be Pareto-composed or prioritized, if there is more than one generator associated with a component.

Assume an ordering of the generators representing the *importance*. For a set of generators $G \subset \mathbb{G}$ we realize this by a function $\pi : \Omega \times G \to \mathbb{N}$, where smaller values stand for a higher importance. π is ω-dependent, which primarily means a dependency of the user type like described at the end of Example 3.

In the following translation from a set of generators to a preference term at first the active generators are selected (Def. 11). With Def. 12 we realize the principle that preferences created by more important generators are prioritized, and Pareto-composed if the according generators are equally important.

Definition 11 (Restriction to Situation, Active Generators). *Consider a set of generators $G \subset \mathbb{G}$. The restriction to a situation $s \in \mathbb{S}$ is defined as*

$$G|_s := \{(S, p) \in G \mid s \in S\}$$

and with $G|_{\mathrm{sit}(\omega)}$ we restrict G to the set of active generators.

Definition 12 (Preference Generation). *For a set of generators $G \in \mathbb{G}$, an ordering $\pi : \Omega \times G \to \mathbb{N}$, and $\omega \in \Omega$ we define the preference generation:*

$$\mathrm{pref}(G, \pi, \omega) := \underset{i \geq 1}{\&} \bigotimes \{p(\omega) \mid (p, S) \in G|_{\mathrm{sit}(\omega)}, \pi(\omega, g) = i\}$$

where $\bigotimes \{p_1, ..., p_n\} := p_1 \otimes ... \otimes p_n$ and $\&_{i=1}^m q_i := q_1 \& ... \& q_m$ for $p_1, ..., p_n$, $q_1, ..., q_m \in \mathbb{P}$. The configuration of a component is described by the tuple (G, π).

Preference terms generated with the pref-function from Def. 12 – also known as *p-skylines* [MC09] – could be very long if many generators are active. But there may be generators which are more "appropriate" for the current situation than others. The measure for this is the inclusion order in the set of associated situations. Formally $g = (S, p)$ is more appropriate than $g' = (S', p')$ for the current situation t if and only if $t \in S \subsetneq S'$. Hence generators which are more specialized to the current situation are preferred to less specialized ones. This unveils a nice analogy to the "best matches only" principle, but now we look out for the *best matching generators*.

Definition 13 (Best Matching Generators). *For $G \in \mathbb{G}$ and $\omega \in \Omega$ we define*

$$\mathrm{best\_gen}(G, \omega) := \{(S, p) \in G|_{\mathrm{sit}(\omega)} \mid \nexists (S', p') \in G|_{\mathrm{sit}(\omega)} : S' \subsetneq S\}$$

The translation from components with configuration (G, π) to a preference term using the best-matching-generators principle is given by $\mathrm{pref}(\mathrm{best\_gen}(G, \omega), \pi, \omega)$.

A similar concept also occurs in [SPV07]. But we use this concept only to *reduce* the number of active generators, we are still allowed to have more active generators. They may also produce contradicting preferences or more than one preference on the same attribute.

Example 8 (Running Use Case). In Example 7 using Def. 13 only one generator is in the set best_gen($\{g_{\text{ws},i} \mid i = 1, 2, 3\}$) because of $\{\text{weather} = x\} \subsetneq \mathbb{S}$ for $x \in \{\text{bad}, \text{warning}\}$.

3.4   The Context Model in Practice

To apply the context model to Example 1 from the introduction we are still missing some definitions. We have generator sets for all the components of the ⟨context⟩ and ⟨profile⟩ part except ⟨weather_conv⟩. We will omit this component for brevity, which is intended to generate preferences on altitude ranges with convenient temperatures. We also omit a formal definition of the generators for ⟨user_input⟩ and ⟨default_input⟩, informally described in Example 3. In our implementation of this use case the concept of generators is "reused" for this components: By allowing that the input from the search mask is also stored in ω and by defining context variables which tell us whether the user filled out the fields for *length*, *duration*, and *ascent* in the search mask, we are able to define generators for ⟨user_input⟩ and ⟨default_input⟩.

Example 9 (Running Use Case). According to the previous examples and the assumption that convenient weather conditions imply a minimal altitude (attribute *alt_min*) higher than 1200 meters (because it is too hot below), we present here the entire preference term created by our model:

$$\langle\textbf{pref\_term}\rangle = \underbrace{\text{AROUND}_{0.5}(duration, 5)}_{P_1 := \langle\text{user\_input}\rangle} \& \underbrace{\text{LESS\_THAN}_{d_a}(alt\_max, 2400)}_{P_2 := \langle\text{weather\_safety}\rangle} \&$$

$$\underbrace{(\text{POS}(recommended, \text{'yes'})}_{P_3 := \langle\text{social\_net}\rangle} \otimes \underbrace{\text{NEG}(visited, \text{'yes'}))}_{P_4 := \langle\text{history}\rangle} \&$$

$$\underbrace{\text{MORE\_THAN}_{d_a}(alt\_min, 1200)}_{P_5 := \langle\text{weather\_conv}\rangle} \&$$

$$\underbrace{(\text{AROUND}_{1.2}(length, 12) \otimes \text{AROUND}_{60}(ascent, 600))}_{P_6 := \langle\text{default\_input}\rangle}$$

Thereby we use the rough structure from Eq. (1)–(3), P_1 and P_6 are generated according to Example 3, P_3 and P_4 are explained in Example 4, and P_2 is from Example 7. P_5 is mentioned above in this example.

With this we have defined a model which is applicable for our problem of designing a hiking tour recommender. But we have still one drawback with the assumption that all the external information in ω is available for the recommender. This may not always be the case.

For example, we assumed we have access to the complete weather forecast to determine the values of our context variables. But what if the user specified no region for the activity or a too big region to get a reliable weather forecast? Or what shall the recommender do if the weather service is unavailable? This can be modeled by replacing the current situation $sit(\omega) \in \mathbb{S}$ by a set of current situations $sit(\Omega_0) \subseteq \mathbb{S}$ with $\Omega_0 \subseteq \Omega$, where all weather states are contained, formally $\{weather = x\} \subsetneq sit(\Omega_0)$ for all $x \in \{good, bad, warning\}$. In the formal model of the generators one must only change all occurrences of the expression $sit(\omega) \in S$ by $sit(\Omega_0) \subseteq S$, which is straightforward but will be not worked out for brevity. The implication of this is that all the weather dependent generators are not active, the ones for good weather as well as the ones for bad weather.

4   Context-Aware Preference Query Evaluation

The complex preference terms generated in Section 3 must be evaluated efficiently to retrieve the best-matching objects for the user. All preference terms can be transformed into *Preference SQL*, an extension of standard SQL for preference query evaluation on database systems, cp. [KEW11]. Syntactically, the SELECT statement of SQL is extended by an optional PREFERRING clause. It selects all interesting tuples, i.e., tuples that are not dominated by other tuples. While the first prototype [KK02] used query rewriting to standard SQL, the current implementation of *Preference SQL* (since 2005) is a *middleware* component between client and database which performs the algebraic and cost-based optimization of preference query evaluation. This allows for a seamless, flexible and efficient integration with standard SQL back-end systems using a *Preference SQL* JDBC Driver. This trend is followed by [AK12], which suggests to implement numerical preferences tightly in the relational database engine. According to [SKP11] currently *Preference SQL* is the only comprehensive approach to support a general preference query model.

The final preference term from Example 9, e.g., can be formulated in *Preference SQL* as demonstrated in Fig. 1. The query is based on a real-world database provided by ALPSTEIN Tourismus GmbH (http://www.alpstein-tourismus.com). It returns all hiking tours in the Bavarian Alps which are best concerning the preference term from Example 9. All hiking tours are stored in the relation tour. There is a join with the relation region, because it holds additional information about recommendations and visited tours, cp. Example 4. A Prioritization is expressed using PRIOR TO, wheres AND in the PREFERRING clause denotes a Pareto preference. The preference constructors AROUND, LESS THAN and MORE THAN are already known from Section 2 and 3. Note that the second argument in these preference constructers denotes the d-parameter. The P_i on the right side of the query correspond to the preference terms in Example 9.

```
SELECT  t.id, t.name
FROM    tour t, region r
WHERE   t.main_region = 'Bavarian␣Alps'    AND
        t.subregion_id = r.subregion_id
PREFERRING    t.duration AROUND 5, 0.5                    -- P₁
    PRIOR TO  t.alt_max LESS THAN 2400, 200               -- P₂
    PRIOR TO  (r.recommend = 'yes'         AND
              r.visited   != 'yes')                       -- P3 ⊗ P4
    PRIOR TO  t.alt_min MORE THAN 1200, 200               -- P₅
    PRIOR TO  (t.length AROUND 12, 1.2   AND
              t.ascent AROUND 600, 60);                   -- P₆
```

Fig. 1. *Preference SQL* query for the preference term in Example 9

4.1   Optimization of Preference Queries

A first view on the query in Fig. 1 leads to the assumption that the evaluation of this query on a database system is very inefficient. Executing such queries on large data sets makes an optimized execution necessary to keep the run-times small. For this, *Preference SQL* implements the accumulated vast query optimization knowhow from relational databases as well as additional optimization techniques to cope with the preference selection operator for complex strict partial order preferences.

For example, the *Preference SQL* rule based query optimizer transforms the query from Fig. 1 into the operator tree depicted in Fig. 2.

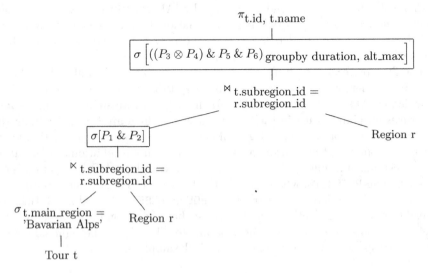

Fig. 2. Optimized operator tree of the *Preference SQL* query in Fig. 1

The optimizer applies among well-known optimization rules the law *L8: Split Prioritization and Push over Join* published by [HK05]. This law splits the large Prioritization preference from Example 9 and pushes $\sigma[P_1 \text{ \& } P_2]$ over the join ($\bowtie$). Since the preference selection $\sigma[P_1 \text{ \& } P_2]$ may eliminate tuples from the relation **tour** which are necessary for the join on the subregions, a semi-join ($\ltimes$) on **tour** and **region** is necessary. Finally, a *grouped preference selection* after the join operation leads to an optimized operator tree of this context-aware preference query. Note that we omit the *Push-Projection* optimization law for a better reading. Afterwards, a cost-based algorithm selection is applied for efficient Pareto and Prioritization evaluation. Such a preference query processing leads to a very fast retrieval of the best-matching objects concerning a users' preference. For further details we refer to [KEW11].

4.2 Practical Performance Tests

We now present some practical runtime benchmarks for the evaluation of context-aware preference queries. We do not look at the quality of the results, which will be future work. But for an useful empirical evaluation the runtime must be small.

In our benchmarks *Preference SQL* operates as a Java framework on top of a PostgreSQL 8.3 database. It stores the already mentioned real-world database of Alpstein Tourismus GmbH. The relations **tour** and **region** used in this paper contain about 48.000 rows and 200 rows, respectively. We build groups of three queries each with a specified *main region* as hard constraint, e.g. *Bavarian Prealps*, *Bavarian Alps*, and *Alps*. For each query in a group we generated different preference terms to show the influence of context-aware preference generation. Furthermore, the hard constraint on the main region leads to a different number of tuples (the basic set) for preference evaluation. We present two context-aware preference queries in detail. The runtimes are depicted in Fig. 3.

Query 4 corresponds to the preference query used in this paper, cp. Fig. 1. It can be evaluated in less than 1 second. The hard selection on the main region

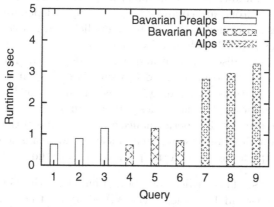

Fig. 3. Runtime for different context-aware preference queries

of the Bavarian Alps as well as the early computation of the P_1 & P_2 preference (cp. Fig. 2) leads to a preference selection on 1650 tuples after the join. Query 5 and 6 are restricted to the same main region but have different preference terms.

The next query, Query 7 depicted below, is similar to Query 1. However, the basic set (about 16.000 tuples) is changed because of a different main region, namely the complete Alps. The runtime is about 3 seconds, which is fast enough for context-aware recommender applications, e.g. in the domain of hiking tours.

```
SELECT  t.id, t.name
FROM    tour t, region r
WHERE   t.main_region = 'Alps' AND
        t.subregion_id = r.subregion_id
PREFERRING  r.duration AROUND 1.5, 0.15
        PRIOR TO  t.alt_max LESS THAN 2000, 200
        PRIOR TO  (r.recommend IN ('yes') AND
                   r.visited NOT IN ('yes'))
        PRIOR TO  t.alt_min MORE THAN 1600, 200
        PRIOR TO  (r.length AROUND 1.5, 0.15 AND
                   r.ascent AROUND 150, 15);
```

Fig. 4. Query 7 of our benchmark

The evaluated queries are only a small selection of the queries occurring in our use case, but they are representative for our performance benchmarks. The runtime of these queries show that our approach of a context-aware preference query composition is not only an intuitive and inductive preference construction but also very efficient in result computation on a real-world database.

5   Summary and Outlook

In the paper we present a new approach to context-aware preference query composition which covers all the way from context modeling to generating the preference query in a well known preference query language. Preferences, after being retrieved from a repository, are composed – on the fly – depending on context. The construction is performed according to an explicitly available set of expressive rules, which easily can be adapted by domain experts to meet their specific requirements. By using discretization and context dependent functional mappings, we achieve an easy to understand (no surprises) system that is suitable for commercial applications. A running use case from the field of commercial e-business platforms for outdoor activities is used to motivate and illustrate the various steps in the process. The performance evaluation shows that the answers for the generated queries can be computed efficiently for realistic data sets. As future work we are planning to perform qualitative user studies.

Acknowledgements. This project has been funded by the German Federal Ministry of Economics and Technology (BMWi) according to a resolution of the German Bundestag, grant no. KF2751301.

References

[AK12]   Arvanitis, A., Koutrika, G.: Towards Preference-aware Relational Databases. In: To appear in International Conference on Data Engineering, ICDE 2012 (2012)

[BD03]   Barkhuus, L., Dey, A.K.: Is Context-Aware Computing Taking Control away from the User? Three Levels of Interactivity Examined. In: Dey, A.K., Schmidt, A., McCarthy, J.F. (eds.) UbiComp 2003. LNCS, vol. 2864, pp. 149–156. Springer, Heidelberg (2003)

[BFA07]   van Bunningen, A.H., Feng, L., Apers, P.M.G.: A Context-Aware Preference Model for Database Querying in an Ambient Intelligent Environment. In: Bressan, S., Küng, J., Wagner, R. (eds.) DEXA 2006. LNCS, vol. 4080, pp. 33–43. Springer, Heidelberg (2006)

[Cho03]   Chomicki, J.: Preference Formulas in Relational Queries. ACM Transactions on Database Systems 28, 427–466 (2003)

[Cho07]   Chomicki, J.: Database Querying under Changing Preferences. Annals of Mathematics and Artificial Intelligence 50(1-2), 79–109 (2007)

[GY02]   Gibson, H., Yiannakis, A.: Tourist Roles: Needs and the Lifecourse. Annuals of Tourism Research 29(29), 358–383 (2002)

[HK05]   Hafenrichter, B., Kießling, W.: Optimization of Relational Preference Queries. In: Australasian Database Conference (ADC 2005), pp. 175–184 (2005)

[KEW11]   Kießling, W., Endres, M., Wenzel, F.: The Preference SQL System – An Overview. IEEE Data Eng. Bull. 34(2), 11–18 (2011)

[Kie02]   Kießling, W.: Foundations of Preferences in Database Systems. In: Very Large Databases (VLDB 2002), pp. 311–322 (2002)

[Kie05]   Kießling, W.: Preference Queries with SV-Semantics. In: International Conference on Management of Data (COMAD 2005), pp. 15–26 (2005)

[KK02]   Kießling, W., Köstler, G.: Preference SQL – Design, implementation, Experiences. In: Very Large Databases (VLDB 2002), pp. 990–1001 (2002)

[KSH+11]   Kießling, W., Soutschek, M., Huhn, A., Roocks, P., Wenzel, F., Zelend, A.: Context-Aware Preference Search for Outdoor Activity Platforms. Technical Report 2011-15, Universität Augsburg

[LKM11]   Levandoski, J.J., Khalefa, M.E., Mokbel, M.F.: An Overview of the CareDB Context and Preference-Aware Database System. IEEE Data Eng. Bull. 34(2), 41–46 (2011)

[MC09]   Mindolin, D., Chomicki, J.: Discovering relative importance of skyline attributes. In: Very Large Databases (VLDB 2009), pp. 610–621 (2009)

[PSV11]   Pitoura, E., Stefanidis, K., Vassilidis, P.: Contextual Database Preferences. IEEE Data Eng. Bull. 34(2), 20–27 (2011)

[SKP11]   Stefanidis, K., Koutrika, G., Pitoura, E.: A Survey on Representation, Composition and Application of Preferences in Database Systems. ACM Transactiopns on Database Systems 36(3), 19:1–19:45 (2011)

[SPV07]   Stefanidis, K., Pitoura, E., Vassiliadis, P.: Adding Context to Preferences. In: International Conference on Data Engineering (ICDE 2007), pp. 846–855 (2007)

An Automaton-Based Index Scheme
for On-Demand XML Data Broadcast⋆

Weiwei Sun[1], Peng Liu[1], Jingjing Wu[1], Yongrui Qin[2], and Baihua Zheng[3]

[1] School of Computer Science, Fudan University, Shanghai, China
{wwsun,liupeng,wjj}@fudan.edu.cn,
[2] School of Computer Science, The University of Adelaide, SA, 5005, Australia
yongrui.qin@adelaide.edu.au
[3] School of Information Systems, Singapore Management University, Singapore
bhzheng@smu.edu.sg

Abstract. XML data broadcast is an efficient way to deliver semi-structured information in wireless mobile environment. In the literature, many approaches have been proposed to improve the performance of XML data broadcast. However, due to the appearance of wildcard "*" and double slash "//" in queries, their performance deteriorates. Consequently, in this paper, we propose a novel air indexing method called Deterministic Finite Automaton-based Index (abbreviated as DFAI) on the XPath queries. Different from existing approaches which build index based on XML documents, we propose to build the index based on the queries submitted by users. The new index treating the XPath queries with "*" or "//" as a DFA actually improves the efficiency of broadcast system significantly. We further propose an efficient compression strategy to reduce the index size of DFAI as well. Experiment results show that our new index method achieves a much better performance in terms of both access time and tuning time when compared with existing approaches.

Keywords: air indexing, deterministic finite automaton, on-demand XML data broadcast.

1 Introduction

In recent years, there is a sharp increase of mobile subscriptions. The international Telecommunication Union estimates there would be 5.3 billion mobile subscriptions by the end of 2010 and the mobile application market will be worth $25 billion in 2014. Mobile devices play an important role in sending and accessing information, which means information can be retrieved by anyone, anywhere, at any time. The data retrieved via various wireless technologies becomes a substantial part of information in our daily life.

⋆ This research is supported in part by the National Natural Science Foundation of China (NSFC) under grant 61073001.

S.-g. Lee et al. (Eds.): DASFAA 2012, Part II, LNCS 7239, pp. 96–110, 2012.

Point-to-point and broadcast are two ways of information access via wireless technologies. In point-to-point access, mobile clients submit queries to the server via a logical channel established between them, and retrieve results returned in a similar way as in a wired network. It is suitable for systems with adequate bandwidth resources and server processing capacity. On the other hand, in broadcast access, data is sent to all the mobile clients. Mobile clients listen to the wireless channel and download the data they are interested in and hence a single broadcast of a data item can simultaneously satisfy an arbitrary number of mobile clients. Compared with point-to-point access, broadcast is suitable for heavily loaded systems in which the number of mobile clients is very large and the bandwidth resources are limited.

There are two critical performance metrics in data broadcast, namely access time and tuning time [1]:

- Access Time (AT): the time elapsed between the moment when a request is issued and the moment when it is satisfied, which is related to access efficiency.
- Tuning Time (TT): the time a mobile user stays active to receive the requested data items, which is related to the power consumption.

Air indexing technique is one of the most important research fields of data broadcast. As the limited battery power heavily restricts the usage of mobile devices, the tuning time metric is more important in wireless data broadcast system. Without any assistance, the mobile clients have to continuously monitor the broadcast data to make sure all interested data items are retrieved. The air indexing technique is used to improve the tuning time. The main strategy is to build a small size index based on broadcast data and interleave indices with broadcast data. Then, the mobile clients are aware of the arrival time of the data items they need according to the index. They can tune to *doze mode* to save the power when waiting and tune into the channel only when the data items of their interest arrive.

Concluded by [1], there are two typical broadcast modes: *push-based* broadcast and *on-demand* broadcast. For *push-based* broadcast, the server disseminates data periodically. The mobile clients listen to the channel and find the requested data items of their queries. For *on-demand* broadcast, the mobile clients submit their queries to the server via an uplink channel and data are sent based on the submitted queries via a downlink channel.

In the past few years, XML has become more and more popular and XML data broadcast has become a hot research topic recently. A wealth of work has focused on indexing and scheduling XML data in wireless environment under different broadcast modes [2,7,8,9,10,12].

In on-demand XML data broadcast, the broadcast XML documents in each cycle are determined by the pending requests dynamically and the broadcast contents of these cycles are different from each other. Moreover, during a broadcast cycle, some queries are satisfied while some queries are being submitted to the server and the server builds the index scheme of the next cycle according to the new query set. All the above features bring more complexity in air indexing technique of on-demand XML data broadcast.

[2] proposed an efficient two-tier index scheme for on-demand XML data broadcast. It first builds DataGuides [5] for each XML document, and then integrates these DataGuides by adopting the RoXSum [6] technique. Also, a pruning strategy and a two-tier structure are designed which can significantly reduce the index size. The main idea of the pruning strategy is to eliminate all the dead nodes that are not accessed by any query.

However, there are some drawbacks of PCI technique in [2]. The efficiency of pruning would be deteriorated when the probability of "*" and "//" in queries increases (users mostly don't quite understand the structures of the documents they need, so fuzzy queries with "*" and "//" are always issued). Firstly, the size of index scheme increases as more branches are satisfied by queries which also means less dead nodes are pruned. Secondly, the tuning time of index scheme will increase as the clients have to probe more index nodes to locate the arrival time of the entire required document set.

Motivated by the drawbacks of existing approaches, in this paper, we propose a novel air indexing method called Deterministic Finite Automaton based Index, abbreviated as DFAI.

To summarize, our main contributions include:

- We are the first to introduce an index scheme built on user queries rather than data in wireless broadcast environments.
- We treat the XPath queries with "*" or "//" as a Deterministic Finite Automaton (DFA) as other queries, and integrate all the XPath queries into one DFA which greatly reduces the access time and tuning time.
- We propose a compression strategy to further reduce the index size of DFAI.
- We compare the DFAI with PCI and provide a detailed analyze about experiment results.

The remainder of the paper is organized as follows. First, Section 2 presents the related work. The details of our index approach DFAI as well as the access protocol are explicitly depicted in Section 3. Section 4 shows the Compression Strategy which further reduces the index size of DFAI. Then, experiments and evaluation are demonstrated in Section 5. Finally, Section 6 concludes our work.

2   Related Work

Air indexing and scheduling methods in XML data broadcast have been extensively investigated in the past few years, we only outline the most related work in the following.

[7] first studies the XML data broadcasting in wireless environment. An XML stream organization and the corresponding event-driven stream generation algorithms are provided. The stream structures support simple XPath queries. [8] provides a novel XML data streaming method in wireless broadcasting environments. It separates the structure information from the text values. The former is

used as index scheme in data broadcast. Mobile clients access the index scheme selectively to process their queries. The total size of the index scheme is further reduced by using the path summary technique. [9] constructs two tree structures to represent the index scheme and the XML data. Three data/index replication strategies are also provided. The main idea of the replication is to repeatedly place part of the index/data tree to improve the efficiency of broadcast.

The work described above all focuses on the push-based broadcast. While in on-demand broadcast mode, as mentioned in Section 1, the queries submitted to the server should be taken into consideration and the broadcast content of each cycle could be changed dynamically. Thus, [2,10] study the air indexing technique for on-demand XML data broadcast .

[2] provides a novel two-tier air index scheme. The basic structure of the index scheme is the combination of all the DataGuides of each XML document. RoX-Sum technique is used to integrate these DataGuides efficiently. In this basic index scheme, every unique-label path of a document appears exactly once for supporting simple XPath queries. Thus it contains the entire unique-label paths in all of the XML documents. The pruning strategy is provided to reduce the size of the index scheme by eliminating the nodes not accessed by any queries. The two-tier structure enables efficient access protocol at the client which facilitates the index access. The mobile clients download the first-tier (i.e. the major index scheme to answer the query which XML documents satisfy) only once and the second-tier (presenting the arrival time of the XML documents in the current cycle) several times.

[10] studies the air index scheme for twig queries. Similar to the DataGuides technique adopted in [2], [10] uses the Document Tree structure as the basic index structure, which keeps only the tree structure information of an XML document. Pruning and combining strategies are also provided to reduce the size of index scheme.

Scheduling techniques are also related to our work. [11] introduces two heuristic algorithms called MPFH and MPLH. The two algorithms choose the candidate item to be broadcasted by the popularity of the item requested by user queries. The popularity of one XML document which can be considered as an item in multi-items traditional broadcast environments, is also an important factor in scheduling XML documents. [12] provides two effective fragment methods for skewed data access. It uses the sub-tree level data organization scheme and fragments the XML document according to the access probability of each element node.

3   Deterministic Finite Automaton-Based Index (DFAI)

The main idea of DFAI is to locate all the XML documents matched with each submitted query based on XML stream filtering techniques (e.g. YFilter [4] and AFilter [3]), build a Deterministic Finite Automaton (DFA) for each query and finally combine all the DFAs into an integrated DFA to share common prefixes among the queries.

DFAI has a better improvement in terms of both access time and tuning time compared with other approaches for listed reasons.

- For access time, DFAI combines all DFAs built based on queries into an integrated DFA to share common prefixes which greatly reduces the index size of DFAI. An improvement strategy is proposed to further reduce the index size which will be introduced in Section 4.
- For tuning time, users mostly submit a fuzzy query with "*" and "//" as they always don't know the detailed structures of the documents they need. DFAI takes "*" and "//" as normal element nodes in building DFA of each query which greatly reduces the tuning time of queries. Besides, matched documents IDs are directly attached to the accept state node. When queries are satisfied by the accept state node, the matched documents IDs can be downloaded immediately.

A running example is used to illustrate indexing scheme, as shown in Fig.1. It contains 5 documents, i.e. d_1, d_2, d_3, d_4 and d_5, with their detailed documents structures depicted in Fig.1(a). We assume there are six user queries submitted, i.e. q_1, q_2, q_3, q_4, q_5 and q_6, as shown in Fig.1(b). Fig.2 shows an example of a DFA for the query set in Fig.1(b). The nodes with double circles represent the *accept states* of DFA.

PCI in [2] uses the structure of DataGuides [5] to build a compact index structure capturing the path information of all the documents. Using the same running example (i.e. Fig.1), a PCI index is shown in Fig.3.

Built based on the users' queries and treating "*" and "//" as normal element nodes, DFAI has its advantages in reducing tuning time. When coming across "//", PCI has to traverse nearly all the subtrees of the current node to find the matched accept state nodes. In contrast with PCI, DFAI just treats it as a normal element node and gets to the next state directly. When coming across "*", PCI has to traverse all the child nodes of current node. The same as "//", DFAI simply jumps to the next state without bringing any extra cost.

For example, in Fig.1(b) $q_4 = $ "/a//c", users have to traverse n_1,n_2,n_3,n_4,n_5 and n_6 in PCI index as shown in Fig.3 to get all the matched documents IDs. However, users just need to traverse n_1,n_3 and n_8 in DFA Index as shown in Fig.2. For query $q_3 = $ "/a/b/*", users have to traverse n_1,n_2,n_4 and n_5 in PCI index, while only n_1,n_2 and n_7 are traversed in DFA index. We use the two-tier structure in implementing the index of DFA. Two-tier structure can not only further reduce the index size but also achieve a much better tuning time [2].

When we put the index into practice, we will partition it into packets with fixed size such as 64 bytes/packet. The minimal unit for user's data retrieval is a packet. Because of the feature of user queries, we sort the nodes according to the depth-first order. When the current packet doesn't have enough space for the next node, a new packet is used to store the whole node. According to the depth-first order mentioned above, the broadcast order of nodes is n_1, n_2, n_5, n_6, n_7, n_3, n_8, n_4 and n_9. The structure of the node is shown in Fig.4. We assign 2 bytes to *flag*, 8 bytes to *entry*, 4 bytes to *pointer* and 2 bytes to *doc*. The partition result is depicted in Fig.5 (assuming packet size is 64 bytes). For query

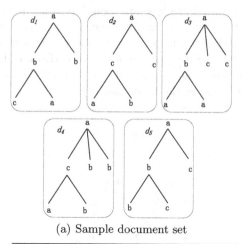

(a) Sample document set

Q	Matched documents IDs list
q_1 : /a/b	d_1, d_3, d_4, d_5
q_2 : /a/b/c	d_1, d_5
q_3 : /a/b/*	d_1, d_3, d_5
q_4 : /a//c	d_1, d_2, d_3, d_4, d_5
q_5 : /a/c/a	d_2, d_4
q_6 : /a/b/a	d_1, d_3

(b) Sample query set

Fig. 1. A running example

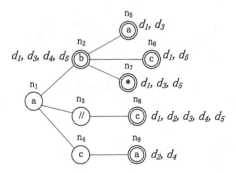

Fig. 2. DFA for the query set

$q_5 = $ "/a/c/a", users need to download P_1 and P_3 in Fig.5 to get all matched documents IDs.

Although the concept of finite automaton has been used in other XML index structures such as YFilter, DFAI is unique because of the following differences. First, in DFAI, the "//" and "*" operators are treated as normal element nodes.

Consequently, they do not introduce any non-determinism to DFAI. However, in YFilter, "//" and "*" can match any elements at any/current level in the current document, which introduces non-determinism. Second, DFAI is designed for XPath queries while YFilter is designed for XML data documents. Third, DFAI is designed for sequential access on the broadcast channel while YFilter is designed for random access in the main memory.

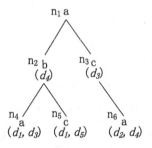

Fig. 3. A PCI index

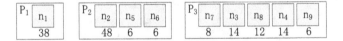

Fig. 4. Node structure of DFAI

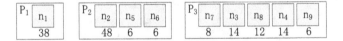

Fig. 5. Partitioning of DFAI

4 Compression Strategy

In this section, first we proof our observation and then we propose our improvement strategy based on the observation.

In Fig.2, we observe that, the matched documents IDs list of an accept state node is a subset of that of its ancestor accept state nodes, which can be proofed in the following part.

Theorem 1. *Assume that Anc is an accept state node in DFA, Des is its descendant accept state node and that Lanc and Ldes are the matched documents IDs lists of Anc and Des respectively. Then Ldes $\subseteq$ Lanc.*

Proof. Anc and *Des* are accept state nodes in DFA, so there are queries q_{anc} and q_{des} satisfied by *Anc* and *Des* respectively. Because of the determinism of DFA, there is only one path from the root of DFA to *Anc* and it's the same with *Des*. Hence, the path from root to *Des* must pass *Anc* definitely, otherwise, *Des* isn't a descendant node of *Anc*. So, q_{anc} (i.e. the path from root to *Anc*) is just a front part of q_{des} (i.e. the path from root to *Des*). The matched documents IDs list of q_{des} (i.e. *Ldes*) is included in that of q_{anc} (i.e. *Lanc*) according to XPath definition. So *Ldes* $\subseteq$ *Lanc*. $\square$

For example, the matched documents IDs list of node n_5 in Fig.2 is $\{d_1, d_3\}$. Its parent node n_2's matched documents IDs list is $\{d_1, d_3, d_4, d_5\}$. The documents IDs lists of nodes n_6 and n_7 also prove our observation. If an accept state node is a leaf node, its matched documents IDs list will repeat many times in its ancestor accept state nodes which results in a bigger index size, even though two-tier index method is adopted.

Based on our observation above, we propose an improvement strategy called Compression Strategy to reduce the index size. As the child accept node's matched documents IDs list is always a subset of that of its ancestor nodes, we use bit to represent the position of the child node's documents ID in its first ancestor accept state node. The pseudo-code of the strategy is shown in Algorithm 1. A new example with XML documents' size 12 is depicted in Fig.6.

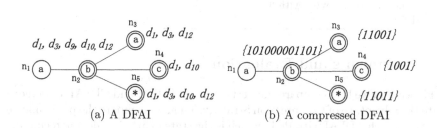

(a) A DFAI   (b) A compressed DFAI

Fig. 6. An example of compressed DFAI

Take n_3 in Fig.6(a) for example, its first ancestor accept state node is n_2 whose matched documents IDs list is $\{d_1, d_3, d_9, d_{10}, d_{12}\}$. d_1, d_3, d_{12} of n_3 are in the first, second and fifth position of $\{d_1, d_3, d_9, d_{10}, d_{12}\}$ respectively. So $\{d_1, d_3, d_{12}\}$ can be represented as '11001' as shown in Fig.6(b) with bits. In reality, we should extend it to a full byte with 0, that is '11001000'. For the first accept state node from root (i.e. the first accept state node in a single path. e.g. n_2 in path $/n_1/n_2$), we need to express the matched document IDs with positions in IDs of XML documents set like n_2 in Fig.6(b).

DFAI with Compression Strategy is named Compressed Deterministic Finite Automaton Index (abbreviated as CDFAI) in next sections.

Algorithm 1 Compression Algorithm

Require: DFAI, XML documents IDs as D
Ensure: Compressed DFAI
 1: initialize r to represent the root node of DFAI
 2: **for** each node $n \in$ DFAI **do**
 3:     initialize the matched documents IDs bit array of n: $B \leftarrow \emptyset$;
 4:     **if** n is the first accept state node from root to n **then**
 5:         **for** each $i \in D$ from smaller document ID to bigger document ID **do**
 6:             **if** $i \in n$'s documents IDs list **then**
 7:                 append a bit '1' to B;
 8:             **else**
 9:                 append a bit '0' to B;
10:             **end if**
11:         **end for**
12:     **else**
13:         find n's first ancestor accept state node from n to r and get its matched documents IDs list L;
14:         **for** each $i \in L$ from smaller document ID to bigger document ID **do**
15:             **if** $i \in n$'s documents IDs list **then**
16:                 append a bit '1' to B;
17:             **else**
18:                 append a bit '0' to B;
19:             **end if**
20:         **end for**
21:     **end if**
22:     extend B to full bytes with '0';
23: **end for**

---

5   Experiments and Evaluation

In this section, we first compare the performance of DFAI and CDFAI, to demonstrate the advantage of Compression Strategy. Then we compare the performance of our approach CDFAI with PCI , which is the state of art air indexing technique in on-demand broadcast, in terms of tuning time and access time.

A.Environment Setup

Similar as existing work [2,3,4], simple XPath queries are used in our experiments. A simple query has the path expression format of

$$P = /N|//N|PP, N = E| * \tag{1}$$

Here, "E" is the element label; "/" means the child axis; "//" means the descendant axis; and "*" is the wildcard.

In our simulation, we use News Industry Text Format (NITF) DTD which is a synthetic data set, and XML Documents generated by the IBM's Generator tool [13]. We generate synthetic XPath queries without predicates by implementing the modified version of the generator [4].

In our experiments, we use the same scheduling algorithm for CDFAI and PCI approaches in order to maintain fairness of all comparisons between them.

In other words, the time cost to retrieve the documents under a scheduling algorithm is independent of the index structures, and hence it remains the same under different index structures. As a result, in our simulation, we only consider the tuning time and access time of index without mentioning the documents retrieval cost. Besides, we assume the bandwidth is constant; the tuning time and access time of CDFAI and PCI can be evaluated by the number of bytes retrieval/broadcasted, and hence, the unit of tuning time and access time is byte. In order to express the figure well, our minimal unit is 1k bytes in the experiments. As mentioned in Section 3, we divide the index into blocks (i.e. block is the minimal unit to be broadcasted. In the experiments, we set the block size to 4K bytes). When users retrieve the index, they have to download a whole block each time. System default parameters are listed in Table. 1.

Table 1. Experimental Setup

Variable	Description	Default value
$N(q)$	The number of queries submitted by users during the broadcast cycle	500
$N(d)$	Total relevant documents number in server	1000
$D(q)$	Maximal depth of queries	8
$Prob$	The probability of wildcard "*" and double slash "//" in queries	0.2

B. Comparison of DFAI and CDFAI

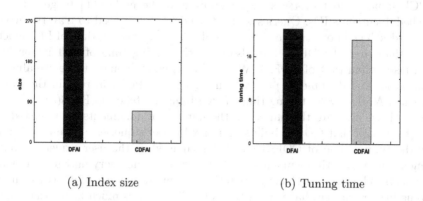

(a) Index size   (b) Tuning time

Fig. 7. Comparison of index size and tuning time

First, we evaluate the improvement of CDFAI with default value in terms of access time and tuning time. As the access time is dependent on the index size, we use the size of index to represent the performance. It's observed that CDFAI significantly reduces the index size and slightly decreases the tuning time compared with DFAI as shown in Fig.7 because of adopting the Compression Strategy.

C. Comparison of CDFAI and PCI

Next, we evaluate access time and tuning time of CDFAI and PCI. We consider the impacts of four parameters, namely the number of queries denoted as $N(q)$, total number of relevant documents denoted as $N(d)$, maximal depth of queries denoted as $D(q)$ and the probability of "*" and "//" in queries denoted as $Prob$. In each set of experiments, we change one parameter only and fix the others at their default values described in Table 1.

1) Tuning Time

In the second set of experiments, we study the tuning time performance presented in Fig.8.

For CDFAI, the tuning time to retrieve relevant index can be approximated by the depth of the query and the block size. Recall that "*" and "//" are treated as normal element nodes.

For PCI, the tuning time for queries with no "*" and "//" is almost the same as that under CDFAI, i.e. the depth of the query multiplying the index block size. However, for queries with "*" and "//", PCI incurs much longer tuning time since the client has to access more packets to get the matched documents.

First, we evaluate the tuning time under different $N(q)$, as depicted in Fig.8(a). Although $N(q)$ increases, the tuning time of CDFAI almost remains the same. The reason is that tuning time of CDFAI depends on the average depth of user queries (i.e. the steps the CDFA needs to jump to the matched state). The increasing $N(q)$ doesn't change the average depth of user queries which results in a steady tuning time. However, with a increasing $N(q)$, fewer and fewer branches and documents are pruned by PCI. For an accept state node of PCI, it needs to traverse a bigger subtree of the node [2,6] to get all the matched documents IDs. Queries with "*" and "//" also need to traverse much more nodes. The above two reasons not only make the tuning time of PCI much longer than that of CDFAI, but also push the tuning time of PCI increasing mush faster than that of CDFAI. When $N(q)$ is large enough, there are almost no branches and documents pruned. tuning time of PCI will remain the same (i.e. when $N(q)>= 800$, tuning time of PCI remains steady in Fig.8(a)).

Second, we evaluate the impact of the number of documents, as reported in Fig.8(b). Recall that CDFAI builds its index by user's queries, so we may think that the tuning time of CDFAI stays the same when the user's queries don't change. However, with documents number increases, one query may match more documents. The matched documents IDs in accept state nodes of index tree may become bigger. Notice that tuning time of PCI increases much faster than that of CDFAI, because PCI is built based on the documents whose number is increasing, while CDFAI is based on the fixed user queries.

Third, we evaluate the impact of $Prob$, as depicted in Fig.8(c). The tuning time of PCI increases as $Prob$ grows. This is because the client has to access

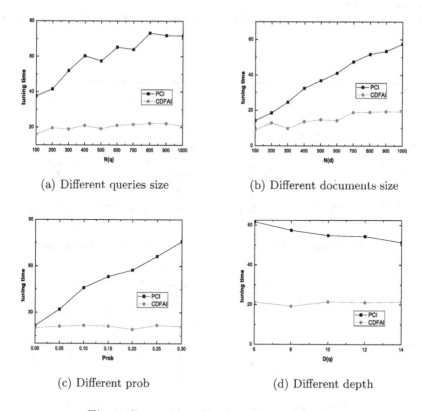

(a) Different queries size   (b) Different documents size

(c) Different prob   (d) Different depth

Fig. 8. Comparison of tuning time performance

much more packets to find all the matched documents which prolongs the performance. However, the tuning time of CDFAI is almost unchanged under different *Prob* as it is mainly determined by the average depth of the query set and less impacted by the selectivity of the queries (i.e. *Prob*).

Finally, we evaluate the impact of the maximal query depth $D(q)$, as reported in Fig.8(d). As the maximal depth of queries increases, more packets have to be accessed in order to get all the matched documents. On the other hand, a deeper query will have fewer matched documents. As a result, increasing $D(q)$ will bring not only positive but also negative influences on the tuning time performance. It can be found in the figure that tuning time of PCI decreases first, remains almost the same in the middle and decreases in the end. It means, initially the reduced documents retrieval has the dominant impact; in the middle, the increasing of the index packets and the reduced documents retrieval almost have the same impact; and later the reduced documents retrieval plays the dominant role again. On the contrary, CDFAI demonstrates a relatively stable performance with the interaction of depth and selectivity (i.e. *Prob*) with tuning time just about 30% of PCI.

2) Access Time

Then, we compare their access time performance, with results presented in Fig.9.

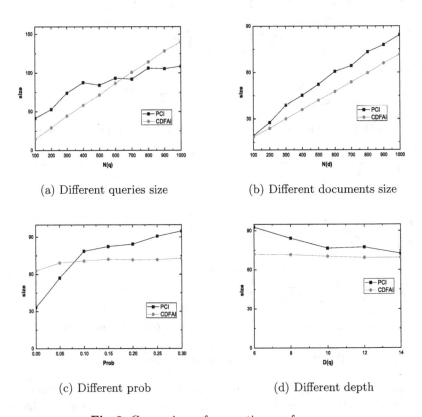

(a) Different queries size   (b) Different documents size

(c) Different prob   (d) Different depth

Fig. 9. Comparison of access time performance

First, the access time of CDFAI and PCI under different $N(q)$ is depicted in Fig.9(a). It is observed that both index sizes of CDFAI and PCI increase with the growing query number. Obviously, when query number increases, the index size of CDFAI will increase proportionally to the query number for the index is built by user queries. For PCI, with more queries, more branches of more documents are matched. The size of pruned branches decreases resulting in the index size increasing with the growing query number. We also can see that the size of CDFAI exceeds that of PCI when the query number is between 600 and 700. That's because as the query number increases, more and more branches of the documents are retained for PCI. The room for the index size of PCI to rise is limited as the size of all documents which is the upper bound of the index size, is fixed. However, for CDFAI with increasing query number, more different queries are added to the growing index.

Second, when the number of documents set increases, the performance of CD-FAI and PCI can be seen in Fig.9(b) . Both of them have the same trend as $N(d)$ increases. A larger documents set means more documents are matched by user queries, so CDFAI and PCI's index sizes both grow with the increasing $N(d)$.

Third, Fig.9(c) shows the change of the CDFAI and PCI index size with $Prob$. It can be found that with the $Prob$ increases, both sizes of CDFAI and PCI grow. The reason is that, with bigger $Prob$, "*" and "//" appear more often in queries and more branches of documents are retained. In other words, fewer branches of documents are pruned by PCI, so the index size increases with $Prob$. For CDFAI, more "*" and "//" means more matched documents. Therefore, the accept state nodes in CDFAI are associated with more matched documents IDs lists which lead to a bigger index. When $Prob$ is less than 0.1, index size of CDFAI is larger than that of PCI, as pruning strategy of PCI behaves well with a small $Prob$.

Finally, the performance under different $D(q)$ is depicted in Fig.9(d). As maximal depth increases, the index size of CDFAI decreases first and then remains the same. The reason is that when depth becomes bigger and bigger, fewer and fewer documents are matched by queries. Recall that CDFAI is built by user queries. A deeper query means CDFAI needs more states to build the index. This is the negative influence. The positive influence is that a deeper query has fewer matched documents and hence the corresponding accepting state node in CDFAI has a shorter matched documents IDs list, resulting in a smaller index. Then we can understand why at the very beginning the index size of CDFAI decreases and then later, it remains the same. That's because positive factor plays a more significant role than negative factor first and then cancels each other out. However, for PCI, a deeper query means more chances for "*" and "//" to appear resulting in more branches retained. At the same time, a deeper query also has fewer matched documents. So at the very beginning, PCI decreases and then slightly increases. Both factors are interact with each other with the increasing of $D(q)$.

From the above analysis we can draw the conclusion that CDFAI is better in its tuning time especially when the probability of wildcard "*" and double slash "//" is high (see fig.8(c)), while the access time is almost the same as that of PCI. Even though the probability of "*" and "//" is very low (e.g. 0.05), the tuning time is just about 1/2 of that of PCI. Of course, when the probability is ZERO, they will fall to the same level. While the queries increase, CDFAI will gradually lose its advantage in access time, but always keeps its advantage in tuning time. So for time-critical and energy-conservation on-demand wireless environments, CDFAI has its incomparable advantages. For other wireless environments that need to gather a large number of queries per cycle and don't pay much attention to limited power, PCI can be considered.

6   Conclusion

In this paper, we propose a novel indexing method, namely DFAI to reduce the tuning time and access time for users in on-demand XML data broadcast.

Existing approaches are typically sensitive to the probability of "*" and "//" while our approach DFAI is not. A compression strategy is proposed to further reduce the index size of DFAI, namely CDFAI. The simulations show us that CDFAI achieves a better performance compared with PCI. In the near future, we plan to find the relationships between document IDs sets in different accept state nodes to further reduce access time and tuning time of DFAI. Promoting DFAI to support twig queries is another future work of us.

References

1. Xu, J., Lee, D.-L., Hu, Q., Lee, W.-C.: Data Broadcast. In: Handbook of Wireless Networks and Mobile Computing. John Wiley & Sons (2002)
2. Sun, W., Yu, P., Qin, Y., Zhang, Z., Zheng, B.: Two-Tier Air Indexing for On-Demand XML Data Broadcast. In: ICDCS 2009, pp. 199–206 (2009)
3. Selcuk Candan, K., Hsiung, W.-P., Chen, S., Tatemura, J., Agrawal, D.: AFilter: Adaptable XML Filtering with Prefix-Caching and Suffix-Clustering. In: VLDB 2006, pp. 559–570 (2006)
4. Diao, Y., Altinel, M., Franklin, M.J., Zhang, H., Fischer, P.M.: Path sharing and predicate evaluation for high-performance XML filtering. ACM Trans. Database Syst. (TODS) 28(4), 467–516 (2003)
5. Goldman, R., Widom, J.: DataGuides: Enabling Query Formulation and Optimization in Semistructured Databases. In: VLDB 1997, pp. 436–445 (1997)
6. Vagena, Z., Moro, M.M., Tsotras, V.J.: RoXSum: Leveraging Data Aggregation and Batch Processing for XML Routing. In: ICDE 2007, pp. 1466–1470 (2007)
7. Park, C.-S., Kim, C.S., Chung, Y.D.: Efficient Stream Organization for Wireless Broadcasting of XML Data. In: Grumbach, S., Sui, L., Vianu, V. (eds.) ASIAN 2005. LNCS, vol. 3818, pp. 223–235. Springer, Heidelberg (2005)
8. Park, S.-H., Choi, J.-H., Lee, S.: An Effective, Efficient XML Data Broadcasting Method in a Mobile Wireless Network. In: Bressan, S., Küng, J., Wagner, R. (eds.) DEXA 2006. LNCS, vol. 4080, pp. 358–367. Springer, Heidelberg (2006)
9. Chung, Y.D., Lee, J.Y.: An indexing method for wireless broadcast XML data. Inf. Sci. (ISCI) 177(9), 1931–1953 (2007)
10. Qin, Y., Sun, W., Zhang, Z., Yu, P., He, Z., Chen, W.: A Novel Air Index Scheme for Twig Queries in On-Demand XML Data Broadcast. In: Bhowmick, S.S., Küng, J., Wagner, R. (eds.) DEXA 2009. LNCS, vol. 5690, pp. 412–426. Springer, Heidelberg (2009)
11. Su, T.-C., Liu, C.-M.: On-Demand Data Broadcasting for Data Items with Time Constraints on Multiple Broadcast Channels. In: Yoshikawa, M., Meng, X., Yumoto, T., Ma, Q., Sun, L., Watanabe, C. (eds.) DASFAA 2010. LNCS, vol. 6193, pp. 458–469. Springer, Heidelberg (2010)
12. Wu, J., Liu, P., Gan, L., Qin, Y., Sun, W.: Energy-Conserving Fragment Methods for Skewed XML Data Access in Push-Based Broadcast. In: Wang, H., Li, S., Oyama, S., Hu, X., Qian, T. (eds.) WAIM 2011. LNCS, vol. 6897, pp. 590–601. Springer, Heidelberg (2011)
13. Diaz, A., Lovell, D.: XML Generator, http://www.alphaworks.ibm.com/tech/xml-generator

Colored Range Searching on Internal Memory

Haritha Bellam[1], Saladi Rahul[2], and Krishnan Rajan[1]

[1] Lab for Spatial Informatics, IIIT-Hyderabad, Hyderabad, India
[2] Univerity of Minnesota, Minneapolis, MN, USA

Abstract. Recent advances in various application fields, like GIS, finance and others, has lead to a large increase in both the volume and the characteristics of the data being collected. Hence, general range queries on these datasets are not sufficient enough to obtain good insights and useful information from the data. This leads to the need for more sophisticated queries and hence novel data structures and algorithms such as the *orthogonal colored range searching (OCRS)* problem which is a *generalized* version of orthogonal range searching. In this work, an efficient main-memory algorithm has been proposed to solve $OCRS$ by augmenting k-d tree with additional information. The performance of the proposed algorithm has been evaluated through extensive experiments and comparison with two base-line algorithms is presented. The data structure takes up linear or near-linear space of $O(n \log \alpha)$, where α is the number of colors in the dataset ($\alpha \leq n$). The query response time varies minimally irrespective of the number of colors and the query box size.

1 Introduction

Multi-attribute queries are becoming possible with both the increase in kind of attributes the datasets are able to store and also the processing power needed to do so. In addition to specific queries across the multiple attributes, there is an increasing need for range queries across these attributes especially in fields like GIS, business intelligence, social and natural sciences. In most of these one needs to identify a set of classes that have attribute values lying within the specified ranges in more than one of these attributes. For instance, consider a database of mutual funds which stores for each fund its annual total return and its beta (a real number measuring the fund's volatility) and thus can be represented as a point in two dimensions. Moreover, the funds are categorized into groups according to the fund family they belong to. A typical two-dimensional orthogonal colored range query is to determine the families that offer funds whose total return is between, say 15% and 20%, and whose beta is between, say, 0.9 and 1.1 [8]. An another example can be to identify potentially suitable locations for afforestation programs and the choice of right vegetation types for these locations. The spatial database can contain multiple themes like soil class, weather related parameters (temperature, relative humidity, amount of precipitation), topography and ground slope, while for each vegetation type there exists a suitable range of values in these themes/attributes. In such cases, one is interested

S.-g. Lee et al. (Eds.): DASFAA 2012, Part II, LNCS 7239, pp. 111–125, 2012.
© Springer-Verlag Berlin Heidelberg 2012

in finding the best match between the two, leading to a need to answer a set of colored range queries. Generally speaking, in many applications, the set S of input points come aggregated in groups and we are more interested in the groups which lie inside q rather than the actual points. (a group lies inside q if and only if at least one point of that group lies inside q.) For convenience, we assign a distinct color to each group and imagine that all the points in the group have that color. This problem is known as *colored range searching*. In this paper, we specifically consider the problem of *orthogonal colored range searching* (see Figure 1) where the query is an orthogonal box in d-dimensional space (i.e $q = \Pi_{i=1}^{d}[a_i, b_i]$) and propose a algorithm to effectively deal with it.

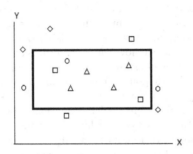

Fig. 1. A set of ten colored points lying in a plane. For the given orthogonal query rectangle, a standard orthogonal range query would report all the points lying inside the query rectangle. However, an orthogonal colored range searching (OCRS) query would report the unique colors of the points lying inside the rectangle.In this figure, each color is represented with a different symbol.The output for this query would be △, □ and ○. Note that ◇ is not reported.

Orthogonal colored range searching ($OCRS$) can be described in an SQL-like language (basically a GROUP-BY operation). However, the current DBMS's do not provide with efficient query processing techniques for $OCRS$. $OCRS$ query will be executed by first executing the standard orthogonal range searching query. This will report all the points lying inside q. Then all the reported points are scanned to filter out the unique colors. In many real-life applications, there could be a huge difference between the number of points lying within q and the number of unique colors among them. This can lead to poor query time.

$OCRS$ and the general class of colored intersection problems [8] have been extensively studied in the field of Computational Geometry. Efficient theoretical solutions for $OCRS$ though have been provided on internal memory (or main memory) with the objective of providing worst-case optimal performance. However they suffer from the following shortcomings:

1. Most of these solutions are good in theory but are extremely difficult to implement.
2. The space of the data structures proposed increase exponentially in terms of the dimension size. Building linear or near-linear size data structures is a requirement in In-memory database systems (IMDS) that are steadily growing

[1]. In fact we implemented one of these theoretical solutions [7] but had to be discarded since it was not fitting into the main-memory for the datasets we tested on (see Figure 3).

3. For most of the practical applications we need not come up with solutions which need to be optimal even in the worst case scenario. In practice the worst case scenario occurs *extremely rarely*.

We seek to to build a solution which works well for *most of the cases/scenarios*. The problem that will be addressed in this paper is formally stated as-

- Preprocess a set S of n colored points lying in d-dimensional space (or $\mathbb{R}^d$) into an index structure, so that for any given orthogonal query box $q{=}\Pi_{i=1}^d[a_i, b_i] \subseteq \mathbb{R}^d$, all the distinct colors of the points of S lying inside q need to be reported efficiently.

Since we are dealing with main-memory, the main focus in this paper is on building space-efficient data structures for answering orthogonal colored range searching query. The following are the main contributions of the paper:

- We come up with non-trivial techniques for building the data structure and for answering the query algorithm of orthogonal colored range searching ($OCRS$) on main-memory. The objective is to *use minimal space* while maintaining efficient query time.
- Detailed experiments and theoretical analysis have been carried out to show the performance of our technique against two base-line algorithms which can be used to solve the orthogonal range searching problem in relation to the distribution that the data exhibits, the query size and the number of colors in the data. These two base-line algorithms are described later in the paper.

2   Related Work

Orthogonal range searching is one of the most popular and well studied problem in the field of computational geometry and databases. The formal definition is the following: "Preprocess a set S of n points lying in d-dimensional space (or $\mathbb{R}^d$) into an index structure, so that for any given orthogonal query box $q{=}\Pi_{i=1}^d[a_i, b_i] \subseteq \mathbb{R}^d$, all the points of S lying inside q need to be reported/counted efficiently". There have been a numerous data structures proposed in the computational geometry literature to handle orthogonal range searching query. Traditionally, the researchers in computational geometry have been interested in building solutions for this problem which aim at coming up worst-case efficient query time solutions, i.e., ensuring that the query time is low for *all* possible values of the query. The most popular among them are the range-trees [6] and the k-d tree [6,5]. The original range-tree when built on n points in d-dimensional space took $O(n \log^d n)$ space and answered queries in $O(\log^{d-1} n + k)$ query time, where k are the number of points lying inside the query region q. By using the technique of fractional cascading we can reduce the query time by a log factor [6].

In a dynamic setting, the range-tree uses $O(n \log^{d-1} n)$ space, answers queries in $O(\log^{d-1} n \log \log n + k)$ time and handles updates in $O(\log^{d-1} n \log \log n)$ time. A k-d tree when built on n points in d-dimensional space takes up *linear space* and answers a range-query in $O(n^{1-1/d} + k)$ time. Updates can be handled efficiently in it. A detailed description of k-d tree's construction and query algorithm shall be provided later.

In the field of databases there have been a significant number of index structures proposed to answer an orthogonal range-query. These are practical solutions which are optimized to work well for the average case queries (with the assumption that the worst-case query will occur rarely). Perhaps no other structure has been more popular than the *R-tree* proposed in 1984 by Guttman [9]. It's main advantage arises from the fact that it is versatile and can handle various kinds of queries efficiently (range-queries being one of them). K-d tree also enjoys similar benifits of being able to handle different kinds of queries efficiently. Varaints of R-tree such as R* tree [4], R+ tree [13], Hilbert tree and Priority R-tree are also quite popular index structures for storing multi-dimensional data.

Orthogonal colored range searching (OCRS) happens to be a *generalization* of the orthogonal range searching problem. Janardan et al. [10] introduced the problem of OCRS. Gupta et al. [7] came up with dynamic (both insertions and deletions) solutions to this problem in 2-dimensional space and a static solution in 3-dimensional space. The best theoretical solution to this problem in $d=2$ has been provided by Shi et al. [14] which is a static data structure taking up $O(n \log n)$ space and $O(\log n + k)$ time, where 'k' is the number of colors lying inside the query rectangle. For $d=3$, the only known theoretical solution takes up $O(n \log^4 n)$ space and answers query in $O(\log^2 n + k)$ time [7]. For $d > 3$, there exists a data structure which answers queries in $O(\log n + k)$ time but takes up $O(n^{1+\epsilon})$ space which from a theoretical point of view is not considered optimal ($O(n$ polylog $n)$ is desirable). As stated before, the objective in the computational geometry field has been to come up with main-memory algorithms which are optimal in worst-case scenario. A good survey paper on OCRS and other related colored geometric problems is [8].

Agarwal et al. [3] solved *OCRS* on a 2-dimensional grid. Recently, there has been some work done on range-aggregate queries on colored geometric objects. In this class of problems, the set S of geometric objects are colored and possibly weighted. Given a query region q, for each distinct color c of the objects in $S \cap q$, the tuple $\langle c, \mathcal{F}(c) \rangle$ is reported where $\mathcal{F}(c)$ is some function of the objects of color c lying inside q [11,12]. Examples of $\mathcal{F}(c)$ include sum of weights, bounding box, point with maximum weight etc. In [11,12], theoretical main-memory solutions have been provided.

3   Existing Techniques to Solve OCRS

In the database community OCRS problem is solved by using the *filter and prune* technique. Two base-line algorithms are described which follow this paradigm. Then we shall discuss about the theoretical solutions which emerged from computational geometry for answering OCRS. Based on these discussions we shall motivate the need for a practical data structure and algorithm which will work efficiently in a main-memory or internal memory environment.

3.1   Base-line Algorithms

Here we shall describe two base-line algorithms with which we shall compare our proposed method. These two algorithms solve the standard orthogonal range searching problem. We will describe how they can be modified to answer the orthogonal *colored* range searching problem.

1. **Plain Range Searching (PRS)** In this method, we build a k-d tree on all the points of S. Given an orthogonal query box q, we query the k-d tree and report all the points of S lying inside q. Then we filter out the unique colors among points that are lying inside q. The space occupied by a k-d tree is $O(n)$. The query time will be $O(n^{1-1/d} + |S \cap q|)$, where $S \cap q$ is the set of points of S lying inside q. If the ratio of $|S \cap q|$ and the number of unique colors is very high, then this technique will not be efficient.

2. **Least Point Dimension (LPD)** A set S of n points lie in a d-dimensional space with some color associated to each point. For each dimension i we project S onto dimension i, then we sort the points of S w.r.t coordinate values in dimension i and store them in an array A_i. Given an orthogonal query $q = \Pi_{i=1}^{d}[a_i, b_i]$, we decompose the query q into intervals corresponding to each dimension (interval corresponding to dimension i will be $[a_i, b_i]$). Then we do a binary search on all the arrays A_i ($\forall\ 1 \leq i \leq d$) with $[a_i, b_i]$ to find the number of points of S lying inside it. The array A_l, in which the total number of points is least is chosen. Each point of A_l which occurs within $[a_l, b_l]$ is also tested for its presence inside q. Among all the points which pass the test, the unique colors of these points are filtered out. Query time for LPD is $O(d \log n + \beta)$ where β is the number of points in A_l which lie inside $[a_l, b_l]$. Note that the performance of LPD algorithm is dependent on the value of β. If the ratio of β and the number of unique colors is high, then the performance of LPD will not be good.

3.2   Existing Theoretical Solutions

As mentioned before in Section 2 (related work), there have been theoretical solutions proposed to answer OCRS. Only a few of these solutions can be implemented and tested. Most of them are meant for theoretical interest and are impossible to implement. We implemented the solution proposed by Gupta et al. [7] for $d = 2$ (semi dynamic solution which handles only insertions). Theoretically it takes up $O(n \log^2 n)$ space, $O(\log^2 n + k)$ query time (k is the number of unique colors reported) and $O(\log^3 n)$ amortized insertion time. We observe that for large datasets $O(n \log^2 n)$ space is not acceptable as that would mean storing $O(\log^2 n)$ copies of each data item in the internal memory. Experiments on real-life and synthetic datasets confirmed our intuition that this data structure takes up a lot more space than the base-line algorithms and the proposed solution in this paper (see figure 3). Therefore, *this solution was discarded.* For higher dimnesions ($d > 2$), the theoretical solutions get too complicated to be implemented and tested.

4   Proposed Algorithm

As we are trying to build main-memory structures, the highest priority is to minimize the space occupied while trying to keep the query time competitive.

4.1   Data Structure

In this section we construct a data structure named BB k-d tree, which is based on a regular k-d tree but is augmented with additional information at each internal node. We chose k-d tree as our primary structure since it takes up linear space for indexing points lying in any dimension and though it has a poor theoretical query performance, it does well in practice [15,5].

We shall first describe the structure for a 2-dimensional scenario The primary structure of BB k-d tree is a conventional k-d tree. A splitting line $l(v)$, here choosen as a median, is stored at every node (v) which partitions the plane into two half-planes. We denote the region corresponding to a node v by $region(v)$. The region corresponding to the left child and right child of the node v are denoted as $region(lc(v))$ and $region(rc(v))$, where $lc(v)$ and $rc(v)$ denote the left and right children of v, respectively.

At each internal node v, apart from the regular information two height-balanced binary search trees $\mathcal{T}_l$ and $\mathcal{T}_r$ are stored. $\mathcal{T}_l$ is built as follows: Let c be one of the colors among the distinct colors of the points lying in the subtree rooted at $lc(v)$. The bounding box of the points of color c in the subtree of $lc(v)$ is calculated and kept at a leaf in $\mathcal{T}_l$. A bounding box for a point set in the plane is the rectangle with the smallest measure within which all the points lie. This process is repeated for each color c which lies in the subtree of $lc(v)$. The colors stored in the leaves of $\mathcal{T}_l$ are sorted lexicographically from left to right. To make navigation easier, each leaf in $\mathcal{T}_l$ is connected to its adjacent leaves. This forms a doubly linked list among the leaf nodes of $\mathcal{T}_l$. Also, the root node of $\mathcal{T}_l$ maintains a special pointer pointing to the leftmost leaf of $\mathcal{T}_l$. Similarly, $\mathcal{T}_r$ is constructed based on the points stored in the subtree of $rc(v)$.

BB k-d tree can also be built for point sets in 3 or higher-dimensional space. The construction algorithm is very similar to the planar case: At the root, we split the set of points into two subsets of roughly the same size by a hyperplane perpendicular to the x-axis. In other words, at the root the point set is partitioned based on the first coordinate of the points. At the children of the root the partition is based on the second coordinate, at nodes of depth two on the third coordinate, and so on, until depth $d - 1$ where we partition on the last coordinate. At depth d we start all over again, partitioning on first coordinate. The recursion stops when there is only one point left, which is then stored at a leaf. The binary search trees ($\mathcal{T}_l$ and $\mathcal{T}_r$) at each internal node are constructed in the same manner as described above.

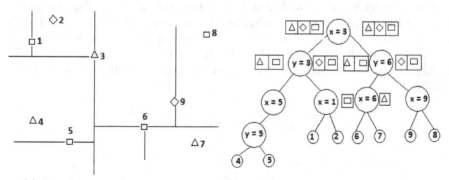

(a) Set of 9 points lying on a plane   (b) BB-tree corresponding to these nine points. Note that all the arrays are not represented for the sake of clarity

Fig. 2. An example of a BB-tree. At every node there are two arrays. The left/right array contains the unique colors present in the left/right subtree along with their respective bounding boxes.

In the example shown in Fig 2 there are 9 points and a color has been assigned to each point. The construction of the K-d tree on these points has been showed in Fig 2(a). At the root we split the point set P with a vertical line $x = 3$ into two subsets of roughly equal size. At the left and right children we split the point set using horizontal median lines. In this way we build a traditional k-d tree first. At each internal node the bounding box of all the colors lying in the left (resp. right) subtree are stored in $\mathcal{T}_l$ (resp. $\mathcal{T}_r$). For example, the left child of the root node has the bounding box of $\triangle$ points 3 and 4, bounding box of $\square$ points 1 and 5, bounding box of the $\diamond$ point 2, are computed and stored at the left binary search tree ($\mathcal{T}_l$) of the of root node of k-d tree. Similarly $\mathcal{T}_r$ is also constructed.

Theorem 1. *A BB-tree with n points and α colors takes $O(n\log\alpha)$ storage*

Proof. To store a set of n points in d-dimensional space, a normal k-d tree uses $O(n)$ space. The space occupied by the internal binary search trees $\mathcal{T}_l$ and $\mathcal{T}_r$ dominates the overall space complexity. Let the number of distinct colors of the points in S be α. We want to find the space occupied in the worst case. The height of a k-d tree will be $O(\log n)$. The height of a node is the length of the longest downward path to a leaf from that node. Consider a node v at height h. If $h \leq \lfloor \log \alpha \rfloor$ (i.e. $2^h \leq \alpha$), then the number of leaves in the subtree rooted at v will be $\leq 2^h \leq \alpha$. Then the size of $\mathcal{T}_l$ and $\mathcal{T}_r$ will be bounded by 2^h (worst case being each point having a unique color). The total space occupied by all the nodes at a particular level h ($\leq \lfloor \log \alpha \rfloor$) will be $O((2^{\log n - h}) \times 2^h) \equiv O(n)$, where $O(2^{\log n - h})$ is the number of nodes at level h. Then the total space occupied by all the internal binary search trees stored at primary nodes having height $\leq \lfloor \log \alpha \rfloor$ will be $\Sigma_{h=0}^{\lfloor \log \alpha \rfloor} O(n) \equiv O(n \log \alpha)$.

Now consider a node v at height $h > \lfloor \log \alpha \rfloor$ (i.e. $2^h > \alpha$). The number of leaves in the subtree at v is bounded by 2^h. However in this case the size of $\mathcal{T}_l$ and $\mathcal{T}_r$ is bounded by $O(\alpha)$. If each node at level h has array size $O(\alpha)$, then the total size of all the arrays at level l will be $O(2^{\log n - h} \times \alpha). \equiv O(\frac{n\alpha}{2^h})$. The overall size of the internal binary search trees added over all levels $h > \lfloor \log \alpha \rfloor$ will be

$$O(\Sigma_{h=\lfloor \log \alpha \rfloor}^{\log n} \frac{n\alpha}{2^h}) \equiv O(n\alpha \Sigma_{h=\lfloor \log \alpha \rfloor}^{\log n} \frac{1}{2^h})$$

$$\equiv O(n\alpha(\Sigma_{h=0}^{\log n} \frac{1}{2^h} - \Sigma_{h=0}^{\lfloor \log \alpha \rfloor - 1} \frac{1}{2^h})) \equiv O(n\alpha(\frac{1}{\alpha} - \frac{1}{n})) \equiv O(n)$$

Therefore, the total size of the BB k-d tree will be $O(n \log \alpha) + O(n) \equiv O(n \log \alpha)$.

□

4.2   Query Algorithm

We now turn to the query algorithm. We maintain a global boolean array *outputSet* of size α, which contains an entry for each distinct color of set S. Given a query box q, an entry of *outputSet*[c] being *true* denotes that color c has a point in the query box and we need not search for that color any more inside our data structure. The query algorithm is described in the form of a pseudo-code in Algorithm 1. The input parameters are the current node being visited (v), the orthogonal query box q and a boolean array A. A has a size of α and $A[c]$, $1 \leq c \leq \alpha$, being *true* denotes that we need to search if color c has a point in the subtree of currently visited node (i.e. v), else we need not.

The processing of the query box q commences from the root of the BB k-d tree. Initially, all elements of A are set to *true* and all elements in *outputSet* are set to *false*. If the current node v is not a leaf, then we intialize two arrays A_l and A_r of size α to *false* (lines $4 - 5$). A_l (and A_r) are updated later in the procedure and will be used when the children of the current node will be visited.

If $region(lc(v))$ is fully contained in q, then all the colors stored in the leaves of secondary tree $\mathcal{T}_l$ are set to *true* in *outputSet* (lines 6–8). However, if $region(lc(v))$ partially intersects q, then we do the following: Using the special pointer at the root of $\mathcal{T}_l$, we will go to the leftmost leaf of $\mathcal{T}_l$. Then we will check if q contains any bounding box b corresponding to each leaf of $\mathcal{T}_l$ (line 11). The adjacent pointers of each leaf help in navigating through the list of leaves. If a bounding box b (of color c) is fully contained in q then there exists at least one point of color c in q. In this case we will update the *outputSet*[c] to *true* and we need not search for this color anymore (lines 11–12). If a bounding box b partially intersects q, then we need to search for that color c in the subtree of v. So, we will update A_l to *true* (lines 13-14).

The last case is when the query box q and bounding box b of all the points in the subtree of v do not intersect at all. In this case, we need not search for any color in the subtree of v. This is automatically reflected in the arrays A_l (or A_r)

Algorithm 1. SearchBBTree(v, q, A)

Input   : A node in BB k-d tree (v) , Query box (q), an array of colors which
need to be searched (A)

Output: An array 'OutputSet' which contains the colors lying inside q

1 **if** v is a leaf and point p stored at v lies in q **then**
2 $\quad$| outputSet[$p.color$]= $true$
3 **else**
4 $\quad$| **forall** colors i from $1 \to \alpha$ **do**
5 $\quad$| $\quad$| $A_l[i] = false$; $A_r[i] = false$
6 $\quad$| **if** $region(lc(v))$ is fully contained in q **then**
7 $\quad$| $\quad$| **forall** colors c in leaves of $\mathcal{T}_l$ **do**
8 $\quad$| $\quad$| $\quad$| outputSet[c]= $true$
9 $\quad$| **else if** $region(lc(v))$ partially intersects q **then**
10 $\quad$| $\quad$| **foreach** bounding box b of color c in leaves of $\mathcal{T}_l$ where $A[c] = true$ **do**
11 $\quad$| $\quad$| $\quad$| **if** q contains b **then**
12 $\quad$| $\quad$| $\quad$| $\quad$ outputSet[c] = $true$
13 $\quad$| $\quad$| $\quad$| **else if** q partially intersects b and outputSet[c] = $false$ **then**
14 $\quad$| $\quad$| $\quad$| $\quad$ $A_l[c] = true$
15 $\quad$| **if** $region(rc(v))$ is fully contained in q **then**
16 $\quad$| $\quad$| **forall** colors c in leaves of $\mathcal{T}_r$ **do**
17 $\quad$| $\quad$| $\quad$| outputSet[c] = $true$
18 $\quad$| **else if** $region(rc(v))$ partially intersects q **then**
19 $\quad$| $\quad$| **foreach** bounding box b of color c in leaves of $\mathcal{T}_r$ where $A[c] = true$ **do**
20 $\quad$| $\quad$| $\quad$| **if** q contains b **then**
21 $\quad$| $\quad$| $\quad$| $\quad$ outputSet[c] = $true$
22 $\quad$| $\quad$| $\quad$| **else if** q partially intersects b and outputSet[c] = $false$ **then**
23 $\quad$| $\quad$| $\quad$| $\quad$ $A_r[c] = true$
24 $\quad$| **if** any $A_l[c] = true$ **then**
25 $\quad$| $\quad$ SearchBBTree ($lc(v)$, q, A_l)
26 $\quad$| **if** any $A_r[c] = true$ **then**
27 $\quad$| $\quad$ SearchBBTree ($rc(v)$, q, A_r)

---

as they are intialized to *false* in the beginning itself. Similar steps are applied for right subtree (lines 15 – 23). If an entry in A_l (or A_r) is *true*, then there is a possibility of existence of that color in the left (or right) subtree of v. So, we shall make a recursive call by passing $lc(v)$ and A_l (lines 24–25). Similarly, if required a recursive call is made by passing $rc(v)$ and A_r (lines 26–27).

4.3   Handling Updates

Insertion and deletion of points can be efficiently handled in the proposed structure. When a point p (having color c) is inserted into BB k-d tree, then an appropriate leaf node is created by the insertion routine of k-d tree [5]. Then we will update all the secondary structures existing on the path (say Π) from the newly created leaf node to the root, in the following manner: At each node $v \in \Pi$, we search for color c in the secondary structures ($\mathcal{T}_l$ and $\mathcal{T}_r$). If no entry of color c exists, then an appropriate leaf is created (in $\mathcal{T}_l$ and $\mathcal{T}_r$) and the bounding box of color c will be the point p. The adjacency pointers are also set appropriately. If the new node happens to be the leftmost leaf (of $\mathcal{T}_l$ or $\mathcal{T}_r$), then the special pointer from the root (of $\mathcal{T}_l$ or $\mathcal{T}_r$) is set to the newly created node. If an entry of color c already exists, then the bounding box of color c is updated. To delete a point p (having color c) from BB k-d tree, we first delete the appropriate leaf node in the primary structure by using the deletion routine of a k-d tree [5]. Before that, at each node v on the path from the leaf node of p to the root we do the following: Search for the color c (in $\mathcal{T}_l$ and $\mathcal{T}_r$). Then update the bounding box of color c. If the bounding box of color c becomes *null*, then that leaf node is removed from ($\mathcal{T}_l$ or $\mathcal{T}_r$). The adjacency pointers are also appropriately adjusted. If the leaf node being removed was the leftmost entry, then the special pointer from the root of ($\mathcal{T}_l$ or $\mathcal{T}_r$) is set to the new leftmost leaf. Next we shall summarize the update time in our structure in a lemma. In the lemma, by random we mean that for each point the coordinate value in each dimension is an independently generated random real number.

Lemma 1. *(Using [5]) The average time taken to insert a random point into the BB k-d tree is $O(\log^2 n)$. The average time taken to delete a random point from a BB k-d tree is $O(\log^2 n)$. In the worst case, the time taken to delete a point from BB k-d tree is $O(n^{1-1/d} \log n)$.*

5   Experimental Setup

All techniques were implemented in C using the same components. The system used is a 1.66 GHz Intel core duo Linux machine with 4 GB RAM. Programs were compiled using cc. We used both synthetic and real life data sets. The following datasets (both real and synthetic) have been used for our experiments (n will denote the number of data points and d denotes the dimensionality):

a) *Uniform Synthetic Dataset (D1)*. n=1,000,000 and d=2.

b) *Gaussian Synthetic Dataset (D2)*. n=100,000, d=2 and σ as 0.2% of the number of points.

c) *Gaussian Skewed Synthetic Dataset (D3)*. $n=100,000$ and $d=2$. The x-coordinate of these points have been assigned using a gaussian function with σ as 10% of the total points and the y-coordinates are produced using a gaussian function with σ 1% of the total points. This helped us in generating a skewed dataset.

d)*Forest Cover real dataset (R1)*. The Cover data set contains 581,012 points and is used for predicting forest cover types from cartographic variables. We used the spatial information to get the 2-d points corresponding to the locations. We used the soil type information, which contains 40 distinct category values to assign colors to the points.

e)*U.S Census real dataset (R2)*. The US Census Bureau data contains 199,523 records, from which we derived two separate sets: (i) the Census3d, having as dimensions the age, income and weeks worked; and (ii) the Census4d, having as additional dimension dividends from stocks. As categorical attribute we selected, in both cases, the occupation type,that has 47 distinct values in total. We went to [2] to obtain the datasets.

6 Results and Performance Evaluation

In this section we shall look at different kinds of factors which effect the query output time. This section describes the effect of each factor on the performance of the three techniques. In all the datasets, queries are generated using each data point as the center of the query box. The output time per query is obtained by averaging all the query times. We also look at the space occupied by these techniques.

	Forest Cover (in MB)	US Census (3d) (in MB)	US Census (4d) (in MB)
PRS	96	38	45
LPD	152	100	160
BB	510	211	249
Gupta et al.[7]	> 4 GB	> 4 GB	> 4 GB

Fig. 3. Comparision of space occipied by various techniques. We implemented the semi-dynamic solution of Gupta et al.[7] for $d=2$. For real-life datasets, the size of this data structure exceeded our main memory capacity.

6.1 Comparision of Space Occupied

Theoretically, both *PRS* and *LPD* techniques take up $O(n)$ space. However, notice that in *LPD* we project all the points to each of the d dimensions. There-fore, it is expected to occupy slightly higher space than the *PRS* technique. This was also observed while testing them on real-life datasets as shown in Figure 3. BB k-d tree occupies $O(n \log \alpha)$ space in any dimensional space. The secondary information stored at each internal node of a k-d tree leads to a *slight* blow up in the space occupied. In contrast, the data structure proposed by Gupta et al. [7] for $d=2$ (semi-dynamic solution) could not be loaded to main memory. This was expected as the space complexity of the data structure is $O(n \log^2 n)$. Hence, we could not compute the query time of it and hence discarded this solution.

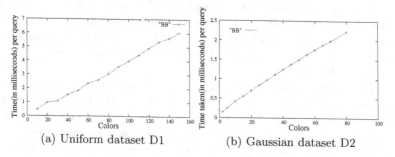

(a) Uniform dataset D1   (b) Gaussian dataset D2

Fig. 4. Performance of BB k-d tree w.r.t the number of colors

6.2   Number of Colors (α)

Query response time of PRS and LPD techniques are independent of the number of colors in the dataset. In a BB k-d tree, as the number of colors start increasing in the dataset (while dataset size remains constant), the size of the secondary structures increase. Consequently, the query time increases. In real life datasets, number of colors remain constant. So we used synthetic datasets $D1$ and $D2$ for observations. Average query time for BB k-d tree for both the datasets is increasing with the increase in the number of colors as shown in the Fig 4(a) and 4(b). However, in real life scenerio, the ratio of the number of colors in a dataset and the size of the dataset is generally very low.

6.3   Size of Query Box

In general, for a given dataset, as the size of the query box increases, the number of points lying inside it also increases. So, naturally the query time of PRS and LPD techniques are expected to increase with increase in size of the query box. Interestingly, *BB k-d technique is minimally affected by the variation in the size of query box which is highly desirable.* As the query box size keeps increasing, the *depth* of the nodes being visited in the BB k-d tree decreases; since the existence or non-existence of the bounding box a color inside the query box becomes clear at an early stage. At the same time when the size of the query box is very small, then the number of primary structure nodes visited will also be less. Hence the query time is minimally varied w.r.t. the query box size. Experimental results have shown that with increase in size of the query box, initially the query time off BB k-d tree increaes slightly and then decreases or stays flat (see Fig 5). In 5(a) and 5(b), we observe the same pattern even when we vary the number of colors in the synthetic datasets $D1$ and $D2$.

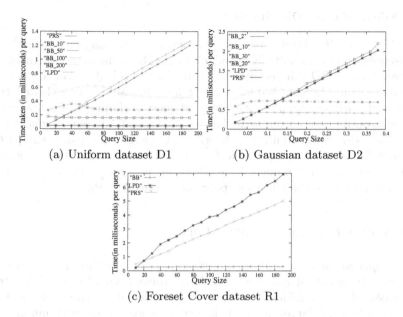

Fig. 5. Effect of size of query box on the query time. Two synthetic datasets and one real-life dataset have been used. In Fig 5(a) and 5(b) BB_x represents that the dataset used has x number of distinct colors. Note that the lines corresponding to BB k-d technique are almost flat.

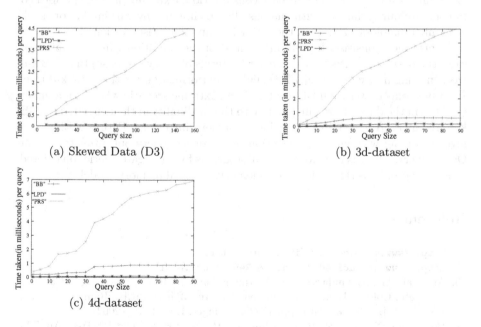

Fig. 6. Effect of skewed datasets on query time. (b) and (c) are based on US census dataset (R2)

6.4   Effect of Data Distribution

Both PRS and BB k-d techniques use k-d tree as their primary structure. Hence, the distribution of data will have similar effect on both the data structures. On the other hand, LPD will perform well if the dataset is highly skewed w.r.t one of the dimensions. As described in Section 2, if the dataset is highly skewed, then the value of β will be small, resulting in good query performance. Fig 6(a) shows the results on the skewed synthetic dataset (D3) to validate our arguments. US Census data (Fig 6(b) and 6(c)) is skewed w.r.t age dimension and hence, LPD does well. However, the query time for BB k-d technique is always within a bounded factor of LPD's query time. However, in Fig 5(a) the datasets have uniform distribution which leads to poor perfomance of LPD.

7   Conclusions and Future Work

In this paper we came up with a main-memory data structure, BB k-d tree, to solve the orthogonal colored range searching problem (OCRS). In BB k-d tree, we augmented the traditional k-d tree with secondary data structures which resulted in significant improvement of the query response time (with minimal increase of $O(\log \alpha)$ factor in space). Comparision of this data structure was done with two base-line brute-force algorithms which solved the traditional orthogonal range searching problem. An existing theoretical solution was implemented but found unsuitable due to its high space consumption. Experiments were performed to compare our technique with the base-line techniques by varying factors such as number of colors (α), size of the query box and data distribution. The BB k-d tree performed consistently well under most of the conditions. In some minimal cases the base-line brute-force do better (in terms of query response) than BB k-d tree: In a highly skewed data LPD slightly performed better than BB k-d tree, PRS performs better than BB k-d tree in an extreme scenario where the number of colors in the dataset are almost close to the cardinality of the dataset. This BB k-d tree method is equally applicable to higher dimensions. Future work would involve coming with efficient execution plans for the query optimizer to answer OCRS query. There is need for practical solutions for aggregate queries on colored geometric problems [11,12] for main-memory, external memory models etc.

References

1. http://www.mcobject.com/in_memory_database
2. http://www.ics.uci.edu/mlearn/MLRepository.html
3. Agarwal, P.K., Govindarajan, S., Muthukrishnan, S.: Range Searching in Categorical Data: Colored Range Searching on Grid. In: Möhring, R.H., Raman, R. (eds.) ESA 2002. LNCS, vol. 2461, pp. 17–28. Springer, Heidelberg (2002)
4. Beckmann, N., Kriegel, H.-P., Schneider, R., Seeger, B.: The R*-Tree: An Efficient and Robust Access Method for Points and Rectangles. In: ACM SIGMOD Conference, pp. 322–331 (1990)

5. Bentley, J.L.: Multidimensional binary search trees used for associative searching. Communications of the ACM 18, 509–517 (1975)
6. de Berg, M., van Kreveld, M., Overmars, M., Schwarzkopf, O.: Computational geometry: algorithms and applications. Springer, Heidelberg (2000)
7. Gupta, P., Janardan, R., Smid, M.: Further results on generalized intersection searching problems: counting, reporting, and dynamization. Journal of Algorithms 19, 282–317 (1995)
8. Gupta, P., Janardan, R., Smid, M.: Computational geometry: Generalized intersection searching. In: Mehta, D., Sahni, S. (eds.) Handbook of Data Structures and Applications, ch. 64, pp. 64-1–64-17. Chapman & Hall/CRC, Boca Raton, FL (2005)
9. Guttman, A.: R-Trees: A Dynamic Index Structure for Spatial Searching. In: SIGMOD Conference, pp. 47–57 (1984)
10. Janardan, R., Lopez, M.: Generalized intersection searching problems. International Journal on Computational Geometry & Applications 3, 39–69 (1993)
11. Rahul, S., Gupta, P., Rajan, K.: Data Structures for Range Aggregation by Categories. In: 21st Canadian Conference on Computational Geometry (CCCG 2009), pp. 133–136 (2009)
12. Rahul, S., Bellam, H., Gupta, P., Rajan, K.: Range aggregate structures for colored geometric objects. In: 22nd Canadian Conference on Computational Geometry (CCCG 2010), pp. 249–252 (2010)
13. Sellis, T.K., Roussopoulos, N., Faloutsos, C.: The R+-Tree: A Dynamic Index for Multi-Dimensional Objects. In: 13th International Conference on Very Large Data Bases (VLDB 1987), pp. 507–518 (1987)
14. Shi, Q., JáJá, J.: Optimal and near-optimal algorithms for generalized intersection reporting on pointer machines. Information Processing Letters 95(3), 382–388 (2005)
15. Kumar, Y., Janardan, R., Gupta, P.: Efficient algorithms for reverse proximity query problems. In: GIS 2008, p. 39 (2008)

Circle of Friend Query in Geo-Social Networks*

Weimo Liu[1], Weiwei Sun[1,**], Chunan Chen[1], Yan Huang[2],
Yinan Jing[1], and Kunjie Chen[1]

[1] School of Computer Science, Fudan University
Shanghai 201203, China
{liuweimo,wwsun,ynjing,chenkunjie}@fudan.edu.cn, huangyan@unt.edu
[2] Department of Computer Science and Engineering, University of North Texas
Denton, TX 76203-5017, USA
chenchunan@fudan.edu.cn

Abstract. Location-Based Services (LBSs) are becoming more social
and Social Networks (SNs) are increasingly including location components. Geo-Social Networks are bridging the gap between virtual and
physical social networks. In this paper, we propose a new type of query
called *Circle of Friend Query (CoFQ)* to allow finding a group of friends
in a Geo-Social network whose members are *close* to each other both
socially and geographically. More specifically, the members in the group
have tight social relationships with each other and they are constrained
in a small region in the geospatial space as measured by a "diameter"
that integrates the two aspects. We prove that algorithms for finding the
Circle of Friends (CoF) of size k is NP-hard and then propose an ε-
approximate solution. The proposed ε-approximate algorithm is guaranteed to produce a group of friends with diameter within ε of the optimal
solution. The performance of our algorithm is tested on the real dataset
from Foursquare. The experimental results show that our algorithm is
efficient and scalable: the ε-approximate algorithm runs in polynomial
time and retrieves around 95% of the optimal answers for small ε.

Keywords: Circle of friend query, geo-social networks.

1 Introduction

The wide availability of wireless communication and positioning enabled mobile
devices allows people to access Location-Based Services (LBSs) whenever and
wherever they want. On the other hand, Social Networks (SNs) are penetrating
people's daily lives. For example, by January 2011, Facebook has more than
600 million active users. Half of these users login everyday and more than 150
million active users access Facebook through mobile devices. LBS and SN are
increasingly integrated together into Geo-Social Networks (GSNs). In GSNs,
e.g. Foursquare, users form a social network, can communicate, and share their

---

* This research is supported in part by the National Natural Science Foundation of
China (NSFC) under grant 61073001.
** Corresponding author.

locations. Recently many SNs are adding support of locating friends nearby. In fact, there is a trend of increasingly bridging the gap between virtual and physical worlds. Aligning with this trend, this paper proposes a new query type called Circle of Friend Query (*CoFQ*) in a GSN. A *k-CoFQ* allows a user q with location $q.l$ to identify k "circle of friends" in "spatial proximity". Here "circle of friends" refers to the group of $k + 1$ users including q with small pairwise social distances in a social network. And "spatial proximity" is to constrain the diameter of the point set formed by the locations of the $k + 1$ users. The *k-CoFQ* is the integration of social and physical world distances. This query allows a circle of friends with the following two properties to be found: (1) The friends are mutual which is ensured by the small pairwise social distance; (2) The friends are in spatial proximity. There are many challenges in answering *k-CoFQ* in a large GSN. The three major ones are:

1. There are two types of distances in GSNs: social network distance and spatial network distance. The integration of these two requires careful balance of the different metrics;
2. The *k-CoFQ* requires to find a subgraph with the smallest diameter. The NP-hard max-clique problem can be reduced to the *k-CoFQ* problem which makes *k-CoFQ* NP-hard and computational expensive. With spatial proximity considered, it is even more challenging to design a scalable algorithm;
3. It is known that obtaining real social network dataset for validation is very difficult in social network research especially in academic settings.

The paper addresses these challenges and makes the following contributions:

1. We formulate a new and useful query in GSNs, namely, *k-CoFQ*. This query enables many applications in a GSN. Examples include group sports, social deal, friend gathering, and community service. To bridge the gap between virtual and physical world, this paper uses a normalization method with a empirical parameter to combine the two metric values.
2. We prove that the *k-CoFQ* is NP-hard. We propose an ε-approximate algorithm to *k-CoFQ*. We identify an upper-bound and lower-bound when searching k-circle of friends in a social network. This allows the the ε-approximate algorithm prunes most of the search space. For the candidate groups remained, the ε-approximate algorithm narrows the range of the optimal group's diameter gradually by checking the candidates in a binary search fashion, tightening the bounds along the way until the distance between bounds is less than ε. The suboptimal result is guaranteed to be within ε of the optimal solution.
3. We algebraically analyze the cost of the algorithms using small world assumption [16]. We test the performance of our algorithm on real dataset from Foursquare. Experimental results demonstrate efficiency and scalability of the algorithms proposed.

The rest of this paper is organized as follows: Section 2 overviews the research on GSNs. Section 3 defines the circle of friend query and proves it is NP-hard.

Section 4 proposes an ε-approximate algorithm in social networks. Section 5 introduces the ε-approximate algorithm in GSNs. Experimental results are provided in section 6. In section 7, we conclude our work and discuss the future work.

2   Related Work

The research on location-based social networks is attracting much attention recently. A main activity is to mine user's location and social network data to retrieve the relationships between the locations and the users. It has been shown that users with short-distance links are often geographically close [15][14]. C. Hung, et al. [8] store users' trajectory profiles by using a probabilistic suffix tree structure. They formulate distance of profiles and uses clustering algorithms to identify communities. M. Lee, et al. [11] focus on the semantic information of location. Their paper proposes a structure that organizes the location information into categories to compute the similarity between users. M. Ye, et al. [18] recommend locations to users based-on their social networks. Their paper proves that there is a co-relationship between friends' ratings on the same location and discusses how to build a location recommendation system based on the co-relationship. The above works attempt to discover some useful Geo-Social information from the history information. In this paper we propose a new kind of query which is very useful in real life applications such as activity planning.

C. Chow, et al. [3] present GeoSocialDB which supports location-based social query operators such as location-based news feed, location-based news ranking, and location-based recommendation. Y. Doytsher, et al. [4] propose a socio-spatial graph model and introduce a query language formed by a set of socio-spatial operators. The socio-spatial operators mentioned in this paper are different from ours. The circle of friend query in our paper aims to find a group based on the social relationship and the location of users.

D. Yang, et al. [17] discuss the problem of finding a group of given number of members from a social network. The group should satisfy the hop and unfamiliar member constraints and make the sum of closeness between the query and each member of the group smallest. Our work aims to find a group with smallest diameter and considers both social network distance and geo-distance.

Finding k points in Euclidean space with minimum diameter or minimum sum of all distance has been studied. It has been shown that polynomial time algorithm is available [1]. However, the problem of finding a k-vertex group with the smallest diameter in a graph has not been studied yet.

Another query to find a group in spatial databases is proposed recently called collective spatial keyword query [2]. It is to find a group of objects that cover all the keywords in the query and minimize the maximal distance between the objects and the sum of distances from the objects to the query location.

In the area of spatial databases, the problem of group nearest neighbor query, also named aggregate nearest neighbor query, is studied in [13][19]. Given a group of k members, the group nearest neighbor query is to find an aggregate point to minimize the total distance from all the members to the aggregate point. This differs from our work because we aim to find a group to minimize the diameter of a group containing a given point. Furthermore, we consider not only spatial distance but also social network distance.

3   Problem Definition

A social network is modeled as an undirected weighted graph $G(V, E)$ with vertex set V and edge set E. For $u, v \in V$, $(u, v) \in E$ if u and v are friends, and the weight of (u, v), denoted as $w(u, v)$, is defined by their direct interactions such as message exchange or co-authorship. For example, in the case of co-authorship $w(u, v)$ can be defined as $1/\Sigma_{i=1}^{n} \frac{1}{x_i}$, where n represents the number of papers u and v co-write, x_i is the number of the ith paper's authors [12]. The closeness $closeness(u, v)$ between u and v is defined as[10][9]:

$$closeness(u, v) = \begin{cases} w(u, v), & \text{if } (u, v) \in E \\ \text{the accumulative weights of the} \\ \quad \text{shortest path between } u \text{ and } v \text{ in } G, \text{otherwise} \end{cases}$$

In a GSN, every vertex $v \in V$ is geo-tagged and associated with a location. The geographical distance between u and v is denoted as $dist(u, v)$. This distance can be the common distance measures such as Euclidean or network distances. We use Euclidean distance in this paper.

3.1   Circle of Friend Query (*CoFQ*) in Social Networks

Definition 1 (diameter). *Given a subset of vertices S in a graph G(V, E), the diameter of S, denoted as dia(S), is the distance between the farthest pair in S:*

$$dia(S) = max_{a,b \in S} closeness(a, b). \tag{1}$$

Problem Statement [*k*-circle of friend query in social networks]. Given a query point q in a social network $G(V, E)$ and the size k, the *k-circle of friend query* (*CoFQ(q, k)*) is to find a set $V' \subseteq V$, satisfying that:

$$q \in V'; \tag{2}$$
$$|V'| = k + 1; \tag{3}$$
$$dia(V') \text{ is the smallest}; \tag{4}$$

The set V' is called the *k-circle of friends* of q, denoted as $CoF(q, k)$.

3.2   Geo-Social Circle of Friend Query (*gCoFQ*)

We denote the diameters of a vertex set S with respect to geometric distance and social closeness as $dia_{geo}(S)$ and $dia_{social}(S)$:

$$dia_{geo}(S) = max_{a,b \in S} dist(a, b), \tag{5}$$
$$dia_{social}(S) = max_{a,b \in S} closeness(a, b). \tag{6}$$

The Ranking Function. Because we want the $dia_{geo}(S)$ and $dia_{social}(S)$ to be small, the combining function of $dia_{geo}(S)$ and $dia_{social}(S)$ needs to be monotonic with respect to both. The linear combination of the two factors is a reasonable method:

$$dist_{gs}(u,v) = \lambda \frac{dist(u,v)}{dia_{geo}(V)} + (1-\lambda)\frac{closeness(u,v)}{dia_{social}(V)}. \tag{7}$$

And the ranking function is

$$f(S) = dia_{gs}(S) = max_{a,b \in S} dist_{gs}(a,b). \tag{8}$$

Here, λ varies according to the users' different demands. Even for the same user, the demand may change at different moment. For example, if a user wants to invite some friends to play football, the geo-distance weights more in the function since the friends nearby are convenient to come together. However if the user wants to hold a party, the closeness is more important. The user probably invites a very close friend although the friend lives a little far. A user can pick up a reasonable λ based on her demand possibly through taking suggestion from the system who learns the value from similar queries.

Problem Statement [*Geo-Social Circle of Friend Query*]. Given a query point q in a GSN $G(V, E)$ and the size k, the *Geo-Social Circle of Friend Query ($gCoFQ(q, k)$)* is to find a set $V' \subseteq V$, satisfying that:

$$q \in V'; \tag{9}$$
$$|V'| = k+1; \tag{10}$$
$$dia_{gs}(V') \text{ is the smallest}; \tag{11}$$

and the set V' is called the Geo-Social circle of friends of q, denoted as $gCoF(q,k)$.

3.3   NP-Hard Proof

The max-clique problem is to find the largest complete subgraph in an unweighted undirected graph and is NP-Hard. We reduce the max-clique problem to *CoFQ* problem.

Theorem 1. *Finding* CoF *(q, k) is NP-hard.*

Proof. Assume that the circle of friend problem's running time is T. Suppose that all the edges' weights of a weighted undirected graph G are one, we iterate all the vertices in G and let k be 1, 2, ..., n ($n = |G|$). We can get different k_0 in the following condition: if G is a complete graph, $k_0 = n$; if G is not a complete graph, there exists q that makes the CoF's diameter equals to 1 when $k=k_0$ and for any $k>k_0$ there is no q makes CoF problem has a solution that satisfies that the diameter equals to 1. The CoF problem's solution for $k=k_0$ is the max-clique problem's solution. The algorithm complexity is $O(k*n)*T$, so MAX-CLIQUE $\leq_P$CIRCLE OF FRIEND. So *CoFQ* is also an NP-hard problem.

Theorem 2. *Finding the Geo-Social Circle of Friends is NP-hard.*

Proof. Assume that the running time of *gCoFQ* is T. Then set the value of λ as 0. We have

$$dist_{gs}(u,v) = \lambda \frac{dist(u,v)}{dia_{geo}(V)} + (1-\lambda)\frac{closeness(u,v)}{dia_{social}(V)} = \frac{closeness(u,v)}{dia_{social}(V)} \quad (12)$$

Since $dia_{social}(V)$ is constant, the *gCoFQ* is equivalent to *CoFQ*. CIRCLE OF FRIEND $\leq_P$ GEO-SOCIAL CIRCLE OF FRIEND. So *gCoFQ* is NP-hard.

4   Algorithm for Circle of Friend Query (*CoFQ*)

In this section, by exploring the upper bound and lower bound of the diameter, we propose an ε-approximate solution to *CoFQ*. We then optimize the algorithm and analyze its complexity.

4.1   Find the Upper Bound and the Lower Bound

The circle of friend result is different from the kNN of the query point. However, the upper bound and the lower bound of the diameter of the circle of friends can be computed based on the kNN result:

Lemma 1 (Upper bound property). *Let N be the union of q and the set of q's kNNs, D_{max} = dia(N), M = $\{u \mid u \in V$ and closeness(q, u) $\leq D_{max}\}$, then CoF(q, k) is a subset of M.*

Proof omitted due to space constraint.

Lemma 2 (Lower bound property). *Let r be the distance between q and its kth NN, then we have dia(CoF(q, k)) $\geq$ r.*

Proof omitted due to space constraint.

   Based on Lemma 1, a very straightforward solution to the *CoFQ* is to compute the diameter of every possible group in M and find the one with the smallest diameter. This method needs $\binom{|M|-1}{k}$ times of diameter computation, once for each possible group. The computation time is very expensive.

4.2 ε-approximate Algorithm

We introduce our first algorithm to *CoFQ*. The main idea of this algorithm is to process a binary search on the $\binom{|M|-1}{k}$ groups generated by M. Firstly, we check if there is a group with diameter smaller than $1/2 \, D_{max}$. If such a group exits, then the diameter of $CoF(q, k)$ falls in the range $[1/4 D_{max}, 1/2 D_{max}]$. Then we continue the binary search in the subspace until we get an ε-approximate result. By ε-approximate result, we mean that we are sure that the optimal result $dia(CoF(q, k))$ falls in the range $[\delta^-, \delta^+]$ and $\delta^+ - \delta^- < \varepsilon$. Lemma 3 shows that with a tight upper bound and special access order, the binary-search-based algorithm will prune most of the groups and avoid scanning the whole space.

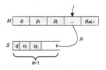

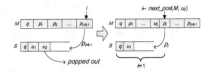

Fig. 1. prune in the ε-approximate algorithm

Fig. 2. pop in the ε-approximate algorithm

Lemma 3. *Suppose that we are currently checking if there is a group with diameter smaller than* D_{cur}. *If there are two points* u, v *in M satisfying that* closeness(u, v) $> D_{cur}$, *then any group containing* u *and* v *can be discarded.*

Proof omitted due to space constraint.

Based on Lemma 3, we develop an ε-approximate algorithm. Suppose that the points in M is stored in a list and q is the first element of the list, as shown in Fig. 1. We use a stack S to store the currently checked group. S is initialized as q. Then we add the elements in M to S in sequence. Before adding a element p to S, we check whether or not the diameter of S will become larger than D_{cur} after p is added. Based on Lemma 3, we just have to check the distance between p and the elements in S. For example in Fig. 1, S currently has three elements: q, u_1 and u_2. Before adding p_i to S, we check that if the distance between p_i and q(or u_1, u_2) is larger than D_{cur}. If the above condition is satisfied, p_i is pruned for the current set in S. Otherwise p_i is added to S.

If for the elements in S, adding any other element will lead to a diameter larger than D_{cur}, the top element of S is popped out. In Fig. 2, u_2 is popped out and the element p_j after u_2 in M is the next element to be added to S.

When the size of S becomes $k+1$, we output the group generated by the elements in S as the currently suboptimal result, denoted as CoF_{cur}. Then we check if the difference between CoF_{cur} and the optimal result is smaller than ε. If so, CoF_{cur} is output as the suboptimal result. Otherwise, the upper bound d_{UP} of $dia(CoF(q, k))$ is set to D_{cur} and we continue to check if there is a group with diameter smaller than $1/2$ $(D_{cur} + d_{LB})$. If all the groups' diameter is larger than D_{cur}, then we continue to check if there is a group with diameter smaller than $1/2$ $(D_{cur} + d_{UP})$.

4.3   Optimized ε-approximate Algorithm

As mentioned above, we use an $|M| \times |M|$ matrix to store the distance between every pair of M in order to compute the diameter efficiently. However, as the value of M becomes larger, both computation time and space cost will increase rapidly. In this section, we propose an optimized algorithm by cutting down the size of the distance matrix. The optimization is based on the following observation.

Observation 1. Suppose that there exists a group whose diameter is smaller than D_{cur}, for a element u in M, if $closeness(u, q) > D_{cur}$, then u will not appear in the suboptimal result. So the columns or rows containing u will not be used in our binary search process. As shown in Fig. 3, only the sub matrix at the top left corner is useful to our ε-approximate algorithm. We call it "useful space".

Observation 1 can be proved by Lemma 3. As Fig. 3 shows, at the first step of the binary search, we don't have to compute the distance matrix of all the elements in M. Instead, we only compute the *useful space* and check if there is a group whose diameter is smaller than D_{cur}. If such a group doesn't exist, we would have to extend the *useful space* according to the new D_{cur}. The check procedure based on the distance matrix is the same as ε-approximate Algorithm. The only difference is the search space. In ε-approximate Algorithm, the search space is all the elements whose distance to q is smaller than $dia(N)$, while in our second algorithm, the search space depends on D_{cur} and may be extended in the following steps.

5   Algorithm for Geo-Social Circle of Friend Query ($gCoFQ$)

In this section, we explore the problem in **Geo**-Social network environment where both social distance and geometric distance are considered. Firstly, we propose a kNN algorithm in GSNs based on the ranking function. Then a binary search based algorithm to $gCoFQ$ is proposed.

The ε-approximate algorithm described in section 4.2 can be applied to $gCoFQ$, too. The first step is to find the kNNs of q ordered by the value of the ranking function. This a multi-objective top-k problem (i.e., the top-k objects are ranked according to the social network distance and spatial proximity). To deal this problem, we propose a Geo-Social kNN algorithm, which is inspired by the NRA algorithm in [5].

5.1   The Geo-Social kNN Algorithm

Definition 2 (Geo-Social kNN query). *Given a query point* q *in a GSN* $G(V, E)$ *and the size* k, *the Geo-Social kNN query (GSkNN(q, k)) is to find a set* V'$\in$ V *with* $| V' |= k$, *and for any point* $u \in V'$ *and* $v \notin V'$,

$$dist_{gs}(q, u) < dist_{gs}(q, v). \tag{13}$$

The main idea of our algorithm is to search the kNN in the social network and space separately and synchronously. Fig. 4 shows the framework of the algorithm. The kNNs in social networks are searched on the graph $G(V, E)$ (see section 4). While the NNs in geo-space are searched using the R-Tree index [6]. The incremental NN search algorithm [7] are used in this paper.

Like NRA, we append the NNs retrieved in the social network or geo-social space with two attributes: W and B (short for *worst* and *best* score). W means the upper bound of the ranking function and B means the lower bound. We use a priority queue Q to store the NNs. The elements in Q are sorted by W in ascending order, if W is the same, ties are broken using B, the lower B wins.

Notice that every time a nearest neighbor in social NN_{social} is retrieved, the Euclidean distance can be computed as well using the coordinates of NN_{social}. So we set the value of $NN_{social}.W$ and $NN_{social}.B$ as the exact Geo-Social distance between q and NN_{social}. Meanwhile, when a nearest neighbor in geo-space NN_{geo} is retrieve, the social network distance is unknown. So we compute the value of $NN_{geo}.W$ and $NN_{geo}.B$ with the lower and upper bound of the social network distance. The social network upper bound is set as ∞, while the lower bound $closeness_{LB}$ is initialized as 0 and increases when a new NN is found. For geo-distance, a lower bound $distance_{LB}$ is also maintained. When the lower bound of the ranking function $f(distance_{LB}, closeness_{LB})$ is larger than the last element's in Q, the algorithm ends and the first k elements in Q are output as the top-k result.

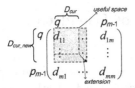

Fig. 3. Useful space in the matrix

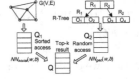

Fig. 4. The Geo-Social kNN algorithm

5.2   The Algorithm for *gCoFQ*

Recall the upper bound property in Lemma 1, the kNN result can be used to prune most of the search space in the social network. This property holds for the *gCoFQ* as well. Suppose that q and its kNNs are stored in N. We compute $dia_{gs}(N)$ and use it as the upper bound. Then we process a range query on the social network with the distance bound of $dia_{gs}(N)$. For every point u lying in this range, we compute $dist_{gs}(u, q)$ using the coordinate of u. If $dist_{gs}(u, q)$ is larger than $dia_{gs}(N)$, u is pruned. Otherwise, u is added to the set M, sorted by $dia_{gs}(N)$. Then we process a binary search on M to get the ε-approximate result, as described in section 4.2.

6   Experiment

We used a real social network from Foursquare, which is a world famous location-based social networking application. The graph consists of 20,550 nodes and 586,424 edges in our experiments. The relation between users and their check-in locations with geographic were crawled from Foursquare by using the open API. Because the API requests' number per hour is limited by the server, we only achieved a subset of the total social network. All experiments were carried out on a Genuine 1.8GHZ CPU desktop computer with 3GB memory and our implementation source code of the algorithms aforementioned were written in C++.

We conducted several experiments for $CoFQ$ and $gCoFQ$ to: 1) compare the performance of the baseline, the ε-approximate and the optimized ε-approximate algorithm with the increase of k; 2) evaluate the effect of ε on the correctness and the response time.

6.1 $CoFQ$ on Social Networks

The Effect of k. In this section, we discuss the effect of k when all algorithms use both upper bound aforementioned as the prune condition. As Fig. 5 shows, the ε-approximate and the optimized ε-approximate algorithm significantly outperform the baseline for all value of k. The gap between them even increases considerably as k increase. As our analysis above, the CPU time of the ε-approximate and the optimized ε-approximate algorithm grows in a polynomial way while the Baseline increases in an exponential way due to its huge amount of computation of diameter for every possible group. The Baseline can hardly return any results when k is bigger than 7 in our experiments, which will be considered unacceptable because in real applications such as *groupon*, k is often a relative large number. We also compared the ε-approximate and the optimized ε-approximate algorithm in a relative large extent of k, as is shows in Fig. 6. The optimized ε-approximate algorithm's CPU time still keeps in a low level but the ε-approximate's becomes bigger with increasing k. This is because the ε-approximate use a $|M| \times |M|$ matrix to store the distance between every pair of M. When k becomes larger which will result in the increase of M, the computation time of the distance matrix will increase dramatically. But for the optimized ε-approximate algorithm, only part of the distance matrix (useful space) will be computed, thus the CPU time greatly reduced.

The Effect of ε. Finally, we explore the effect of ε on the correctness and the response time of our algorithm. We used our two ε-approximate algorithms to process 1,000 CoF queries whose query points are randomly chosen in the social network. k is set as 6 and ε ranges from 0.001 to 0.1. For each of the queries, we compute the actually CoF using the baseline algorithm. We then compare ε-approximate CoF with the actual CoF and record the number of the correct answers. We define the correct rate $R_{correct}$ and the approximate rate R_{appro} as follows, and the effect of R_{appro} on $R_{correct}$ is shown in Fig. 7.

$$R_{correct} = \frac{num(correctanswers)}{num(totalqueries)} \times 100\% \tag{14}$$

$$R_{appro} = \frac{\varepsilon}{dia(G)} \times 1000\text{\textperthousand} \tag{15}$$

6.2 Geo-Social CoF Query

The Effect of k. Fig. 8 measures the effect of the k (ranging from 3 to 6) on the running time for $gCoFQ$. When k equals to three, three methods make little difference in CPU time. As k increase, the optimized ε-approximate's CPU time still keep in a low level due to its effective exploring way despite the addition

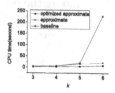

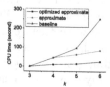

Fig. 5. Small k on the CPU time

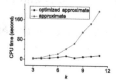

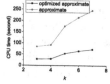

Fig. 6. Large k on the CPU time

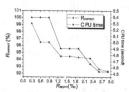

Fig. 7. The effect of ε

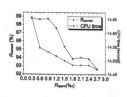

Fig. 8. Small k on the CPU time

Fig. 9. Large k on the CPU time

Fig. 10. The effect of ε

computation on the spatial distance. On the other hand, baseline's performance degrades very fast. Again, we still get no result in a reasonable time using baseline, since the method exhausts all possible combinations of the candidate set. And the ε-approximate performs similar as before with using addition 50 seconds mostly because the algorithm need to get the Euclidean space distance.

Fig. 9 illustrates the effect of k on two approximate algorithms. The optimized ε-approximate's CPU time increases in a different way as shown in Fig. 6. With k increase, the method takes more time to calculate the spatial distance for every candidates. The same holds for the ε-approximate algorithm.

The Effect of ε. Fig. 10 shows the effect of ε on the CPU time and the correctness. The approximate ratio is redefined as

$$R_{appro} = \frac{\varepsilon}{dia_{gs}(G)} \times 1000\%_0. \tag{16}$$

With small ε, most of the results retrieved by the ε-approximate algorithm are the same as the optimal solution. For example, with $R_{appro} = 0.3\%_0$, all the results got are correct. When R_{appro} gets bigger, the response time becomes less, and the correctness ratio decreases.

7 Conclusion

In this paper, we define a new query in GSNs, namely, the *Geo-Social Circle of Friend Query*. Given a query point in a GSN, the $gCoFQ$ is to obtain a group including q and the members are close to each other with respect to Euclidean distance and social relationship closeness. We proved that the problem is NP-hard. After exploring the upper bound and lower bound property of the $gCoF$, we proposed two ε-approximate algorithms to find a suboptimal solution. We tested the performance of our algorithms on real dataset with different parameters. The response time was reduced from exponential to polynomial. By experimental results, we showed that with $\varepsilon < 0.01$, the ε-approximate algorithms yielded 95% of the correct answers.

An interesting direction for future work is to process Geo-Social queries based on the trajectories of the mobile users. The main challenge is how to calculate the geo-distance between users based on the history of the locations, not only the current locations.

References

1. Aggarwal, A., Imai, H., Katoh, N., Suri, S.: Finding k points with minimum diameter and related problems. Journal of Algorithms 12(1), 38–56 (1991)
2. Cao, X., Cong, G., Jensen, C.S., Ooi, B.C.: Collective spatial keyword querying. In: SIGMOD, pp. 373–384. ACM (2011)
3. Chow, C.Y., Bao, J., Mokbel, M.F.: Towards location-based social networking services. In: LBSN, pp. 31–38. ACM (2010)
4. Doytsher, Y., Galon, B., Kanza, Y.: Querying geo-social data by bridging spatial networks and social networks. In: LBSN, pp. 39–46. ACM (2010)
5. Fagin, R., Lotem, A., Naor, M.: Optimal aggregation algorithms for middleware. In: PODS, pp. 102–113. ACM (2001)
6. Guttman, A.: R-trees: a dynamic index structure for spatial searching, vol. 14. ACM (1984)
7. Hjaltason, G.R., Samet, H.: Distance browsing in spatial databases. TODS 24(2), 265–318 (1999)
8. Hung, C.C., Chang, C.W., Peng, W.C.: Mining trajectory profiles for discovering user communities. In: LBSN, pp. 1–8. ACM (2009)
9. Jøsang, A., Gray, E., Kinateder, M.: Simplification and analysis of transitive trust networks. Web Intelligence and Agent Systems 4(2), 139–161 (2006)
10. Jøsang, A., Pope, S.: Semantic constraints for trust transitivity. In: APCCM, vol. 43, pp. 59–68. Australian Computer Society, Inc. (2005)
11. Lee, M.-J., Chung, C.-W.: A User Similarity Calculation Based on the Location for Social Network Services. In: Yu, J.X., Kim, M.H., Unland, R. (eds.) DASFAA 2011, Part I. LNCS, vol. 6587, pp. 38–52. Springer, Heidelberg (2011)
12. Newman, M.E.J.: Scientific collaboration networks. ii. shortest paths, weighted networks, and centrality. Physical Review E 64(1), 016132 (2001)
13. Papadias, D., Shen, Q., Tao, Y., Mouratidis, K.: Group nearest neighbor queries. In: ICDE, p. 301. IEEE Computer Society (2004)
14. Scellato, S., Mascolo, C., Musolesi, M., Latora, V.: Distance matters: Geo-social metrics for online social networks. In: Proceedings of the 3rd Conference on Online Social Networks, p. 8. USENIX Association (2010)
15. Singla, P., Richardson, M.: Yes, there is a correlation:-from social networks to personal behavior on the web. In: WWW, pp. 655–664. ACM (2008)
16. Watts, D., Strogatz, S.: Collective dynamics of small-world networks. Nature 393(6684), 440–442 (1998)
17. Yang, D.N., Chen, Y.L., Lee, W.C., Chen, M.S.: On social-temporal group query with acquaintance constraint. VLDB 4(6), 397–408 (2011)
18. Ye, M., Yin, P., Lee, W.C.: Location recommendation for location-based social networks. In: SIGSPATIAL GIS, pp. 458–461. ACM (2010)
19. Yiu, M.L., Mamoulis, N., Papadias, D.: Aggregate nearest neighbor queries in road networks. TKDE, 820–833 (2005)

A Power Saving Storage Method That Considers Individual Disk Rotation

Satoshi Hikida, Hieu Hanh Le, and Haruo Yokota

Department of Computer Science, Tokyo Institute of Technology
2-12-1 Ookayama, Meguro-ku, Tokyo 152-8552, Japan
{hikida,hanhlh}@de.cs.titech.ac.jp, yokota@cs.titech.ac.jp

Abstract. Reducing the power consumption of storage systems is now considered a major issue, alongside the maintenance of system reliability, availability, and performance. In this paper, we propose a method named as the Replica-assisted Power Saving Disk Array (RAPoSDA) to reduce the electrical consumption of storage systems. RAPoSDA utilizes a primary backup configuration to ensure system reliability and it dynamically controls the timing and targeting of disk access based on individual disk rotation states. We evaluated the effectiveness of RAPoSDA by developing a simulator that we used for comparing the performance and power consumption of RAPoSDA with Massive Arrays of Inactive Disks (MAID), which is a well-known power reduction disk array. The experimental results demonstrated that RAPoSDA provided superior power reduction and a shorter average response time compared with MAID.

Keywords: storage, power saving, performance, large scale, reliability.

1 Introduction

Ongoing increases in the total electricity consumed by data centers present significant problems that must be solved to reduce the running cost of centers and to keep the global environment green. A governmental report estimated that the electricity consumption of US data centers in 2011 would be double that of 2006 [11].

Servers and storage systems are the two main data center components that consume electricity. They tend to grow at a numerically faster rate, because of the increasing requirements for processing load and the targets of processes. The amount of data stored in storage systems is increasing particularly rapidly, because of the explosive increase in the volume of data generated by the Internet. The IDC reports that the growing amount of digital data created and replicated will surpass 1.8 zettabytes in 2011 [7]. Many services are leveraged by cloud computing, which is becoming widespread in our daily life and business, including social network services, data sharing services, and movie sharing services. These services require large-scale storage, which is characterized by writing new data and reading recent data, while only a few read old data. In this paper, we focus on a power saving method for large-scale storage systems that involve time-skewed data access.

Many methods have been proposed to reduce the power consumption of storage systems. A typical approach is to spin down some of the hard disk drives in a storage

S.-g. Lee et al. (Eds.): DASFAA 2012, Part II, LNCS 7239, pp. 138–149, 2012.

system, because rotation of the platters in a disk drive accounts for most of the electricity consumption by a storage system. Thus, the electricity consumption of a disk is decreased by spinning it down.

A well-known system that uses this approach is MAID (Massive Arrays of Inactive Disks) [3]. MAID keeps a small number of disk drives rotating at all times, which are used as a cache (known as cache disks), and this allows other disk drives to spin down (known as data disks). The MAID approach is effective for reducing power consumption when data access is time skewed, because numerous data disk drives store infrequently accessed old data and this allows disk rotation to be suspended. However, the original version of MAID did not consider system reliability. Thus, data stored on a disk are lost if one of data disks fails in the MAID system.

If we introduce a simple replica-based fault-tolerant configuration into MAID, synchronization between replicas during update operations would reduce the effect of saving power consumption because an increased number of write operations on disks elevates the number of spinning disks. In contrast, if we choose to improve timing when applying replica updates, we can reduce power consumption and maintain system reliability. In this paper, we propose a Replica-assisted Power Saving Disk Array (RA-PoSDA) as a method for dynamically controlling the timing of disk write operations by considering the rotation status of disks storing replicas.

RAPoSDA employs a duplicated write cache memory to maintain reliable data and it waits for a suitable time to write to an appropriate disk, thereby avoiding unnecessary disk spin-ups. We assume that there is an independent power supply (UPS) for the duplicate cache memory to ensure tolerance of power shortages or a UPS failure. Data on data disks are replicated in RAPoSDA as a primary backup configuration to ensure system reliability. RAPoSDA also has cache disks that provide large read cache spaces, as found with MAID, but they do not need to be replicated. Cache disks are an optional feature in RAPoSDA if there is sufficient cache memory.

The main contribution of this approach is the introduction of careful control over timing and targeting of disk access by replicas, which is achieved by dynamically checking the rotation status of individual disk drives. To ensure system reliability and reduce power consumption, RAID-based low power consumption storage systems [9,8] and power proportional disk arrays using replicas [14,13,1] have also been proposed. However, those other methods do not check the rotation status of each disk drive.

A simulation was more appropriate than empirical experiments and disks when evaluating the performance and power consumption of a storage system with changing configurations, including various sizes of cache memory, disk capacity, number of disks, and particularly with large numbers of disks. Therefore, we developed a dedicated simulator to compare RAPoSDA with MAID and a simple disk array with no power saving mechanism. We also prepared synthetic workloads with skewed data access based on a Zipf distribution and different ratios of read/write requests. The experimental results using the simulator indicated that RAPoSDA provided superior power reduction and a shorter average response time compared with MAID.

The remainder of this paper is organized as follows. Section 2 describes the power consumption model of a disk drive and a general approach to power reduction for a storage system with disk drives. Section 3 describes the details of our proposed storage system, RAPoSDA. Section 4 outlines the simulator we developed to evaluate the performance and power consumption of a storage system. A detailed discussion of our evaluation of RAPoSDA and MAID in terms of power saving and performance is reported in Section 5. Section 6 presents a number of related studies. Finally, Section 7 provides the conclusions of this paper.

2   Disk Drive Power Consumption Model

A disk drive is mainly composed of mechanical and control parts. The mechanical part has four components: platters, a spindle motor, read and write heads, and an actuator to move the heads over the platters. The controller is a component of the control part that performs read and write operations to comply with requests. Of these components, the spindle motor, actuator, and controller consume the most electricity in a disk drive, depending on its state.

A disk drive has three states, as shown in Table 1. It consumes the most power when it is in an active state, because the spindle motor, actuator, and controller are all working. It consumes the least in the standby state, because all components are not working, whereas only the spindle motor is working in the idle state. The standby state transfers to the active state after a read or write request. The idle state transfers to the standby state via spin-down, and vice versa via spin-up.

Compared with the active state, a disk drive consumes larger amounts of electricity when it spins up or down, especially at the beginning of spin-up. This means that frequent spin-ups and spin-downs to provide short standby states are not good for power saving. Thus it is important to keep unaccessed disk drives in a standby state for as long as possible. The duration of a standby state where a spin-up and spin-down requiring energy consumption does not exceed that of the idle state is known as the break-even time. A power saving is produced only if the duration of the standby state is longer than the break-even time.

Table 1. Disk-drive Status and Corresponding Power Consumption

State	I/O	RPM	Head location	Power
Active	In operation	max rotation	on disk	Large
Idle	No operation	max rotation	on disk	Middle
Standby	No operation	0	off disk	Small

3   RAPoSDA

One approach for reducing power consumption in a storage system with multiple hard disk drives is to divide the disk drives into two groups. Disk drives in one group are mainly kept in the standby state for longer than the break-even time, while disk drives in the other group are kept in active or idle states by access localization.

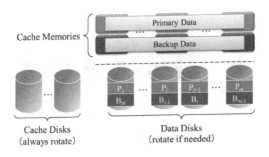

Fig. 1. Configuration of RAPoSDA

MAID (Massive Arrays of Inactive Disks) [3] is a well-known system that takes this approach. MAID ensures that a small number of disk drive are always kept in idle or active state as cache disks, while the remaining large number of disks are kept in a standby state as data disks. All read and write requests are initially dealt with by the cache disks and a replacement algorithm is used to keep the hit ratio high. This approach is effective in reducing the power consumption of a storage system. However, the original MAID method had problems of reliability, which is a major requirement for data centers. There is no means of recovering data in the event of a data disk failure. Direct importation of replicas into MAID might make it more reliable, but it would have detrimental effects on power consumption because the increased disk access required for replicas might violate the break-even time restriction.

To solve this problem, we propose a method for controlling the timing and targeting of replica disk access by dynamically checking the rotation status of individual disk drives. This reduces the power consumption and maintains system reliability. To implement the method, we propose the RAPoSDA storage system configuration.

3.1   RAPoSDA Configuration

In RAPoSDA, we adopt the chained declustering [6] method for data placement on data disks as a primary backup configuration that tolerates disk failures. When writing data onto data disks, RAPoSDA tries to select a disk that is currently rotating or one that has been longer in the standby state than the break-even time. This means that RAPoSDA has to maintain data elsewhere when waiting for the write timing. We use a write cache memory to temporarily maintain data. It is important to maintain reliability when the data are in the volatile cache memory, so we provide the cache memory with a primary backup configuration that corresponds to the data disks. As with MAID, RAPoSDA can use some disks as cache disks to provide larger read cache spaces. Figure 1 shows the configuration of RAPoSDA.

Data Disks. Chained declustering is a simple but effective strategy for data placement that provides good reliability and accessibility when the backup data is logically located in the disk next to the primary. We assume that a data disk spins down and moves to a standby state if the time without access is longer than a predefined threshold time.

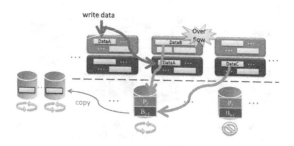

Fig. 2. Data Flow in RAPoSDA During Write Requests

Cache Disks. When the size of cache memories is not sufficient for comparing the total capacity of a disk array, cache disks are effectively used for enlarging the read cache space. However, cache disks are optional in RAPoSDA if there is sufficient cache memory and access localization. The cache contents are copies, so they do not need to be replicated. In contrast to data disks, we assume that cache disks do not spin down. Therefore, a large number of cache disks is not appropriate for reducing power consumption.

Cache Memory. RAPoSDA has two layers of cache memories that correspond to the primary and backup data disks, while one cache memory is shared by more than one disk drive. Each layer is connected to an individual power supply (UPS) to ensure the tolerance of power shortages or UPS failures.

3.2 Handling Write Requests

Write requests are processed in RAPoSDA as shown in Figure 2. Data are initially written into both the primary and backup layer of the cache memory. The written data are gathered in a corresponding buffer location in the cache memory of each individual disk that is responsible for storing the data. Buffered data are written onto their corresponding data disks when the amount of buffered data on the cache memory exceeds a predefined threshold. The rotation status of the data disk is investigated at this point. The data disk will spin up if it is in a standby state. When the disk enters an active state, all data on the location that corresponds to the disk in the cache memory are written to disks and then deleted from the cache memory. When a buffer overflow occurs in the primary layer corresponding to P_i, all data in the backup layer buffer for B_{i-1} are also transferred onto the data disk. However, if the buffer overflow occurs in B_i, the data in P_{i+1} are also transferred. Therefore, the amount of data in the primary layer of the cache memory is different from that in the backup layer.

This collective writing process reduces the frequency of spin-ups and spin-downs. Data are also copied to the cache disks to ensure quick responses for future read requests when there is time-skewed access. The remaining data disks with data still in the buffer stay in a standby state beyond the break-even time.

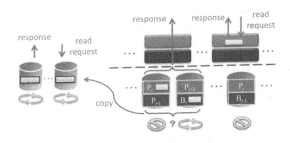

Fig. 3. Data Flow in RAPoSDA During Read Requests

3.3 Handling Read Requests

As shown in Figure 3, read requests initially check the existence of data in the cache memory, followed by cache disks. The cache memory has primary and backup layers, and both layers are searched for data. If the target data are in the cache memory or cache disks, the data are returned without accessing the data disks. If the data do not exist in the cache memory or cache disk, the data are read from a data disk.

Chained declustering is the method used for primary backup data placement on data disks, so RAPoSDA selects an appropriate disk from the two disks that correspond to the primary and backup for data. This selection is also important for reducing power consumption while still maintaining the break-even restriction. We propose the following set of rules for selecting the disk.

– If only one data disk is active, select the one that is active.
– If both data disks are active, select the disk with the largest memory buffer capacity.
– If both data disks are in standby, select the disk with the longer standby duration period.

Data read from data disks are also copied to cache disks to ensure a rapid response in future read requests when there is time-skewed access.

4   Disk Array Simulator

A number of simulators, such as DiskSim [4], have been proposed for evaluating the performance of storage systems. However, many cannot measure power consumption, including DiskSim. Dempsey [15] used an expansion of DiskSim with a function for measuring power consumption, but it needed to measure the power of the actual disk drives in advance and there was a limitation on the number of hard disks that could be used in a simulation. Thus, we developed a new simulator to evaluate the performance and power consumption of RAPoSDA and MAID.

The developed simulator is shown in Figure 4. It simulates behavior of each disk drive in the target storage system with a given workload, including the response time and power consumption, and it can flexibly change its configuration and workload.

The simulator initially sets up the parameters for the workload and the configuration of the cache memories, cache disks, and data disks. Clients inside the simulator then

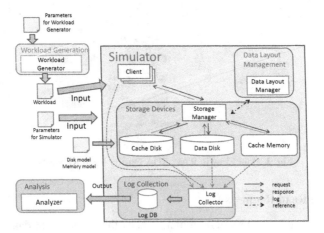

Fig. 4. Simulator Configuration

generate requests at times assigned by the workload. The Storage Manager dispatches the requests to the cache memories, cache disks, and data disks, based on information from the Data Layout Manager. Logs of the operation status for each device are collected by the Log Collector and analyzed by the Analyzer after finishing the simulation.

5   Evaluation

We used the simulator described in the previous section to compare the performance and power consumption of RAPoSDA, MAID, and a Normal system that was composed of a simple disk array with no power saving mechanism. In this paper, we mainly focus on the effect of read/write ratio for their performance and power consumption. We also evaluated other aspects, including their scalability. The evaluation results on the scalability indicate the superiority of RAPoSDA to MAID. Because of the page limits, we will report the details of the scalability evaluation in the other chance.

The original MAID proposed by [3] had no cache memory. However, many practical storage systems have cache memories and the cache memories in RAPoSDA play a very important role in reducing power consumption. To ensure a fair comparison, we introduced cache memory into MAID and evaluated the effect of cache memory with the three storage systems.

To ensure reliability, we also modified MAID so it had two replicas on the data disks to match RAPoSDA. However, MAID had no replica in the cache memory and cache disks because the data in the cache was only a copy. To prevent data loss from the cache during failures, the cache memory and cache disk in MAID used the 'write through' protocol. To handle replicas in the data disks, MAID randomly selected disks for the replicas. Normal also used replicas and it provided a faster response than the other method. RAPoSDA determined the access disk based on the rotation state of individual disks.

parameter	value
Capacity (TB)	2
Number of platters	5
RPM	7200
Disk cache size (MB)	32
Data transfer rate (MB/s)	134
Active power (Watt)	11.1
Idle power (Watt)	7.5
Standby power (Watt)	0.8
Spin-down energy (Joule)	35.0
Spin-up energy (Joule)	450.0
Spin-down time (sec)	0.7
Spin-up time (sec)	15.0

Fig. 5. Parameters for the Hard Disk Drive Used in the Simulation

Workload parameter	value
Time	5 hours
read:write	7:3, 5:5, 3:7
Number of files	1,000,000 (32KB/file)
Amount of file size	64GB (Primary × Backup)
Number of requests	$\lambda \times 3600 \times$ Time
Distribution of access	Zipf distribution
Request arrival distribution	Poisson process
Zipf factor	1.2
Mean arrival rate (λ)	25 (request/sec)

Fig. 6. Parameters of the Synthetic Workload

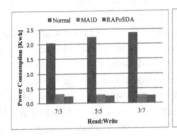 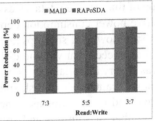

Fig. 7. Power Consumption of Each Storage System and the Power Reduction Ratios of RA-PoSDA and MAID versus Normal

The hard disk drive model used in our simulator was based on the Hitachi Deskstar 7K2000[12] produced by Hitachi Global Storage Technologies. Table 5 shows the parameters of the model. Furthermore, We prepared a synthetic workload for the evaluation. The workload parameters used in this experiment are listed in Table 6. In the workload, the access skew was based on Zipf distribution and the request arrival rate was based on a Poisson distribution. A workload was generated with three different types of read:write ratio, 7:3, 5:5, and 3:7.

In this evaluation, we set the simulation parameter as follows. The number of data disks was 64 with six cache disks and the total capacity of the cache memory was the number of data disks × 1/4 GB.

5.1 Power Consumption and Power Reduction Rate

Figure 7 shows the power consumption of Normal, MAID, and RAPoSDA, and the power reduction ratio of MAID and RAPoSDA when compared with Normal.This

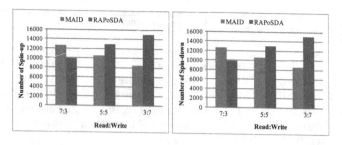

Fig. 8. Spin-up and Spin-down Counts with MAID and RAPoSDA

figure shows that Normal was the highest power consumer and that the power consumption of Normal increased with an increase in the write frequency. In contrast, MAID and RAPoSDA decreased power consumption with an increased write frequency. This shows that MAID and RAPoSDA achieve significant power savings. The power reduction ratio of RAPoSDA was higher than MAID with all read/write ratios. This matched our assumption of time-skewed data access with high write requests.

The disk drive power consumption depended on the rotation state and the number of spin-ups or spin-downs. If spin-ups and spin-downs occurred frequently, the power consumption may exceed that with no disk spin-downs. Control of excessive spin-ups and spin-downs is very important for power saving in the storage system.

Figure 8 shows the number of disk spin-ups and spin-downs with MAID and RA-PoSDA (Normal had no spin-downs). The graph shows that the trend for MAID was decreasing a number of spin-ups and spin-downs with a decreasing read frequency, while the trend for RAPoSDA was an increasing spin-up and spin-down count. From the perspective of the write cache size, MAID had a larger write cache than RAPoSDA because MAID used the cache disks for reads and writes, whereas RAPoSDA only used them for reads and this led to a smaller write cache memory compared with cache disks. This led to the possibility of a higher hit ratio of write requests for MAID and a smaller number of spin-ups and spin-downs with a frequent write workload.

However, the power reduction ratio shown in Figure 7 indicates that RAPoSDA gave a greater power reduction ratio compared with MAID. This shows that the group writing method of RAPoSDA provided a beneficial effect in maintaining the break-even time for individual data disks, even though the total counts of spin-ups and spin-downs with RAPoSDA exceeded those with MAID.

5.2   Average Response Time

Figure 9 shows the average response time for each storage system. The graph shows that Normal had the fastest response time but Normal consumed the most amount of power, because the disks were always spinning. The average response time of RAPoSDA was faster than that of MAID. When the average response time of MAID changed from 2.302 sec to 3.308 sec, the corresponding values for RAPoSDA fell in the range 0.559 sec to 1.296 sec.

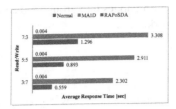

Fig. 9. Average Response Time with Normal, MAID, and RAPoSDA

Based on the previous discussion of hit ratios for the write cache, it seemed possible that the average response time of MAID was shorter than that of RAPoSDA, because the cache hit was effective in reducing the response time. However, Figure 9 demonstrates that RAPoSDA was also superior to MAID in terms of the average response time. One of the main reasons for this was the cache protocol. MAID adopted a write through protocol for cache memories and cache disks to prevent data loss because of failures.

6 Related Studies

This section briefly reviews related systems other than MAID.

In DRPM [5], the power consumption of disk drives can be expressed as a function of the rotation speed (RPM). To achieve good performance and low power consumption, DRPM utilizes multispeed disk drives that can dynamically change the rotation speed of the disk depending on the system's workload. Similarly, Hibernator [16] exploited this concept to implement power saving and performance controlled by a RAID configuration of multispeed disk drives.

The reduction of power consumption by network servers, such as Web servers and Proxy servers, was investigated by Carrera [2]. According to that report, multispeed hard disk drives are necessary to reduce power consumption and maintain server response performance. However, dynamic changes in the frequency of disk drive rotations present many technical challenges. As a result, such hard disk drives are not currently popular in practical use.

PARAID [14] is a powerful power saving technique for targeting RAID-based storage [10]. The controller skews the access to a small number of disk drives. It then creates inaccessible disks and spins down these disks to reduce power consumption. GRAID [9] places an emphasis on ensuring reliability and power savings based on RAID10 disk arrays. Using information at the log disk that is added into the normal disk to store logs, the system only needs to update the mirror disks periodically, so the system can spin down all the mirror disks to a low-power mode for most of the time and save energy. EERAID [8] is a power saving method that is focused in the RAID controller layer. EERAID reduces the power consumption using dynamic I/O scheduling and a cache management policy.

SIERRA[13] and RABBIT[1] are distributed storage systems that implement power proportionality through leveraging cluster-based data placement with data replication. These methods achieve proportional relation between power consumption and system performance by dividing all storing nodes into groups and controlling which groups are to be active to serve certain workloads.

7 Conclusions

Large-scale, high-performance, reliable, and low-power storage systems are required to construct better data centers for cloud computing. In this paper, we proposed RA-PoSDA (Replica-assisted Power Saving Disk Array) for use with such storage systems. RAPoSDA carefully controls the timing and targeting of disk access by dynamically checking the rotation status of individual disk drives. We compared the performance and power saving effects of RAPoSDA with modified MAID in simulations with different ratios of read/write requests in the workloads. The original version of MAID had no cache memory and no replication mechanism, so we added them to MAID to make a fair comparison.

The experimental results showed that RAPoSDA and the modified MAID provide reduced power consumption compared with a simple disk array with no power saving mechanism. However, RAPoSDA was superior to the modified MAID. From the performance perspective, the simulation results showed that the average response time of RAPoSDA was shorter than that of the modified MAID. Thus, consideration of individual disk rotation is an effective method for reducing the power consumption of storage systems, while maintaining good performance.

In future work, we will consider distributed file systems such as Hadoop Distributed File System (HDFS) and Google File System (GFS) which accepts more than two replicas, as power saving storage systems. We also aim to develop an experimental system with actual disk drives and evaluate the system using actual workloads.

Acknowledgment. This work is partly supported by Grants-in-Aid for Scientific Research from Japan Science and Technology Agency (A) (#22240005).

References

1. Amur, H., Cipar, J., Gupta, V., Ganger, G.R., Kozuch, M.A., Schwan, K.: Robust and flexible power-proportional storage. In: Proceedings of the 1st ACM Symposium on Cloud Computing, SoCC 2010, pp. 217–228. ACM, New York (2010)
2. Carrera, E.V., Pinheiro, E., Bianchini, R.: Conserving disk energy in network servers. In: ICS 2003: Proceedings of the 17th Annual International Conference on Supercomputing, pp. 86–97. ACM, New York (2003)
3. Colarelli, D., Grunwald, D.: Massive arrays of idle disks for storage archives. In: Supercomputing 2002: Proceedings of the 2002 ACM/IEEE Conference on Supercomputing, pp. 1–11. IEEE Computer Society Press, Los Alamitos (2002)
4. Ganger, G., et al.: The DiskSim Simulation Environment (v4.0),
   http://www.pdl.cmu.edu/DiskSim/
5. Gurumurthi, S., Sivasubramaniam, A., Kandemir, M., Franke, H.: DRPM: Dynamic Speed Control for Power Management in Server Class Disks. In: International Symposium on Computer Architecture, p. 169 (2003)

6. Hsiao, H.I., DeWitt, D.J.: Chained Declustering: A New Availability Strategy for Multiprocessor Database Machines. In: Proceedings of the Sixth International Conference on Data Engineering, pp. 456–465. IEEE Computer Society, Washington, DC (1990)
7. IDC: Extracting value from chaos (2011), http://idcdocserv.com/1142
8. Li, D., Wang, J.: EERAID: energy efficient redundant and inexpensive disk array. In: EW 11: Proceedings of the 11th Workshop on ACM SIGOPS European Workshop, p. 29. ACM, New York (2004)
9. Mao, B., Feng, D., Wu, S., Zeng, L., Chen, J., Jiang, H.: GRAID: A Green RAID Storage Architecture with Improved Energy Efficiency and Reliability. In: MASCOTS, pp. 113–120 (2008)
10. Patterson, D.A., Gibson, G., Katz, R.H.: A case for redundant arrays of inexpensive disks (raid). SIGMOD Rec. 17(3), 109–116 (1988)
11. Program, U.E.P.A.E.S.: Report to Congress on Server and Data Center Energy Efficiency Public Law 109-431 (2007),
    http://www.energystar.gov/ia/partners/prod_development/
    downloads/EPA_Datacenter_Report_Congress_Final1.pdf
12. Technologies, H.G.S.: Hard disk drive specification, hitachi deskstar 7k2000,
    http://www.hitachigst.com/tech/techlib.nsf/techdocs/
    5F2DC3B35EA0311386257634000284AD/$file/
    USA7K2000_DS7K2000_OEMSpec_r1.4.pdf
13. Thereska, E., Donnelly, A., Narayanan, D.: Sierra: practical power-proportionality for data center storage. In: Proceedings of the Sixth Conference on Computer Systems, EuroSys 2011, pp. 169–182. ACM, New York (2011)
14. Weddle, C., Oldham, M., Qian, J., Wang, A.I.A., Reiher, P., Kuenning, G.: PARAID: A gear-shifting power-aware RAID. Trans. Storage 3(3), 13 (2007)
15. Zedlewski, J., Sobti, S., Garg, N., Zheng, F., Krishnamurthy, A., Wang, R.: Modeling Hard-Disk Power Consumption. In: Proceedings of the 2nd USENIX Conference on File and Storage Technologies, pp. 217–230. USENIX Association, Berkeley (2003)
16. Zhu, Q., Chen, Z., Tan, L., Zhou, Y., Keeton, K., Wilkes, J.: Hibernator: helping disk arrays sleep through the winter. In: Proceedings of the Twentieth ACM Symposium on Operating Systems Principles, SOSP 2005, pp. 177–190. ACM, New York (2005)

ComMapReduce: An Improvement of MapReduce with Lightweight Communication Mechanisms

Linlin Ding, Junchang Xin, Guoren Wang, and Shan Huang

Key Laboratory of Medical Image Computing (NEU),
Ministry of Education, P.R. China
College of Information Science & Engineering, Northeastern University, P.R. China
{linlin.neu,xinjunchang}@gmail.com, wanggr@mail.neu.edu.cn,
milesandnick@163.com

Abstract. As a parallel programming model, MapReduce processes scalable and parallel applications with huge amounts of data on large clusters. In MapReduce framework, there are no communication mechanisms among Mappers, neither are among Reducers. When the amount of final results is much smaller than the original data, it is a waste of time processing the unpromising intermediate data objects. We observe that this waste can be avoided by simple communication mechanisms. In this paper, we propose ComMapReduce, a framework that extends and improves MapReduce for efficient query processing of massive data in the cloud. With efficient lightweight communication mechanisms, ComMapReduce can effectively filter the unpromising intermediate data objects in Map phase so as to decrease the input of Reduce phase specifically. Three communication strategies, Lazy, Eager and Hybrid, are proposed to filter the unpromising intermediate results of Map phase. In addition, two optimization strategies, Prepositive and Postpositive, are presented to enhance the performance of query processing by filtering more candidate data objects. Our extensive experiments on different synthetic datasets demonstrate that ComMapReduce framework outperforms the original MapReduce framework in all metrics without affecting its existing characteristics.

1 Introduction

As a well-known programming model, MapReduce [1] has gained extensive attention in recent years from industry and research communities. MapReduce processes scalable and parallel applications with huge amounts of data on large clusters. This programming model is scalable, fault tolerant, cost effective and easy to use. Query processing applications can be easily solved by MapReduce. In MapReduce framework, the implementations of Mappers and Reducers are independent without communication among Mappers, neither among Reducers. Each Mapper has no processing information of the other ones. When the amount of final results is much smaller than the original data, it is a waste of time processing the unpromising data objects. Combiner of MapReduce framework is only

S.-g. Lee et al. (Eds.): DASFAA 2012, Part II, LNCS 7239, pp. 150–168, 2012.

a local filtering strategy to compress the intermediate results of one Mapper. There is no global information to filter the unpromising data objects of MapReduce. In our opinion, when there is a little communication to filter numerous unpromising data objects, this waste of time can be avoided by simple and efficient communication mechanisms for many query processing applications, such as top-k, kNN and skyline.

In this paper, we propose a new improvement of MapReduce framework named ComMapReduce, with simple lightweight communication mechanisms for query processing of massive data in the cloud. ComMapReduce adds a lightweight communication function to effectively filter the unpromising intermediate output of Map phase and the input of Reduce phase, and then further extends MapReduce to enhance the efficiency of query processing applications without sacrificing its existing availability and scalability. In summary, the primary contributions of this paper can be summarized as follows:

- A framework with simple lightweight communication mechanisms, ComMapReduce, is proposed. The Mappers and Reducers of ComMapReduce use global communication information to enhance the query processing performance without sacrificing the availability and scalability of the original MapReduce.
- Three communication strategies, Lazy, Eager and Hybrid, are proposed to analyze ComMapReduce framework in depth. Two optimization strategies, Prepositive and Postpositive, are also proposed to enhance the query processing performance of ComMapReduce framework.
- Abundant synthetic datasets are adopted to evaluate the query processing performance of ComMapReduce framework. The experimental results demonstrate that ComMapReduce outperforms the original MapReduce in all metrics.

The rest of this paper is organized as follows: Section 2 proposes ComMapReduce framework. Three communication strategies are presented in Section 3. Section 4 proposes two optimization strategies. Section 5 reports the experimental results. Section 6 reviews the related work and the conclusion is given in Section 7.

2   ComMapReduce Framework

2.1   MapReduce Analysis In-depth

MapReduce framework was first presented by Dean et.al. in 2004, with one of its pubic available implementations Hadoop [2]. The nodes of MapReduce are divided into two kinds: one Master node and the other Slave nodes. The Master node is like a brain of the whole framework scheduling a number of parallel tasks to run on the Slave nodes. Non-experts can deploy their applications in parallel simply by implementing a Map function and a Reduce function. In the following, we give a simple example of query processing, top-2 query, to illustrate how query processing applications are executed on MapReduce framework.

EXAMPLE 1. (*Simple Top-k Query*) *There are fifteen data objects from '1' to '15' in the initial dataset and the top-2 query searches the two biggest data objects from '1' to '15'. Obviously, '14' and '15' are the final results.*

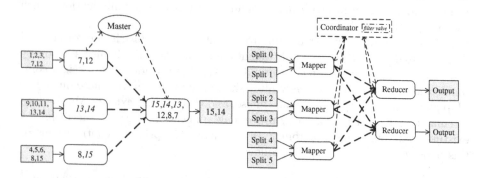

Fig. 1. Top-*k* Query on MapReduce (*k*=2)   **Fig. 2.** Framework of ComMapReduce

Figure 1 shows the simple top-*k* query of EXAMPLE 1 processing on MapReduce framework. There are three Mappers and one Reducer on the platform. The original fifteen data objects are partitioned into three splits as the input of three Mappers and then processed by the Mappers. Each Mapper reads the partition of its input data objects and computes its local top-*k* query results by calling a user-defined Map function. For example, the first Mapper reads data objects '1', '2', '3', '7', '12' and generates '7' and '12' as its intermediate results. The Reducer copies the output of three Mappers and then invokes a user-defined Reduce function to generate the final results of this top-2 query, data objects '14' and '15'.

Obviously, MapReduce framework has the ability of query processing naturally. In addition, Combiner of MapReduce processes the intermediate data objects compression of one Mapper. However, Combiner is a local data compression and has no processing information of the other Mappers. It cannot obtain a global *filter value* to process data compression for further filtering. It is meaningless of filtering unpromising data objects by using Combiner for query processing applications.

Just as this simple example, before the intermediate results of three Mappers are written into their disks, if the Mappers can receive data object '13' as a *filter value* to prune their intermediate results first with simple communication mechanisms, the output of Mappers can be decreased dramatically. The input of Reducer can be decreased from data objects '15', '14', '13', '12', '8', '7' to data objects '15', '14', '13'. Based on this idea, our ComMapReduce is proposed in the following.

2.2   ComMapReduce Overview

In order to further optimize the intermediate data objects of the original MapReduce framework, we propose ComMapReduce, an efficient lightweight communication framework extending MapReduce by pruning more unpromising data objects of query processing applications dramatically.

Figure 2 illustrates the architecture of ComMapReduce framework. ComMapReduce inherits the basic framework of MapReduce and takes HDFS [3] to store the original data objects and the final results of each application. In ComMapReduce framework, the Master node and Slave nodes play the same roles as the original MapReduce, so the Master node is not shown in Figure 2 and the following figures for convenience. The Coordinator node is deployed on ComMapReduce to share global communication information to enhance the performance of ComMapReduce without sacrificing the availability and scalability of the original MapReduce. The Coordinator node communicates with the Mappers and Reducers with simple lightweight communication mechanisms to filter the unpromising data objects. The Coordinator node receives and stores some temporary variables, and then generates *filter values* of the query processing applications. Each Mapper obtains a *filter value* from the Coordinator node to prune its data objects inferior to the other ones. After filtering, the output of Mappers and the input of Reducers both decrease dramatically. Therefore, the query processing performance of massive data on ComMapReduce framework is enhanced.

In Figure 2, when a client submits an application to the Master node, it schedules several Mappers to process this application. In Map phase, the original data objects of the client are divided into several splits. The Mappers on clusters process the splits in parallel and obtain their intermediate results. Then, each Mapper generates its *filter value* and sends it to the Coordinator node with simple communication mechanisms. The Coordinator node gains the most optimal result as a global *filter value* from the results it receives. Simultaneously, the Mappers receive a global *filter value* from the Coordinator node to filter their intermediate results and generate their new ones as the input of Reducer. Finally, in Reduce phase, the final results of this application are generated. In the following, we illustrate that ComMapReduce retains the existing characteristics of the original MapReduce from two aspects: availability and scalability.

2.3   ComMapReduce Availability

The availability of ComMapReduce is dependent on MapReduce. The recovery strategies of the Master node and Slave nodes are the same as MapReduce. For the recovery of the Coordinator node, easy method of periodic checkpoints can be used to check it whether fails or not. If the Coordinator node fails, a new copy can be deployed from the last checkpoint state to implement the filtering of unprocessed Mappers. However, there is only a single Coordinator node of ComMapReduce framework, so its failure is unlikely. Our current implementation stops the ComMapReduce computation if the Coordinator node fails. The clients can check and restart the ComMapReduce operation according to their desire. Therefore, ComMapReduce has its availability the same as MapReduce.

2.4   ComMapReduce Scalability

The scalability of ComMapReduce inherits the scalability of the original MapReduce. The Master node and Slave nodes retain the same functions of MapReduce, so they own the same scalability as MapReduce. To illustrate the scalability of ComMapReduce, we only need to illustrate that the scalability of ComMapReduce isn't affected by the Coordinator node. In the following, we analyze the scalability of the Coordinator node from two aspects: time complexity and memory usage.

Time Complexity: Suppose that the number of system setting Map tasks is M indicating that the number of Map tasks implementing at the same time is no more than M. Each Map task sends its *filter value* to the Coordinator node after it completes. That is to say, the maximum value of sending *filter values* at the same time is M. However, the probability of all Mappers completing at the same time is so low that this situation scarcely happens. As a result, if we can illustrate the scalability of ComMapReduce isn't affected in this extreme case, the scalability isn't affected certainly in the other cases. Further, if we can illustrate the time complexity of the Coordinator node is the same as the Master node, the scalability of ComMapReduce isn't affected by the Coordinator node.

- First, the time complexity of the Coordinator node receiving *filter values* of Map tasks is the same as the time complexity of the Master node receiving Map tasks requirements of locating data chunks.
- Second, the time complexity of the Coordinator node identifying the global *filter value* is the same as the time complexity of the Master node sending the data chunks locations to Map tasks. They both search a value in a Hash table.
- Third, the time complexity of the Coordinator node sending a global *filter value* to Map tasks is equal to the time complexity of the Master node sending the locations of data chunks to Map tasks.

Therefore, the scalability of ComMapReduce isn't affected by the Coordinator node from the aspect of time complexity.

Memory Usage: The memory usage of the Coordinator node is much lower than the Master node for only storing the global *filter value* and other variables of communication strategies. Therefore, the scalability of ComMapReduce isn't affected by the Coordinator node from the aspect of memory usage.

From the above analysis, we draw a conclusion that ComMapReduce has the same availability and scalability of the original MapReduce.

3   ComMapReduce Communication Strategies

Towards scalable and efficient query processing on ComMapReduce, how to identify the global *filter value* with simple communication mechanisms without influencing the correctness of results is an important problem. ComMapReduce doesn't change the functions of the Master node in MapReduce, so we only present the communication strategies of the Coordinator node. We design three communication strategies to illustrate how to communicate with the Coordinator node to obtain a global *filter value*. In this section, a complex top-*k* query example is adopted to illustrate the three communication strategies, Lazy, Eager and Hybrid.

EXAMPLE 2. *(Complex Top-k Query) The original dataset is fifty data objects, from '1' to '50'. There are five Mappers processing ten data objects respectively. Top-5 query returns the five biggest data objects as the final results. Obviously, data objects '50', '49', '48', '47' and '46' are the results.*

3.1 Lazy Communication Strategy

The principle of Lazy Communication Strategy (LCS) is that the Coordinator node generates a global *filter value* after all Mappers complete. It receives the *filter value*s of all Mappers and chooses the most optimal one as the global *filter value*. Then the Mappers filter their intermediate values using the global *filter value* from the Coordinator node and generate the input of Reducer. The Reducer produces the final results with less input than MapReduce.

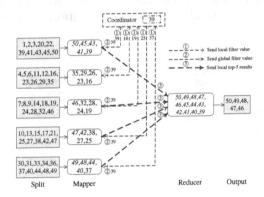

Fig. 3. Top-*k* Query Processing on ComMapReduce with LCS

Figure 3 shows the complex top-*k* query processing on ComMapReduce with LCS. The data objects in original dataset are divided into five splits processed by five Mappers respectively. We use three steps to illustrate the course of LCS. Step 1. Each Mapper processes its own data objects and generates its *filter value*, the *k*th one of its intermediate results, and then sends it to the Coordinator node. For example, the first Mapper generates its intermediate results '50', '45', '43', '41', '39' and sets '39' as its *filter value*, and then sends it to the Coordinator node after it completes. Step 2. The Coordinator node receives all the *filter value*s of Mappers, '39', '16', '19', '25', '37' and chooses the biggest one, '39', as the global *filter value*, and then sends it to all the Mappers. Step 3. After receiving the global *filter value* '39', the intermediate results of Mappers are filtered and sent to the Reducer. For example, after filtering, there are no intermediate results of the second Mapper, because its intermediate results are all smaller than the global *filter value* '39'. The italic type shown data objects in Figure 3 are the intermediate results of Map phase after filtering. The final input of Reducer is data objects '50', '49', '48', '47', '46', '45', '44', '43', '42', '41', '40', '39', which is about 48% of the original dataset without filtering the intermediate results of Mappers. By obtaining the most optimal one from the *filter values* of all Mappers as the final global *filter value*, LCS filters the intermediate data objects which are not the final results sharply for query processing applications of ComMapReduce.

3.2   Eager Communication Strategy

In order to improve the response speed of query processing, Eager Communication Strategy (ECS) is proposed with the principle as follows. After each Mapper completes, it sends its *filter value* to the Coordinator node and receives a global *filter value* simultaneously. If the global *filter value* is null, the Mapper processes its split normally. Otherwise, the Mapper prunes its intermediate results with the global *filter value*. Once receiving a *filter value* of another Mapper, the Coordinator node chooses the more optimal one as a new global *filter value* according to the application by comparing with the original global *filter value*. This course still repeats until all Mappers in the platform complete.

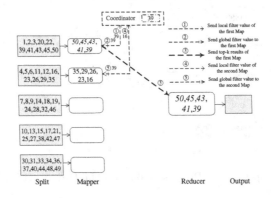

Fig. 4. Top-*k* Query Processing on ComMapReduce with ECS

Figure 4 shows the complex top-*k* query processing on ComMapReduce with ECS. Six steps are taken to illustrate the implementations of the first and the second Mapper. Step 1. After the first Mapper completes, it sends data object '39' to the Coordinator node as its *filter value*. Step 2. After receiving '39', the Coordinator node sets '39' as the global *filter value* and sends it to the first Mapper. Step 3. The final intermediate results of the first Mapper are data objects '50', '45', '43', '41', '39' after filtering. Step 4. The second Mapper sends its *filter value* '16' to the Coordinator node. Step 5. After the Coordinator node receives the *filter value* '16', it chooses the bigger one '39' as the new global *filter value.* Step 6. The second Mapper filters its intermediate results with '39'. All the intermediate results of the second Mapper are smaller than '39', so there are no intermediate results of the second Mapper without showing in Figure 4. This course continues until the fifth Mapper completes and then generates the final results.

The Coordinator node generates the temporary global *filter value* in time with ECS instead of waiting for all the Mappers completing. Mappers can receive a temporary global *filter value* immediately to filter their intermediate results. So ECS can dramatically enhance the efficiency of query processing of ComMapReduce.

3.3   Hybrid Communication Strategy

In order to both effectively filter the unpromising intermediate results with more optimal global *filter value* and improve the response speed of query processing, Hybrid Communication Strategy (HCS) is proposed to enhance the query processing performance of ComMapReduce. The main workflow of HCS is as follows. Suppose that one Mapper completes, it sends its *filter value* to the Coordinator node at once. However, the Coordinator node waits for a preassigned a period of time (t_w) to receive the other Mappers' *filter values* instead of generating a global *filter value* immediately. After receiving the other completing Mappers' *filter values*, the Coordinator node chooses the most optimal one as a global *filter value* according to the application. The Mappers receive a global *filter value* from the Coordinator node to prune their intermediate values and decrease the input of Reducer. The same course repeats until all the Mappers in the system complete. Nevertheless, when the last Mapper completes, it sends its *filter value* to the Coordinator node and receives a global *filter value* immediately.

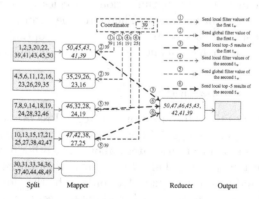

Fig. 5. Top-*k* Query Processing on ComMapReduce with HCS

Figure 5 shows the complex top-*k* query processing on ComMapReduce with HCS. For simple description, we suppose that there are two Mappers completing in each preassigned a period of time t_w. Six steps are taken to illustrate the implementations of the first and the second t_w. Step 1. The first Mapper generates '39' as its *filter value* and sends it to the Coordinator node. After receiving data object '39', the Coordinator node waits for the second Mapper completing in the first t_w and receives its *filter value* '16' instead of generating a global *filter value* in time. Step 2. After receiving '39' and '16', the Coordinator node chooses the larger one, '39', as a global *filter value* and sends it to the first and the second Mapper. Step 3. The first and the second Mapper receive the global *filter value* '39' to filter their intermediate results. Step 4. In the second t_w, the third and the forth Mapper complete and send their *filter values* '19' and '25' to the Coordinator node. Step 5. After receiving *filter values* '19' and '25', the Coordinator node compares '19' and '25' to the original global *filter value* '39' and sets '39' as a global *filter value* again. Step 6. After receiving the global *filter value* '39', the third and the forth Mapper filter their intermediate results and send them to the Reducer.

HCS neither waits for the completion of all the Mappers to generate a global *filter value* nor immediately sends local *filter value* after each Mapper completes. It can both effectively filter the unpromising intermediate results of Mappers and don't need to wait for a long time to generate a more optimal global *filter value*.

4   Optimizations

In practical applications, the number of Mappers is much larger than the number of virtual machines, about hundreds multiples. Even adopting the above communication strategies to filter the unpromising data objects, the data objects which are not the final results can be filtered ulteriorly. Therefore, we propose two optimization strateties to enhance the performance of query processing of massive data on ComMapReduce. The two optimization strategies are orthometric to the communication strategies in Section 3. Therefore, they can be combined with any communication strategy above. In addition, the two optimization strategies can be used together to further enhance the performance of ComMapReduce.

4.1   Prepositive Optimization Strategy

The principle of Prepositive Optimization Strategy (PreOS) is as follows. After a part of Mappers complete, the Coordinator generates a temporary global *filter value* by any communication strategy above. The other unprocessed Mappers can retrieve the temporary global *filter value* in their initial phase from the Coordinator node, and then prune their data objects by the global *filter value*. The number of data objects to be sorted of Mappers decreases dramatically, so the sorting time of Mappers also decreases. PreOS improves the query processing performance of ComMapReduce by further filtering.

Figure 6 shows the complex top-k query processing on ComMapReduce with PreOS. We take five steps to illustrate the workflow of PreOS. For simple description, from Step 1 to Step 3, the Coordinator node generates a temporary

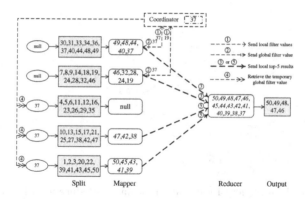

Fig. 6. Top-k Query Processing on ComMapReduce with PreOS

global *filter value* '37' with HCS after the first and the second Mapper complete. Step 4. The other three unprocessed Mappers retrieve the temporary global *filter value* '37' from the Coordinator node in their initial phase and filter their input data objects that cannot be the final results. For example, all the input data objects of the third Mapper are smaller than '37', so it is unnecessary to process these data objects. Step 5. The third to the fifth Mapper send their intermediate results to the Reducer only with PreOS not using any communication strategy above.

In practical applications, when the number of Mappers is much more than the number of virtual machines, the Mappers run in several batches. The unprocessed batches of Mappers can filter the unnecessary data objects in time by retrieving a temporary global *filter value* from the Coordinator node in their initial phase of Mappers with PreOS to enhance the query processing performance.

4.2 Postpositive Optimization Strategy

When all the Mappers in the system are processed in one batch, Postpositive Optimization Strategy (PostOS) is proposed to further filter the input data objects of Reducer. The main idea of PostOS is as follows. After the intermediate results of Mappers are sent to the Reducer, the Reducer first retrieves a temporary global *filter value* from the Coordinator node and filters its input data objects again. As the global *filter value* changes along the running of Mappers, it is necessary to filter the input of Reducer again. Therefore, the Reducer processes the input data objects after filtering to enhance the performance of query processing of massive data in ComMapReduce framework.

Figure 7 shows the complex top-k query processing on ComMapReduce with PostOS. Only four steps are taken to illustrate the implementation of PostOS simply. From Step 1 to Step 2, after the first and the second Mapper complete, the Coordinator node generates a temporary global *filter value* '37' with HCS. Step 3. All Mappers send their intermediate results to the Reducer. Step 4.

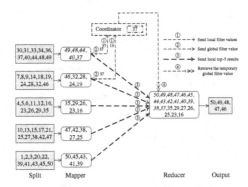

Fig. 7. Top-k Query Processing on ComMapReduce with PostOS

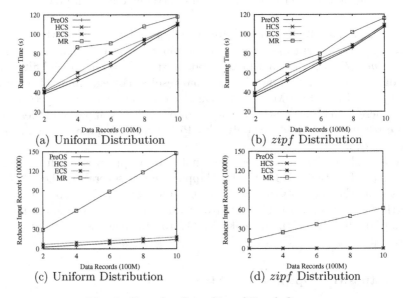

Fig. 8. Changing Data Size of Top-k Query

After receiving the intermediate results of Mappers, the Reducer retrieves the temporary global *filter value* '37' from the Coordinator node and prunes its input data objects. The data objects smaller than '37' are all filtered. Therefore, the input of Reducer decreases sharply so as to relieve the workload of Reducer.

5   Experiments

In this section, three types of query processing are taken as examples to evaluate the query processing performance of our ComMapReduce framework. Two communication strategies of ComMapReduce, ECS, HCS and two optimization strategies, PreOS and PostOS, are taken to compare with MapReduce (MR). For LCS, if some weak Mappers occupy long processing time, the Coordinator node waits for the completion of these Mappers and then generates the global *filter value*, which leads to the increase of running time of query processing. When the number of Mappers is unknown, it is hard to detect the time of all Mappers completing and sending their local *filter value*s. Therefore, LCS is not widely used in actual applications, so we don't evaluate the performance of LCS.

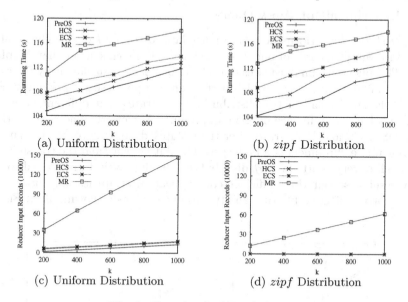

(a) Uniform Distribution   (b) *zipf* Distribution

(c) Uniform Distribution   (d) *zipf* Distribution

Fig. 9. Changing k of Top-k Query

5.1   Experimental Setup

We set up a cluster of 9 commodity PCs in a high speed Gigabit network, with one PC as the Master node and the Coordinator node, the others as the Slave nodes. Each PC has an Intel Quad Core 2.66GHZ CPU, 4GB memory and CentOS Linux 5.6. We use Hadoop 0.20.2 and compile the source codes under JDK 1.6. Two main experimental benchmarks are the running time of query processing and the number of Reducer input records. We compare ECS, HCS and PreOS to MapReduce processing top-k and kNN query. We compare ECS, HCS and PostOS to MapReduce processing skyline query. The experimental parameters of top-k, kNN and skyline query are as follows:

- The top-k query is evaluated in uniform distribution and *zipf* distribution. The default number of data records is 1000M, ranging from 200M to 1000M. The default number of k value of top-k query is 1000.
- The kNN query is evaluated in uniform distribution and clustered distribution. The default number of data records is 1000M, ranging from 200M to 1000M. The default number of k value of kNN query is 1000 and the default number of data dimensions of kNN query is 4.

– The skyline query is evaluated in independent distribution and anti-related distribution. The default number of data records is 1000K, ranging from 200K to 1000K. The default number of data dimensions of skyline query is 4.

5.2   Experiments of Top-k Query

Figure 8 shows the performance of changing data size of top-k query. The number of Reducer input records of our three strategies are much smaller than MapReduce. This phenomenon illustrates that ComMapReduce can effectively filter the unpromising data objects with high performance of processing top-k query. The similar changing trend of two distributions illustrates that different data distributions have no influence on ComMapReduce framework. The running time of PreOS, HCS and ECS are shorter than MapReduce. One reason is that the experimental environment is a high speed Gigabit network and the transmitting time itself is very short. In $zipf$ distribution, the data object are skewed to the query results, so our three efficient communication strategies only have 1000 records for its effective filtering ability without obviously showing in in Figure 8(d).

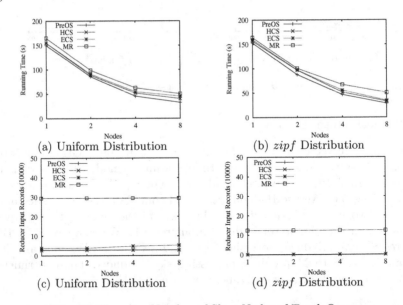

Fig. 10. Changing Number of Slave Nodes of Top-k Query

Figure 9 shows the performance of changing k of top-k query. PreOS, HCS and ECS are more optimal to MapReduce both in uniform distribution and $zipf$ distribution. The similar changing trend of different data distributions also shows the high scalability of ComMapReduce. As the k of top-k query increases, the number of final results also increases, so the number of Reducer input records increases certainly. In Figure 9(a) and Figure 9(b), the unprocessed Mappers can retrieve a temporary global *filter value* to filter their unpromising data objects in the initial phase by PreOS, so its running time is more optimal to the others.

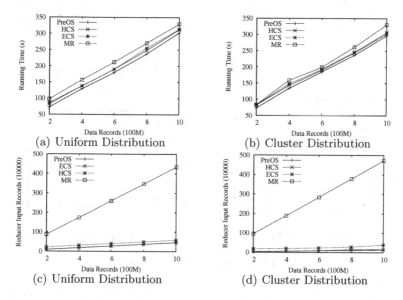

Fig. 11. Changing Data Size of *k*NN Query

Figure 10 shows the performance of changing the number of slave nodes to evaluate top-*k* query. As the number of Slave nodes increases, the running time decreases for enlarging the parallelism of the system. The similar changing trend of different data distributions illustrates that ComMapReduce can reach the same scalability of MapReduce. In addition, the running time and the number of Reducer input records of ComMapReduce are both more optimal to MapReduce.

5.3 Experiments of *k*NN Query

Figure 11 shows the performance of changing data size of *k*NN query. As the data size increases, the running time and the number of Reducer input records both increase sharply. The number of Reducer input records of ComMapReduce is extremely fewer than MapReduce showing the efficient filtering ability of ComMapReduce, about 14% of the number of Reducer input records of MapReduce. PreOS is better than HCS and ECS by obtaining a global *filter value* in the initial phase of Mappers and filtering numerous unpromising data objects.

Figure 12 shows the performance of changing the dimensions of *k*NN query. As the dimensions of *k*NN query increases, the running time and the number of Reducer input records increase accordingly as the computation and transmission cost increase. The number of Reducer input records is much fewer than MapReduce showing the efficient filtering ability of ComMapReduce.

Figure 13 shows the performance of changing *k* of *k*NN query. With increasing *k*, the final results increase, so the running time and number of records also

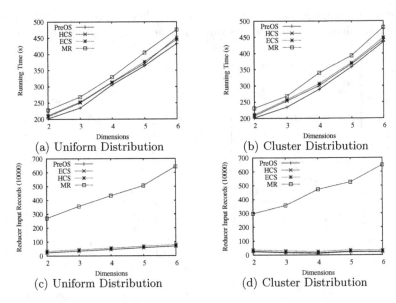

Fig. 12. Changing d of kNN Query

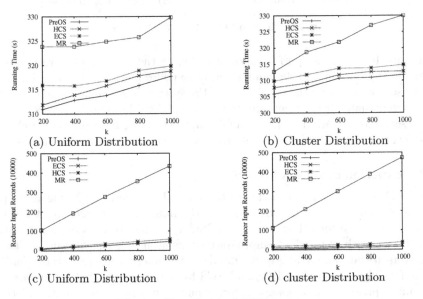

Fig. 13. Changing k of kNN Query

increase. The running time of ComMapReduce and the number of Reduce input records of ComMapReduce are dramatically optimal to MapReduce.

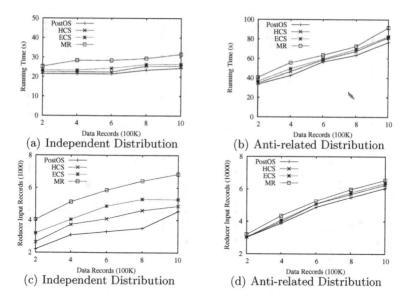

Fig. 14. Changing Data Size of Skyline Query

5.4   Experiments of Skyline Query

The computation cost of skyline query is larger than top-k and kNN, so the Mappers processing skyline query implement in a batch instead of many batches incurring the waste of running time. Therefore, we choose PostOS to compare with the other strategies. Figure 14 shows the performance of changing the data size of skyline query. As the data size increases, the running time and the number of Reducer input records increase dramatically. ComMapReduce is more optimal to MapReduce under different data distributions. In anti-related distribution, although the data objects skew to the final results and the computation cost increases accordingly, the performance of ComMapReduce is still optimal to MapReduce.

Figure 15 shows the performance of changing the number of slave nodes of skyline query. The running time decreases as the number of nodes increases for enhancing the parallelism of the system. The same changing trend of different distributions shows that ComMapReduce has high scalability. ComMapReduce filters the unpromising data objects with efficient communication mechanisms, so the number of Reducer input records of ComMapReduce is fewer than MapReduce. Without using efficient filtering strategies, the number of Reducer input records of MapReduce keeps a constant.

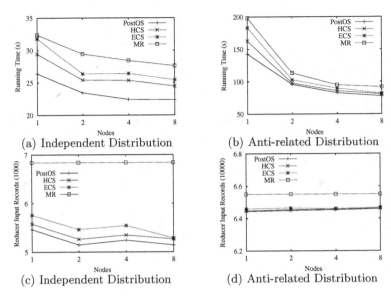

(a) Independent Distribution   (b) Anti-related Distribution

(c) Independent Distribution   (d) Anti-related Distribution

Fig. 15. Changing Number of Slave Nodes of Skyline Query

6   Related Work

Google's MapReduce [1] was first proposed in 2004 for massive parallel data analysis in shared-nothing clusters. Hadoop [2] is an open-source implementation of MapReduce followed by several systems including Hive [4], HBase [5] and Pig [6]. Paper [7] illustrates the limitations and opportunities of data management in the cloud and there has already been several research achievements as follows. Map-Reduce-Merge [8] is a new model improved MapReduce adding a Merge phase to merge data already partitioned and sorted by Map and Reduce models. Azza Abouzeid et. al. proposes HadoopDB [9], using MapReduce as the communication layer above numerous nodes running single-node DBMS instances. PLANET [10] is a scalable distributed framework for learning tree models over large datasets, with implementing each one using MapReduce model. Ariel Cary et. al. [11] applies MapReduce model to solve the problems of bulk-construction of R-trees and aerial image quality computation. Papers [12,13] introduce how to process Join algorithms using MapReduce without the query processing algorithms. Jens Dittrich et. al. proposes Hadoop++ [14], injecting index and join techniques at the right places by Use Defined Functions only to achieve advantages. HaLoop [15] is an improvement of MapReduce supporting iterative applications, which improves their efficiency by making the task scheduler loop-aware and by adding various caching mechanisms. Pergel [16] is a distributed system for processing largesize graph datasets and doesn't support general query processing applications. Contrary to these research, ComMapReduce adopts efficient communication mechanisms to improve the performance of query processing of massive data in the cloud.

7   Conclusions

This paper proposes the design and evaluation of ComMapReduce, an improved version of MapReduce framework with lightweight communication mechanisms, supporting query processing for a large number of data objects. ComMapReduce maintains the main great features of MapReduce, availability and scalability, while filtering the unpromising data objects by communicating with the Coordinator node. Three communication strategies, Lazy, Eager and Hybrid, are presented to effectively implement query processing applications of massive data in ComMapReduce framework. We also propose two improvements, named Prepositive Optimization Strategy and Postpositive Optimization Strategy to enhance the performance of ComMapReduce. Three query processing applications, top-k, kNN and skyline, are taken as examples to evaluate the query processing performance of ComMapReduce. The experimental results demonstrate that the performance of ComMapReduce is beyond of MapReduce dramatically in all metrics.

Acknowledgement. This research is supported by the State Key Program of National Natural Science of China (Grant No. 61033007), the National Science Foundation for Distinguished Young Scholars of China (Grant No. 61025007), the National Natural Science Foundation of China (Grant No. 60973020, 61073063), the National Natural Science Foundation for Young Scientists of China (Grant No. 61100022), and the National Marine Public Benefit Research Foundation (Grant No. 201105033).

References

1. Dean, J., Ghemawat, S.: MapReduce: Simplified Data Processing on Large Clusters. In: Proc.of OSDI, pp. 137–150 (2004)
2. Hadoop, http://hadoop.apache.org/
3. HDFS, http://hadoop.apache.org/common/hdfs/
4. Thusoo, A., Sarma, J.S., Jain, N., et al.: Hive-A Warehousing Solution Over a Map-Reduce Framework. PVLDB 2(2), 1626–1629 (2009)
5. Carstoiu, D., Lepadatu, E., Gaspar, M.: Hbase-non SQL Database, Performances Evaluation. IJACT-AICIT 2(5), 42–52 (2010)
6. Olston, C., Reed, B., Srivastava, U., et al.: Pig Latin: A Not-so-foreign Language for Data Processing. In: Proc.of SIGMOD, pp. 1099–1110 (2008)
7. Abadi, D.J.: Data Management in the Cloud: Limitations and Opportunities. IEEE Data Eng. Bull. (DEBU) 32(1), 3–12 (2009)
8. Yang, H., Dasdan, A., Hsiao, R., et al.: Map-reduce-merge: Simplified Relational Data Processing on Large Clusters. In: Proc. of SIGMOD, pp. 1029–1040 (2007)
9. Abouzeid, A., Baida-Pawlikowski, K., Abadi, D., et al.: HadoopDB: An Architectural Hybrid of MapReduce and DBMS Technologies for Analytical Workloads. PVLDB 2(1), 922–933 (2009)
10. Panda, B., Herbach, J.S., Basu, S., et al.: PLANET: Massively Parallel Learning of Tree Ensembles with MapReduce. In: Proc. of VLDB, pp. 1426–1437 (2009)

11. Cary, A., Sun, Z., Hristidis, V., Rishe, N.: Experiences on Processing Spatial Data with MapReduce. In: Winslett, M. (ed.) SSDBM 2009. LNCS, vol. 5566, pp. 302–319. Springer, Heidelberg (2009)
12. Blanas, S., Patel, J.M., Ercegovac, V., et al.: A Comparision of Join Algorithms for Log Processing in MapReduce. In: Proc. of SIGMOD, pp. 975–986 (2010)
13. Pavlo, A., Paulson, E., Rasin, A., et al.: A Comparison of Approaches to Large-scale Data Analysis. In: Proc. of SIGMOD, pp. 165–178 (2009)
14. Dittrich, J., Quian-Ruiz, J., Jindal, A., et al.: Hadoop++: Making a Yellow Elephant Run Like a Cheetah (Without It Even Noticing). PVLDB 3(1), 518–529 (2010)
15. Bu, Y., Howe, B., Balazinska, M., et al.: HaLoop: Efficient Iterative Data Processing on Large Clusters. PVLDB 3(1), 285–296 (2010)
16. Malewicz, G., Austern, M.H., Bik, A.J.C., et al.: Pregel: A System for Large-scale Graph Processing. Proc. of SIGMOD, pp. 135–146 (2010)

Halt or Continue: Estimating Progress of Queries in the Cloud

Yingjie Shi, Xiaofeng Meng, and Bingbing Liu

School of Information, Renmin University of China, Beijing, China
shiyingjie1983@yahoo.com.cn, {xfmeng,liubingbing}@ruc.edu.cn

Abstract. With cloud-based data management gaining more ground by day, the problem of estimating the progress of MapReduce queries in the cloud is of paramount importance. This problem is challenging to solve for two reasons: i) cloud is typically a large-scale heterogeneous environment, which requires progress estimation to tailor to non-uniform hardware characteristics, and ii) cloud is often built with cheap and commodity hardware that is prone to fail, so our estimation should be able to dynamically adjust. These two challenges were largely unaddressed in previous work. In this paper, we propose PEQC, a Progress Estimator of Queries composed of MapReduce jobs in the Cloud. Our work is able to apply to a heterogeneous setting and provides a dynamically update mechanism to repair the network when failure occurs. We experimentally validate our techniques on a heterogeneous cluster and results show that PEQC outperforms the state of the art.

Keywords: progress estimate, PERT network, MapReduce, cloud.

1 Introduction

As a solution to manage big data with high cost performance, cloud data management system has attracted more and more attentions from both industry and academia, and it is now supporting a wide range of applications. These applications need operations on massive data, such as reporting and analytics, data mining and decision support. A large part of the applications are implemented through MapReduce[5] jobs. MapReduce integrates parallelism, scalability, fault tolerance, load balance into the simple framework, and MapReduce-based applications are very suitable to be deployed in the cloud which is composed of large-scale commodity nodes. Although data queries are sped-up in cloud-based systems, many queries cost a long period of time, even to several days or months[15]. So users want to know the remaining time of such long-running queries to help decide whether to terminate the query or allow it to complete. For them, some early results within a certain confidence interval which can be computed through online aggregation[7] are even better. The accurate feedback and online aggregation both depend on the progress estimate of queries. There are also some applications in which every query requires critical response time, such as advertisement applications, customer profile management of SNS(Social Network Site), etc. Estimating the remaining time of the queries is beneficial for scheduling them to guarantee their deadlines.

S.-g. Lee et al. (Eds.): DASFAA 2012, Part II, LNCS 7239, pp. 169–184, 2012.

In addition, providing the accurate remaining time of queries is also helpful to performance debugging and configuration tuning for applications in the cloud. Progress estimate of queries in the cloud first confronts challenges in parallel environment, such as parallelism and concurrency[11]. The characteristics of cloud raise this problem's complexity , common failures and environment heterogeneity are two main challenges.

In order to achieve high performance cost, cloud system is often constructed with cheap, commodity machines, and the platform is always of large scale (hundreds or thousands of nodes[4]), so failures are very common. Cloud systems support high fault tolerance and consider failure as normal situation. According to the usage statistics in [4], the average worker deaths per MapReduce job in Google is 5.0 with the average worker machines 268. It is reported in[8] that in a system with 10 thousand of super reliable servers (MTBF of 30 years), according to typical yearly flakiness metrics, servers crash at least twice with the failure rate of 2-4%. For commodity nodes in the cloud, the failure rate is even higher. So the challenge is that once a failure happens, the progress indicator has to cope with failures promptly to provide continuously revised estimates.

Scalability is one main advantage of cloud systems. With the development of applications and the increase of data volume, there is a trend that more nodes will be added into the existing system to get more computing capability. So it is difficult to keep all the nodes in the cloud belonging to the same hardware generation. This brings another challenge – heterogeneity. Different software configurations and concurrent workload further reduce the homogeneity of cluster performance [1]. As [14] reports, in the virtual environment, the performance COV (coefficient of variation) of the same application on the same virtual machines is up to 24%. A MapReduce job is always composed of several rounds of tasks, and the tasks during each round are executed parallel on different nodes. The heterogeneity and variation of performance results in that the task scheduling is composed of irregular task rounds, and it raises the complexity of estimating the progress.

We propose PEQC, a progress estimator of queries based on MapReduce jobs in the cloud. The main contributions of our work include:

1. We model every MapReduce task of the query with its duration and failure probability, and transform the query procedure into a stochastic PERT(Project Estimate and Review Technique) network by allowing the task duration to be random vairables;
2. We propose a method to compute the most critical path of the PERT network, which can present the execution of the whole query in the heterogeneous environment;
3. We provide an efficient update mechanism to repair the network and re-compute the critical path when failure happens;
4. We implement PEQC on Pig[19] & Hadoop[18] to verify its accuracy and robustness in presence of failures.

The rest of this paper is structured as follows. Section 2 describes the related work. Section 3 presents the problem modeling, and discusses the uncertainties and stochastic characteristics of task duration. In Section 4, we propose our solution to estimating progress of MapReduce queries, we also present the repair mechanism to react to common failures. Experimental validation of PEQC on Pig & Hadoop is presented in Section 5. We conclude the paper and discuss the future work in Section 6.

2   Related Work

There are two kinds of areas that are related to our work. First is the area of estimating the progress of SQL queries on single-node DBMS. [9] separates a SQL query plan into pipelined segments, which is defined by blocking operators. The query progress is measured in terms of the percentage of input processed by each of these segments. Its sub-sequent work [10] widens the class of SQL queries the estimator supports, and it increases the estimate accuracy by defining segments at a finer granularity. [3] decomposes a query plan into a number of pipelines, and computes the query progress through the total number of getnext() calls made over all operators. [2] characterize the query progress estimation problem introduced in [3] and [9] so as to understand what the important parameters are and under what situations we can expect to have a robust estimation of such progress. The work of this area focuses on partitioning a complex query plan into segments(pipelines) and collecting the statistics on cardinality information to compute the query progress. These techniques are helpful for estimating the progress of queries running on a single node. However, they do not account for the challenges brought by query parallelization, such as parallelism and data distribution.

Our work is also related to progress estimate of MapReduce jobs. We can classify the related work into two categories: estimating the progress of tasks during one MapReduce job, and estimating the progress of MapReduce pipelines. [17] estimates the time left for a task based on the progress score provided by Hadoop. This paper focuses on task scheduling in MapReduce based on the *longest approximate time to end* of every task, so it orders the task by their remaining times. It computes the progress rate through the progress score and the elapsed time of the task execution, and computes the remaining time of a task based on the progress rate. [17] provides a method to estimate the progress of a MapReduce task, however, there are also several challenges to estimate the progress of MapReduce jobs and MapReduce DAGs. Parallax[12] estimates the progress of queries translated into sequences of MapReduce jobs. It breaks a MapReduce job into pipelines, which are groups of interconnected operators that execute simultaneously. Parallax estimates the remaining time by summing the expected remaining time across all pipelines. It addresses the challenges of parallelism and variable execution speeds, without considering concurrent workloads. ParaTimer[11] extends Parallax to estimate queries translated into MapReduce DAGs. It estimates the progress of concurrent workloads through a critical path based on task rounds, which works well in a *homogeneous* environment. ParaTimer handles task failure through comprehensive estimation, which provides an upper bound on the query time in case of failure. However, ParaTimer assumes only one worst-case failure before the end of the execution, and it has to repeat all the steps in the estimate algorithm to adjust to failures. It may become inefficient when failures are very common, which is one characteristic of cloud. None of the above work handles heterogeneity in the estimation, which is also an important characteristic of cloud.

3   Problem Modeling and Stochastic Characteristics

The project evaluation and review technique (PERT) is widely used to estimate completion time of projects with concurrent workloads[6], we formulate the problem into a stochastic PERT network. In this section, we discuss the problem modeling procedure and the stochastic characteristic.

3.1 Why stochastic PERT?

We take a query related to HTML document processing and log-file analysis as an example to discuss the problem. The query is to find the ad revenue of all the web page visitings with both page rank and duration in specific ranges. Table *Rankings* contains basic information of pageURLs, and table *U serVisits* models log files of HTTP server traffic. The SQL command is like:

```
SELECT sourceIP, destURL, adRevenue FROM Rankings, UserVisits WHERE
Rankings.pageRank > 10 AND UserVisits.duration > 1 AND
UserVisits.duration <= 10 AND Rankings.pageURL = UserVisits.destURL;
```

This query can be translated into three MapReduce jobs: job1 and job2 filter tuples according to the query conditions, they can be executed concurrently; job3 joins the tables and executes after job1 and job2. Suppose job1 is composed of 2 mappers and 1 reducer, job2 is composed of 6 mappers and 1 reducer, and job3 includes 1 mapper and 1 reducer. In the cluster there are 5 nodes with different hardware settings, and every node has 1 map slot and 1 reduce slot. The scheduling and execution of tasks in this query can be illustrated by a Gantt chart, as shown in Fig. 1.

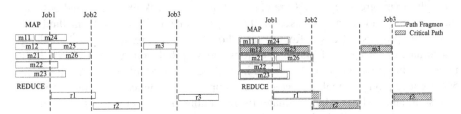

Fig. 1. Task Scheduling and Execution   **Fig. 2.** The Critical Path Detected by ParaTimer

Though map tasks in one job execute the same operations on the same amount of data(one data block), their elapsed time is different because of the cluster hetero-geneity. So even during one task round, time differences between tasks are obvious. ParaTimer[11] assumes that tasks in one round process at approximately the same speed. It detects the *path fragment* based on the task round, then composes them to a critical path. In the heterogeneous environment, ParaTimer does not find the actual critical path. The critical path ParaTimer detects is shown in Fig. 2, it is m12-m25-part of r1-part of r2-m3-r3. Not all the tasks on the path are actually *critical activities* that can directly impact total duration of the query, so monitoring these *fake critical tasks* is redundant. The critical path contains not only total tasks, but also some task fragments, which makes the estimation more complicated.

We can predict the duration of every task with sampling debug run or the execution performance history. However, the tasks do not execute exactly as planned due to the variation of hardware and software, which is called "duration uncertainty". So fixed task durations cannot reflect the real execution of the whole query. During the field of operational research, stochastic PERT[16] is always used to solve planning problems and recognize uncertainty in the activity durations.We formulate every task of the MapReduce query into an edge of a DAG, and model the execution processing into a stochastic PERT network.

3.2   PERT Modeling

In this section, we discuss the problem modeling. First, we give the problem definition:

Definition 1. *[Problem Definition] Given the MapReduce job DAG $D(N,E)$ of Query Q in the cloud, return accurate and real-time estimate of the query progress T.*

The execution plan of a query is represented by a MapReduce job DAG D, during which the nodes N represent the jobs, and the arcs E represent the job's logical relationships. The modeling can be defined as follows:

Definition 2. *[Modeling to PERT network] Given DAG $D(N,E)$ of Query Q, construct a PERT network $G(U,\xi)$. Node set $U = \{u_i\}_{i=1}^n$ represents the completing of tasks; Edge set ξ is composed of two subsets: $EP = \{ep_{ij}\}$ represents execution procedure of a task, with edge weight ω_{ij} denoting the task duration; $ER = \{er_{ij}\}$ only represents logical relationship of tasks,with ω_{ij} equal to zero.*

In $G(U,\xi)$, nodes represent completing of one or more tasks, and there are two kinds of edges: *EP* and *ER*, $EP \cup ER = \xi$. *EP* represents the execution procedures of tasks, the direction of arcs represent precedence relation between two nodes, and the edge weights represent task durations, which are considered as random variables in PEQC; *ER* doesn't represent real task execution, they only represent the precedence relationship of nodes, and the edge weights are always fixed values - 0. The precedence relationship of tasks results from two sequences: logical sequence between jobs and execution sequence of tasks on the same slot. The logical sequence between jobs is from query plan D, if two jobs have the relationship: job1→job2, then all the map tasks of job2 must execute after all the reduce tasks of job1. There can only be one task executing on one slot at the same time, so tasks assigned to the same slot have the the precedence relationships. Given the execute plan and task scheduling of the query example, we can construct the PERT network as Fig. 3 shows. The precedence relationship of m3 and r1 is determined by the job DAG, while the precedence relationship of m11 and m24 is determined by the scheduler that assigns them to the same slot.

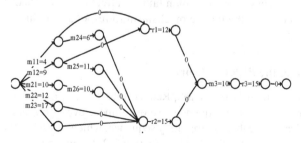

Fig. 3. PERT Network

3.3   The Stochastic Characteristics

There are two kinds of uncertainties during the query execution procedure: duration uncertainty and scheduling uncertainty. The duration uncertainty will change PERT network by alter the weight of arcs, while the scheduling uncertainty will change the structure of PERT network. We discuss duration uncertainty and the stochastic characteristics it brings in this section, the algorithm reacting to scheduling uncertainty will be shown in Section 4.4.

Suppose there are m tasks in the query, let $T = \{T_i\}_{i-1}^m$ represents the duration of every task in query Q. We can estimate T_i from sampling debug or running history of tasks on the node, however, tasks will never execute with the durations we expected. It is unrealistic to associate T_i with a deterministic value, so we treat T_i as a random variable which obeys some distribution. We assume all the duration of tasks have independent distributions. Let $P = \{P_i\}_{i=1}^r$ represents the path between initial node U_1 and terminal node U_n, where r denotes the cardinality of set P. D_i represents the duration of P_i, where $D_i = \sum_{ep_k \in P_i} (T_k)$. According to the central limit theorem (CLT), D_i approximately obeys normal distribution [8]. The duration D* of the PERT network is $D^* = max_{p_i \in P}(D_i)$, then P^* will be the critical path. Let μ_k, σ_k^2 represent the mean and variance of task duration T_k on one path P_i. Then the expected value and variance of D_i are:

$$E(D_i) = E(\sum_{ep_k \in P_i} T_k) = \sum_{ep_k \in P_i} E(T_k) = \sum_{ep_k \in P_i} \mu_k \tag{1}$$

$$\sigma^2(D_i) = \sigma^2(\sum_{ep_k \in P_i} T_k) = \sum_{ep_k \in P_i} \sigma^2(T_k) = \sum_{ep_k \in P_i} \sigma_k^2 \tag{2}$$

The expected value of D* is $E(D^*) = max_{p_i \in P}(E(D_i))$. Given a deadline t for the query, we can compute the probability for completing the query before t according to the distribution characteristic and parameters of D*: $P\{D^* < t\} = \int_{-\infty}^t f(x)dx$.

4   Proposed Solution

In this section, we introduce our three-step solution to estimating query progress in the cloud. First, PEQC constructs the PERT network of tasks depending on environment resources and scheduling strategy. Secondly, PEQC chooses the most critical path from PERT network. Thirdly, it computes the progress of tasks on the critical path to represent the query progress. Since common failures affect the task scheduling and query progress, we also discuss the update mechanism to respond to failures with less time cost and changes.

4.1   Constructing the PERT network

Given the execution plan DAG of MapReduce jobs for one query, we can compute a topological order of the job sequence. We decompose every job into map tasks and reduce tasks and compute the task topology sequence, which is used as the input of constructing PERT network. As discussed in Section 3.2, the precedence relationship between tasks stems from job DAG and task scheduling. Although we demonstrate PEQC with FIFO scheduler in this paper, it can be extended to other schedulers with some manageable modifications. Under the FIFO policy, the scheduler maintains a waiting queue of jobs sorted by arrival time and job priority. The scheduler assigns available slots to tasks from the job at the head of the queue. When allocating slots to tasks from one job, it also follows the *data locality* rule, which means that the available slot is preferentially allocated to the task that operates on data in the same node of this slot. We mimic the schedule policy and compute the relationships between tasks scheduled on the same slot.

Among all the jobs of one query, there are some jobs that can execute concurrently, which we call a *job batch*. More precisely:

Definition 3 (Job Batch). *Given the execution plan DAG $D(N,E)$ of a query Q, a job batch J, is a set of jobs $j \in N$ that can execute concurrently with no precedence relationships among them.*

The DAG $D(N,E)$ can be broken into several job batches, which execute sequentially. The job batches can be computed from $D(N,E)$, or can be gotten directly from the MapReduce job launcher. Algorithm 1 computes the task scheduling using job batch as the unit, because only tasks during the same job batch can execute concurrently and influence the task scheduling of each other. In Algorithm1, every task in the input sequence contains a list of dependencies, which includes its precursor tasks according to the job DAG. We adopt two arrays called *MapSlots* and *ReduceSlots* to store scheduling information for map slots and reduce slots, respectively. Every element of the array denotes a task slot, and it is associated with three attributes: pointer to the currently scheduled task, sum of length of tasks executed on this slot, number of tasks scheduled on this slot. Both *MapSlots* and *ReduceSlots* contain information for one job batch, so they are set to initial status at the beginning of scheduling tasks of every job batch(line4). Take the query in Section 3.1 for example, the task sequence is m11, m12, r1, m21, m22, m23, m24, m25, m26, r2, m3, r3. There are two *job batches*: job1 and job2 compose JB1, and job3 composes JB2. The initial state of MapSlots[] is listed in Table 1.

Table 1. Initial State of MapSlots

	slot1	slot2	slot3	slot4	slot5
pointer	null	null	null	null	null
length	0	0	0	0	0
round	0	0	0	0	0

Table 2. MapSlots State after Round One

	slot1	slot2	slot3	slot4	slot5
pointer	m11	m12	m21	m22	m23
length	4	9	10	12	17
round	1	1	1	1	1

Algorithm 1 schedules jobs from the head of the job sequence. It chooses the slot with the smallest length from MapSlots (line8), then arranges a task to this slot. During the tasks belonging to the scheduling job, the one that operates data on the same node with this slot is scheduled preferentially (line8). If there are no such tasks in the scheduling job, it chooses the task at the head of task queue. Table 2 shows the state of MapSlots[] when all the slots have been arranged with one task. Next the algorithm chooses the slot with the smallest length - slot1, and arrange a task from m24, m25, m26 according to data locality. After arranging a new task to one slot, Algorithm1 first adds a new node and edge of this task into the AdjList(line9-11), then it deals with the task's dependencies.

Algorithm 1. Construct PERT Network

---

input : TaskSequence *T*: task topology sequence, it maintains six variables for each task *t*
　　　　 ∈ *T*: *t.name, t.type, t.length, t.dependency, t.failureprob, t.nodeip.*
output: *AdjList[]*:PERT network

1　Set *AdjList[0].data* = Start; *AdjList[0].out* = null;
2　*current* = 0; /*current represents the processing node in the network*/;
3　**for** all jobbatch *JB* **in** *T* **do**
4　│　*MapSlots*.setInitial(); *ReduceSlots*.setInitial();
5　│　**for** all job *J* **in** *JB* **do**
6　│　│　Slots = MapSlots;
7　│　│　**while** existing unscheduled task *in J* **do**
8　│　│　│　*s* = getShortestSlot(*Slots*); *t* = getLocalTask(*s, J*);
9　│　│　│　*t.length* = getLength(*t, s*); *t.failureprob* = getFailureProb(*t.s*);
10　│　│　│　*AdjList[++current].data* = *t.name*; *AdjList[current].adj* = null;
11　│　│　│　Edge *e* ← new Edge(*t.name, t.length, t.failureprob*, null);
12　│　│　│　**if** t.dependency *!= null* **then**
13　│　│　│　│　**if** *length(* t.dependency *) ==1* && Slots[s].pointer==*null* **then**
14　│　│　│　│　│　addEdge(getEdge(*t.dependency*), *e*); /*add *e* consecutively after *t′s*
　　　　　　　　　　　dependency task*/
15　│　│　│　│　**else**
16　│　│　│　│　│　*AdjList[++current].data* = virtual_*t.name*; *AdjList[current].out* = *e*;
17　│　│　│　│　│　**for** i *in* t.dependency **do**
18　│　│　│　│　│　│　Edge *ed* = new Edge(virtual_*t.name, 0,0*,null);
19　│　│　│　│　│　│　*ed.out* = *AdjList[i].out*; *AdjList[i].out* = *ed*
20　│　│　│　│　**if** Slots[s].pointer*!=null* **then**
21　│　│　│　│　│　Edge *ed* = new Edge(virtual_*t.name, 0,0*, null);
　　　　　　　　　　　addEdge(*Slots[s].pointer, ed*);
22　│　│　│　**else** addEdge(*start, e*); /*make the network's start node tail of *e*/
　　　　　　　　Slots[s].length += T.length; Slots[s].round++; Slots[s].pointer = e ;
23　│　│　└　**if** (all map tasks in *J* are scheduled) **then** *Slots = ReduceSlots*;

24　*AdjList[++current].data* = End; *AdjList[current].out* = null;
25　**for** i *with* AdjList[i].out == *null* **do**
26　│　Edge *ed* = new Edge(End, 0, 0, null); *AdjList[i].out* = *ed*;

The dependency tasks of task *t* stem from two precedence relationships: the logical relationship from job DAG is included in the dependency attribute of *t*, and the precursor task of *t* on slot *s* can be gotten from MapSlots[s] or ReduceSlots[s]. If a task has more than one dependency task, PEQC creates a virtual node which represents the event of all the dependency tasks' completing, the weights of all the arcs pointing to the virtual node are set to 0(line16-19). After scheduling all the tasks in the sequence, Algorithm1 adds edges from the nodes without out edges to the end node (line20-22). PEQC adopts an adjacency list to store the PERT network, because the netwok is not too dense, and adjacency list is more convenient in the network update. After constructing the PERT network for the task execution of the query, the remaining work of PEQC will be all operations on the graph.

4.2 Computing the Critical Path

The *job batches* of a query partition $G(U,\xi)$ into k sub-networks:$SG = \{SG_i(U_i, \xi_i)\}_{i=1}^k$. Let P_i^* represents the critical path of SG_i, D_i^* represents the duration of P_i^*, then the critical path of the whole network is $P^* = \bigcup_{i=1}^k (P_i^*)$, the duration is $D^* = \sum_{i=1}^k (D_i^*)$. PEQC computes the critical path of each sub-network and connects them to form the critical path of the whole network. PEQC adopts the *divide and rule* mechanism for two reasons: computing the critical paths of all the sub-networks parallel;re-computing the critical path of some job batch instead of the whole network when failure happens.

Given a PERT network using the duration expected value as the arc weight, there can be several critical paths whose durations are equal, which we call *candidate critical paths*. How to find the *most critical path* that can represent the execution of the whole query from these candidate critical paths? We discuss how the task failure can influence the path length first. When a task failure happens on one slot, the task duration on this slot is changed to the time interval between its start time and failure time, which is shorter than the expected task duration. So task failure can result in the duration reduction of path where the failed task is. However, the critical path is always the longest path in the network. If there are task failures on the path, its duration may be exceeded by other paths. We use a metrics called *path reliability* to measure the probability that a candidate critical path maintains the longest path.

Definition 4 (Path Reliability). *Suppose a candidate critical path P_c includes m tasks, the failure rate of task i is denoted by F_i, then the reliability of P_c is: $P_c = \prod_{i=1}^m (1 - F_i)$.*

A task failure can be caused by many reasons. However, modelling the failure probability of tasks is beyond the scope of this paper. So we do not address this problem and assume predefined task failure rate to focus on the other factors. Among all the candidate critical paths of each sub-network, PEQC chooses the one with the biggest path reliability, then connects them together as the critical path of the whole query.

4.3 Estimating the Progress

Tasks on the critical path represent the query progress. They can be classified into three types: completed tasks, running task, and pending tasks. Suppose there are n pending tasks of the query, the remaining time of the query is: $T_{remaining} = T_{running} + \sum_{i=1}^n (T_i)$. For the pending tasks that haven't executed, PEQC adopts their duration expected values as the estimate time. The elapsed time of the running task can be evaluated through its actual execution speed to provide more accurate estimate. Existing methods break map or reduce task into pipelines and sum the elapsed time of every pipeline as the task estimate[17][12]. In our experiment, we adopts the finish time estimate method of [17], which estimates the time left for a task based on the progress score provided by Hadoop, as $(1 - ProgreeScore)/ProgressRate$, and the $ProgressRate=ProgressScore/$elapsed time t.

4.4 Reacting to Failures

PEQC constructs a PERT network to identify how the MapReduce query behaves before its execution, we call it *baseline scheduling*. However, the actual jobs do not execute exactly as the baseline scheduling because of common failures. Predicting accurately all the failures to happen in the execution is unrealistic, so PEQC provides reactive update algorithm to adjust to the changes caused by failures.

PEQC has to "repair" the whole network and re-compute the critical path whenever a task failure comes up. An intuitive method is to reconstruct the PERT network and repeat the regular steps of Algorithm1. However, this method involves much unnecessary work and costs much time when failures are very common. We can partition the baseline PERT network into three parts depending on the time when failure occurs, just as Fig. 4 shows(failure happens at 3s): executed part contains tasks that have completed or being executed when failure occurs; scheduling part contains tasks that haven't executed and their scheduling is affected by the failure; pending part contains tasks that have not been executed, and their scheduling is not changed by this failure. Tasks in the scheduling part have two characteristics: they belong to the same *job batch* with the failure task, and they have not been executed when the task failure happens. The update algorithm only changes the scheduling part instead of the whole PERT network.

Algorithm 2. Updating PERT When Task Failure Happens

---

    input : AdjList[]: baseline PERT network; T: TaskSquence; tf: failed task; TP_f: time point when failure happens.

    output: Updated PERT network

1 *Schedule*.setInitial();

2 *start_task* $=T$.*getNext*(*Schedule*.*getLastTask*()); *slot* = *Schedule*.getSlot(tf);

3 $RTP_f = TP_f - ST$; /* ST: start time of job batch; RTP_f: relative failure time to ST */

4 *Schedule*[*slot*].*current_lengh* = *Schedule*[*slot*].*scheduling_lengh* = RTP_f;

5 *CandidateSlots* = getShortestSlot(*Schedule* − *slot*);

6 *ls* = getLocalSlot(*CandidateSlots*, tf'); /* tf' represents the re-executing task of tf*/

7 $AdjList[++length]$.*data* = tf'.*name*; $AdjList[length]$.*adj* = null;

8 Edge e = new Edge(tf'.*name*, tf'.*length*, tf'.*failureprob*, null);

9 addEdge(s.*executing_task*, e);

10 *ls*.*scheduling_task* = tf';

11 *ls*.*scheduling_length*+ = tf'.*length*;

12 **for** *edge in getFollowingEdge*(tf) **do**

13 | **if** *edge.name* $\in$ tf.*dependency* **then**

14 | | moveEdge(*edge*, tf'); /*move *edge* consecutively after tf', update the *scheduling_length* and *scheduling_task* of two corresponding slots*/

15 **for** all job J *between* *start_task* **and** *end_task* **do**

16 | s = getShortestSlot(*Schedule*); t = getLocalTask(s, J);

17 | *edge* = getEdge(t); moveEdge(*edge*, s.*scheduling_task*);

18 | **for** *edge in getFollowingEdge*(t) **do**

19 | | **if** *edge.name* $\in$ t.*dependency* **then**

20 | | | moveEdge(*edge*, t)

The left bound ary of scheduling part can be determined by the event nodes representing the completion of executing tasks. The right bound ary is composed of the event nodes that represent the completion of reduce tasks owing to the same job batch with the

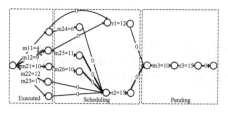

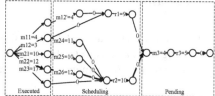

Fig. 4. Task Failure Happens   **Fig. 5.** The Task Scheduling after Failure

failed task. After determining the start task and end task from the task scheduling queue, Algorithm2 reschedules the tasks between the start task and end task, and updates the network. PEQC maintains two arrays called *MapSchedule* and *ReduceSchedule* for every *job batch*, which store the execution and scheduling information for map slots and reduce slots respectively. Every element in the array contains four variables: *executing_task* represents the task that is executing on this slot; *scheduling_task* represents the last scheduled task of this slot; *current_length* represents the duration sum of tasks executed on this slot; *scheduling_length* represents the duration sum of tasks scheduled on this slot. At the beginning of the update algorithm, *scheduling_task* is set to the *executing_task*, and *scheduling_length* is set to *current_length*(line1). The failure task affects the scheduling of tasks of the same type with it, the processing procedure after failed map task and reduce task are the same except the scheduling arrays they used. During the scheduling queue T, the start task is the one following the last task of the executed part (line2), and the end task is the one before the first task of the pending part. For the corresponding slot of tf, its *current_length* should be changed to the relative time of TP_f to the start time of the job batch (line3-4). Algorithm 2 schedules tf' first because failure task always has the highest priority (line5-6). After finding the proper slot ls, it creates a new node and a new edge for tf', the new node is added at the end of Adjlist (line7-8), and the new edge is added after the *executed_task* of ls (line9-11). The edges following tf which represent the logical precedence of the query, are moved consecutively after the new task tf' (line12-14). After processing tf', Algorithm2 reschedules the tasks between *start_task* and *end_task*(line15-20). The re-computed task scheduling after failure is shown in Fig. 5.

The previous method focuses on updating the network when a task failure happens. When node failure happens, all the executing tasks on the failure node have to be re-executed on other nodes. If a map task has completed on the node before failure happens, but the job the map task belongs to hasn't completed, then this map task also has to be re-executed. Because the output results of map task are stored on the local node, and the output results are unavailable after the node failure. The updating algorithm after node failure should take three things into consideration: first, there may be more than one failed task to be re-executed; second, the slots on the failure node should be disabled when re-scheduling tasks; third, the scheduling of tasks in the pending part may be changed because of the disabled slot. The pending part may include several *job batches*, and the updating can execute parallel.

5   Evaluation

In this section we evaluate PEQC, and compare it with ParaTimer and the default estimator of Pig from two aspects: the precision of query estimate and the resilience reacting to failures. The experiment is implemented on Pig 0.8.0 and Hadoop 0.20.2.

5.1   Experimental Setup and DataSet

All the experiments are run on a heterogeneous cluster of 31 nodes connected by a 1Gbit Ethernet switch. One node serves as the namenode of HDFS and jobtracker of MapReduce, and the remaining 30 nodes act as slaves. There are four levels of hardware equipment of all the nodes in the cluster, the setup details are shown in Table 3. The master node which acts as the jobtracker belongs to levelIII.We set the block size to 64M, and configure Hadoop to run 1 mapper and 1 reducer per node.

Table 3. Testbed Setup

	LevelI	LevelII	LevelIII	LevelIV
CPU	Quad Core 2.33GHz	Quad Core 2.66GHz	Quad Core 4GHz	Quad Core 2.13GHz
RAM	7GB	8GB	4GB	4GB
Disk	1.8TB	2TB	2TB	500GB
No. of Node	4	10	14	2
Node Type	PC	PC	PC	Server

In the experiment we adopt the query introduced in Section 3.1, which contains three MapReduce jobs. We perform the tests on two datasets with different data size. In dataset1, the data size of *Rankings* and *UserVisits* are 2G and 24.5G; in dataset2, the data size of the two table are 1.4G and 61G. We adopt the data generation method from[13]. It first generates a collection of random HTML documents, then generates the data of *Rankings* and *UserVisits*. The task number of every job is listed in Table 4. We estimate the distribution parameters of task duration from the task running history of every node, Table 5 shows the distribution parameters of map task in Job2 on four nodes, they belong to four different hardware levels.

Table 4. No. of Tasks

	Mappers Dataset1	Reducers Dataset1	Mappers. Dataset2	Reducers Dataset2
Job1	393	27	970	66
Job2	32	3	21	2
Job3	406	27	998	67

Table 5. Distribution Parameters

	Node1	Node6	Node11	Node30
Mean Value	12.60	8.19	7.30	9.60
Variance	0.77	0.72	0.46	0.99

5.2 Accuracy Evaluate

In this experiment, we evaluate the estimate accuracy and the time overhead of getting the first estimate result in the heterogeneous environment. We obtain progress estimate every 10 seconds and examine three metrics: mean estimate error, max estimate error[3][11], and time overhead. Pig reports the progress as the percentage of the query completed, PEQC and ParaTimer provide the remaining time. In order to compare them with the same metric, we transform the percentage of query f reported by Pig into remaining time: $t_{remaining} = \frac{(1-f)(t_i-t_0)}{f}$. t_0 represents the time when the query is submitted, t_i represents the time when this estimate is reported. Let t_e represents the estimated remaining time at time t_i, t_n represents the time when the query actually completes, then the estimate error at time t_i is: $error_i = \left|\frac{t_i+t_e-t_n}{t_n-t_0}\right|$.

Fig. 6 and Fig. 7 illustrate the remaining query time estimated by the progress indicator over time on two datasets. There are four lines in each figure. The almost straight line represents the actual remaining time. In Fig. 6, there are two turning points on Pig's curve: $t1$(140s) and $t2$(380s). Pig's indicator assumes all the jobs execute sequentially with the same weight. Job1's mappers occupied the slots before $t1$, so the estimate is pessimistic. After that Job1 and Job2 execute concurrently, the estimate changes to be optimistic. Job3 starts approximately at $t2$, and its duration accounts for about one-third of the whole query's time. So after $t2$, the estimate is close to the actual remaining time. ParaTimer's estimate becomes close to the actual line at $t3$(448s), when there are only the reducers of Job3 executing. The critical path ParaTimer detects after $t3$ is close to the actual one. PEQC computes the critical path from PERT network, and it supplies relatively stable estimate. In general, PEQC's estimate is slightly optimistic. This is because PEQC computes the critical path based on the mean value of every task duration, which is different from the task's duration in the actual query execution. The trend of every curve in Fig. 7 is similar to that of Fig. 6. The metrics of tests on two datasets are shown in Fig. 8. Pig fetch the job progress from Hadoop directly without any overhead. PEQC has to construct the PERT network and compute the critical path before estimating, and ParaTimer has to compute the path fragment and compose the critical path, which costs more time than PEQC.

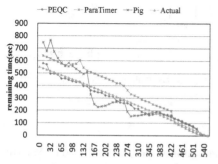

Fig. 6. Estimated Result (Dataset1)

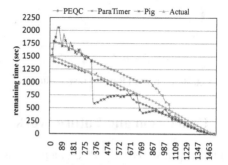

Fig. 7. Estimated Result (Dataset2)

	Dataset1			Dataset2		
	PEQC	ParaTimer	Pig	PEQC	ParaTimer	Pig
Mean Error	4.2%	10.3%	12.3%	4.7%	13.1%	14.1%
Max Error	10.6%	31%	44.9%	11.7%	24.1%	33.5%
Overhead(s)	1.05	6.45	0	3.40	18.52	0

	Task Failure			NodeFailure	
	PEQC	Paratimer	Pig	PEQC	Pig
Mean Error	4%	15.4%	16.2%	4.5%	15.3%
Max Error	8.6%	28.4%	38.2%	23.4%	39.8%
Overhead(s)	1.02,0.87,0.54	6.42,6.2,6.31	0	2.9	0

Fig. 8. Metrics of Accuracy Test   **Fig. 9.** Metrics of Failure Test

5.3   Robustness to Failures

In this section, we conduct two tests to evaluate the estimators' robustness to task failures and node failures. The metrics we adopt are: mean estimate error, max estimate error, the time overhead of reacting to each failure. Fig. 10 shows the results of task failure test. ParaTimer provides an additional estimate called *PessimisticFailureEstimate*, which assumes a single worst-case task failure will occur. We run this test on dataset1 and fail three tasks at 134s, 225s, and 377s. The metrics are shown in Fig. 9. When task failure occurs, PEQC only recomputes part of the network, so it cost less time than ParaTimer.

Fig. 11 shows the results of node failure. ParaTimer does not support estimate in presence of node failures, we conduct this test on PEQC and Pig on dataset2. We make two node failures(two nodes in LevelI) by cutting off their network connections to the cluster concurrently. The node failure happens at about 275s, when Job1 is executing. There are 28 slave nodes left in the cluster after the failure, all the executing tasks and completed map tasks of Job1 have to be re-executed, and the scheduling of the remaining tasks have to be changed. The metrics of this test is shown in Fig. 9. It costs PEQC 2.9s to repair the network and re-compute the critical path. Though tasks in both *JB1* and *JB2* have to be re-scheduled, and their critical paths have to be re-computed, PEQC can do the repair work in parallel.

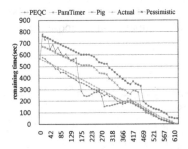

Fig. 10. Estimated Result (Task Failure)

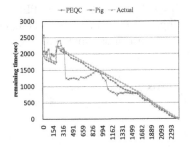

Fig. 11. Estimated Result (Node Failure)

6   Conclusion and Future Work

In this paper, we propose PEQC, a progress indicator of queries composed of MapReduce jobs in the cloud. PEQC focuses on solving challenges brought by two features of cloud: environment heterogeneity and common failures. PEQC models the task execution of a whole query into a stochastic PERT network. It adopts partial update mechanism to react to the task failures. Based on our implementation on Pig & Hadoop on a heterogeneous cluster, PEQC provides promising remaining time estimate of a query in the cloud, and can repair the PERT network when failure happens in acceptable time. There are also some inherent characteristics within MapReduce that give rise to difficulties of the problem, such as speculative execution and data skew in the reduce phase. We will judge the tradeoff between algorithm complexity and processing time, and make PEQC more robust to these challenges in the future work.

Acknowledgements. This research was partially supported by the grants from the Natural Science Foundation of China (No. 91024032, 91124001, 61070055, 60833005), the Fundamental Research Funds for the Central Universities, and the Research Funds of Renmin University of China (No. 11XNL010, 10XNI018), National Science and Technology Major Project (No. 2010ZX01042-002-003).

References

1. Abouzeid, A., Bajda-Pawlikowski, K., Abadi, D., Silberschatz, A., Rasin, A.: HadoopDB: an architectural hybrid of MapReduce and DBMS technologies for analytical workloads. In: 35th ACM Conference of Very Large Databases, pp. 922–933. ACM Press, New York (2009)
2. Chaudhuri, S., Kaushik, R., Ramamurthy, R.: When can we trust progress estimators for SQL queries. In: 25th ACM International Conference on Management of Data, pp. 575–586. ACM Press, New York (2005)
3. Chaudhuri, S., Narassaya, V., Ramamurthy, R.: Estimating progress of execution for SQL queries. In: 24th ACM International Conference on Management of Data, pp. 803–814. ACM Press, New York (2004)
4. Dean, J.: Experiences with mapreduce, an abstraction for large-scale computation. In: PACT, p. 1. IEEE Press, Washington (2006)
5. Dean, J., Ghemawat, S.: Mapreduce: simplified data processing on large clusters. In: OSDI, pp. 137–150. ACM Press, New York (2004)
6. Malcolm, D.G., Roseboom, J.H., Clark, C.E., Fazar, W.: Application of a technique for research and development program evaluation. Operations Research 7(5), 646–669 (1959)
7. Hellerstein, J.M., Haas, P.J., Wang, H.J.: Online Aggregation. In: 17th ACM International Conference on Management of Data, pp. 171–182. ACM Press, New York (1997)
8. Dean, J.: Designs, lessons and advice from building large distributed systems. In: Keynote from LADIS 2009 (2009)
9. Luo, G., Naughton, J.F., Ellmann, C.J., Watzke, M.: Toward a progress indicator for database queries. In: 24th ACM International Conference on Management of Data, pp. 791–802. ACM Press, New York (2004)
10. Luo, G., Naughton, J.F., Ellmann, C.J., Watzke, M.: Increasing the accuracy and coverage of SQL progress indicators. In: 21st IEEE International Conference on Data Engineering, pp. 853–864. IEEE Press, Washington (2005)
11. Morton, K., Balazinska, M., Grossman, D.: ParaTimer: A progress indicator for mapreduce DAGs. In: 30th ACM International Conference on Management of Data, pp. 507–518. ACM Press, New York (2010)
12. Morton, K., Friesen, A., Balazinska, M., Grossman, D.: Estimating the progress of MapReduce pipelines. In: 26th IEEE International Conference on Data Engineering, pp. 681–684. IEEE Press, Washington (2010)

13. Pavlo, A., Rasin, A., Madden, S., Stonebraker, M., DeWitt, D., Paulson, E., Shrinivas, L., Abadi, D.J.: A comparison of approaches to large-scale data analysis. In: 29th ACM International Conference on Management of Data, pp. 165–178. ACM Press, New York (2009)

14. Schad, J., Dittrich, J., Quian-Ruiz, J.: Runtime measurements in the cloud: observing, analyzing, and reducing variance. J. Proc. of VLDB Endowment 3(1), 460–471 (2010)

15. Schatz, M.C.: CloudBurst: highly sensitive read mapping with MapReduce. Bioinformatics 25(11), 1363–1369 (2009)

16. Shogan, A.W.: Bounding distributions for a stochastic pert network. Networks 7(4), 259–381 (1977)

17. Zaharia, M., Konwinski, A., Joseph, A.D., Katz, R., Stoica, I.: Improving MapReduce performance in heterogeneous environments. In: OSDI. ACM Press, New York (2008)

18. The Hadoop Website, http://hadoop.apache.org

19. The Pig Website, http://pig.apache.org

Towards a Scalable, Performance-Oriented OLAP Storage Engine

Todd Eavis[1] and Ahmad Taleb[2]

[1] Concordia University, Montreal, Canada
[2] Najran University, Saudi Arabia,
College of Computer Science and Information Systems

Abstract. Over the past generation, data warehousing and OLAP applications have become the cornerstone of contemporary decision support environments. Typically, OLAP servers are implemented on top of either proprietary array-based storage engines (MOLAP) or as extensions to conventional relational DBMSs (ROLAP). While MOLAP systems do indeed provide impressive performance on common analytics queries, they tend to have limited scalability. Conversely, ROLAP's table oriented model scales quite nicely, but offers mediocre performance at best relative to the MOLAP systems. In this paper, we describe a storage and indexing framework that aims to provide both MOLAP like performance and ROLAP like scalability by essentially combining some of the best features of both. Based upon a combination of R-trees and bitmap indexes, the storage engine has been integrated with a robust OLAP query engine prototype that is able to fully exploit the efficiency of the proposed storage model. Experimental results demonstrate that not only does the framework improve upon more naive approaches, but that it does indeed offer the potential to optimize both query performance and scalability.

1 Introduction

Data warehousing and OLAP have been popular targets for researchers over the past 10-15 years, with papers published on a wide variety of related topics. In the OLAP domain, early work often focused on the development of algorithms for the efficient computation of the data cube. Later, the cube methods were expanded to include mechanisms for the computation or representation of hierarchies derived from the cube's dimensions. For the most part, academics built upon table-based models, as the associated relational systems were well understood. On the positive side, scalability for relational OLAP (ROLAP) was very impressive and was generally limited only by the hardware. Unfortunately, such systems often provided poor query performance as they were ill suited to OLAP's complex, multi-dimensional data model.

For this reason, commercial vendors often developed proprietary array-based server products that were meant to more closely resemble the hyper-cubic nature of the data cube. Performance on these multi-dimensional OLAP (MOLAP) servers was/is indeed impressive as the direct indexing provided by arrays often leads to much improved query response time. Of course, everything comes at a

S.-g. Lee et al. (Eds.): DASFAA 2012, Part II, LNCS 7239, pp. 185–202, 2012.

price and, in the case of MOLAP, scalability remains a concern. Specifically, the sparsity of high cardinality OLAP spaces significantly limits the size of the cube structures in enterprise environments.

In this paper, we discuss the storage architecture — from a systems perspective — for an OLAP-specific server designed from the ground up as a high performance analytics engine. The system is capable of efficiently generating full or partial cubes and subsequently providing complex processing on common cube queries (slice/dice, drill down/rollup, etc.). Until recently, the DBMS essentially relied on the file system for storage services. Some indexing was available but it was limited in nature. Our recent work, as presented in this paper, has significantly extended the original model to include both R-tree and bitmap indexing facilities. Specifically, we have integrated the open source Berkeley DB libraries into the server so as to encapsulate both indexes and cube data within a single data store. Cube dimensions are also efficiently stored as Berkeley DB databases and are, as expected, hierarchy aware. Non-hierarchical attributes, in turn, are stored as a set of FastBit bitmap indexes. Ultimately, the integrated architecture represents a very efficient OLAP storage engine that provides the kind of query performance that one would expect of a MOLAP system, with the scalability typically associated with table-oriented relational servers.

This paper is organized as follows. In Section 2, we discuss a number of related research projects. In Section 3, we introduce the conceptual model upon which the storage engine is based. A detailed look at how data is encoded is provided in Section 4, including both its abstract and physical representation. The integration with the Berkeley DB and Fastbit bitmap libraries is discussed in Section 5. Then, in Section 6, we present an overview of the query processing logic that actually utilizes the relevant storage components. We round out the paper with some final conclusions in Section 8.

2 Related Work

Subsequent to the initial definition of the data cube cube operator [9], a number of researchers proposed techniques for the compact representation of the cube. Both the DWARF cube [18] and QC-trees [11], for example, define compact non-relational tree-based structures that provide efficient data access. However, their complex models were never integrated into practical systems, academic or otherwise. Conversely, the CURE cube [14] supports the representation of cubes and dimension hierarchies and does so with relatively compact table storage. Still, the CURE model lacks the native multi-dimensional indexing schemes that are essential for high performance query functionality.

In terms of OLAP indexing, a number of researchers have proposed methods that would improve warehouse access. In the simplest case, clusters of B-trees have been proposed, though such an approach is neither scalable nor efficient in higher dimensions [10]. A more interesting proposal was perhaps the CUBE Tree, a warehouse-specific mechanism based upon the R-tree [17]. The CUBE tree not only demonstrated that the R-tree was well suited to OLAP access patterns, but also provided an efficient update mechanism.

Recently, column store databases have been investigated as means to minimize IO costs on aggregation queries [16]. It is important to note, however, that column stores are best suited to general purpose warehouses that perform ad hoc, real time querying involving massive amounts of raw data. True OLAP servers, often working in conjunction with a supporting warehouse (possibly a column store), typically use a combination of pre-aggregation, specialized indexing, and query optimization to target the most common query forms. In practice, OLAP severs and column store DBMSs can be seen as complimentary rather than competitive.

A second recent theme has been the exploitation of the increasingly popular MapReduce framework [3] — and its open source implementation Hadoop — as a kind of parallel DBMS subsystem. Integration of Hadoop and traditional relational DBMSs (with their storage and indexing architectures) has also been suggested [1]. While this work is indeed interesting, there remains considerable doubt as to whether such systems can compete directly with the performance offered by purpose-built DW/OLAP servers [19].

Finally, we note that *non-academic* DBM systems are certainly available. The open source Java-based Mondrian server, for example, provides OLAP query functionality [13]. Mondrian, however, is primarily an OLAP query API and actually piggy-backs on top of existing database servers. Commercially, leading vendors such as Microsoft [12] and Oracle [15] also provide warehousing and OLAP applications, often with very rich functionality. Even here, however, users generally have a choice between the scalability of ROLAP or the performance of MOLAP. It is true that Hybrid OLAP (HOLAP) promises the "the best of both worlds" by incorporating relational tables and array-storage into the same repository but, in practice, this is an awkward, complex configuration at best.

3   The Data Cube Model

Before we look at the storage framework, we briefly discuss the model the DBMS is meant to represent. We consider analytical environments to consist of one or more *data cubes*. Each cube is composed of a series of d dimensions — sometimes called *feature* attributes — and one or more *measures*. The dimensions can be visualized as delimiting a d-dimensional hyper-cube, with each axis identifying the members of the parent dimension (e.g., the days of the year). Cell values, in turn, represent the aggregated measure (e.g., sum or count) of the associated members. Figure 1(a) provides an illustration of a very simple three dimensional cube on Store, Time and Product. Here, each unique combination of dimension members represents an aggregation on the measure. For example, we can see that Product OD923 was purchased 53 times at Store CA69 in January (assuming a Count measure). Note, as well, that each dimension is associated with a distinct aggregation hierarchy. Stores, for instance, are organized in Country, Province, and City groupings. There are in fact many variations on the form of OLAP hierarchies [7] (e.g., symmetric, ragged, non-strict). Regardless of the form, however, traversal of these aggregation paths — typically associated with

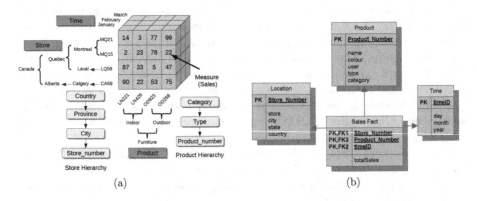

Fig. 1. (a) A three dimensional data cube (b) The Star Schema

rollup and drill down operations — is perhaps the single most common query pattern in the OLAP domain.

In practice, the cube is modeled as what is known as a *Star Schema*, essentially a central *Fact table* surrounded by one or more *Dimension* tables. Figure 1(b) demonstrates how the cube would be logically represented in a relational system. Note that the primary keys of the dimension tables form a composite primary key in the Fact table. Moreover, it is important to keep in mind that the Fact table typically dwarfs the Dimension tables in size. Consequently, we must seek to minimize both its size and processing costs.

4   Encoding the Database

Given the cube model presented above, we now examine how our DBMS actually encodes the contents of the schema. We begin with an overview of Dimension encoding. We note that Dimension attributes can be described as either *hierarchical* or *non-hierarchical*, with each requiring a distinct representation. In short, hierarchical attributes are those found within a dimension aggregation pathway (e.g., Country-Province-City), while non-hierarchical attributes are descriptive elements typically used to restrict a user query (e.g., Product Name). Encoding of a non-hierarchal dimension — a dimension that doesn't contain any hierarchy — is relatively straightforward and is accomplished with a linear pass through the native data set that simply assigns an incremental *surrogate key* to the table (i.e., an artificial, integer-base primary key).

Encoding of hierarchical values is more involved and is based upon the notion of *hierarchy linearity* [6]. Given an a hierarchy A, we define the hierarchy levels $A_1, A_2, \ldots A_k$, for hierarchy depth k, in terms of decreasing *granularity*. So, for example, $A_1 = $ city is more granular than $A_2 = $ province. Briefly, we say that a hierarchy on an attribute A is linear if for all *direct descendants* $A_{(j)}$ of $A_{(i)}$ there are $|A_{(j)}| + 1$ values, $x_1 < x_2 \ldots < x_{|A_{(j)}|}$ in the range $1 \ldots |A_{(i)}|$ such that

$$A_{(j)}[k] = \sum_{l=x_k}^{x_{k+1}} A_{(i)}[l]$$

where the array index notation [ ] indicates a specific value within a given hierarchy level. Informally, we can say that if a hierarchy is linear, there is a contiguous range of values $R_{(j)}$ on $A_{(j)}$ that may be aggregated into a contiguous range $R_{(i)}$ on $A_{(i)}$.

The DBMS exploits hierarchy linearity by using *mapping tables* that represent a sorting of the hierarchical column values of the associated dimension. In effect, values are ordered as $A_k, A_{k-1}, \ldots, A_1$, where A_1 is the *base* attribute in the hierarchical dimension. For each hierarchical attribute level L in the dimension, a sibling column L_ID is added. Values of L_ID are created as consecutive integer IDs and are used to delineate hierarchical group-by levels. Figure 2(a) illustrates the mapping table for the Product dimension with the three-level hierarchy ProductNumber → Type → Category. In the next section, we will see that these mapping tables are used to load hierarchy-aware data structures that provide real-time query value transformations.

4.1 Dimension Table Storage

We must now consider how these dimension tables are physically stored on disk. We address hierarchical attributes first. Once the mapping tables have been defined, they are used to load a data structure called a mapGraph [6]. Figure 2(b) shows a mapGraph sub-structure called an hMap that is used to model a simple Product hierarchy (We note that the mapGraph is typically loaded in a single linear pass when the cube is first accessed). In effect, the mapGraph is a two-way hashing structure for the hierarchy levels above the base. For a sub-attribute

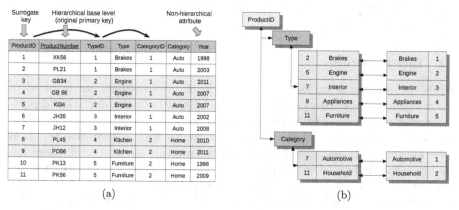

(a)   (b)

Fig. 2. (a) Product Mapping Table (b) mapGraph structure for the Product dimension

(A_j), $j \geq 2$, the associated map is made up of the *maximum encoded value* from the range on (A_1), corresponding to the current encoded value of (A_j). We add the native values of (A_j) to allow conversion of any encoded value of (A_j) to its native value. For example, we can use the `Category` component to retrieve the encoded value (1) of the `Automotive` category in O(1) time. Then, we can use the map associated with `Category` to find all `ProductIDs` $(1 \rightarrow 7)$ that are `Automotive`. Conversely, mapping base level IDs to coarser levels (i.e., base value $7 \rightarrow$ `Interior` at the `Type` level) can be done as a $O(logn)$ binary search, where $n = $ the cardinality of the given level (typically quite small relative to the base level). Because data in the associated Fact structure is physically stored at the base level (i.e., the dimension record's surrogate key), the mapGraph allows extremely efficient run-time mapping in response to the user's query specification. Specifically, constraints can be transformed between levels at run-time, with the final results aggregated as required.

Non-Hierarchical Attributes. While the transparent mapping of hierarchical attribute values is crucial for optimal query performance, non-hierarchical attribute processing must also be efficiently supported. In particular, if a non-hierarchical attribute is used in the restriction of an OLAP query or displayed in an OLAP report (e.g., "All customers older than 40"), then joins between the appropriate group-bys and dimension tables are required. This process can be very expensive and its costs must therefore be minimized.

The DBMS utilizes bitmap indexes for this purpose. For each non-hierarchical attribute we provide one bit string for each distinct value on the dimension. For k-non-hierarchical attributes, with each attribute having m distinct values, we would therefore have *(k*m)* bit strings. In practice, compression techniques (typically some variation of *Run Length Encoding*) significantly minimize storage requirements. Ultimately, the advantage of bitmap indexes for non-hierarchical attributes is that they allow us to identify the surrogate key values (e.g., `ProductID`) matching multiple non-hierarchical column constraints, typically *without* retrieving any records from the dimension table itself. In the current context, the DBMS uses the open source FastBit bitmap indexing libraries [8] as its bitmap subsystem.

4.2 Fact Structures

While the dimension tables, and their indexes in particular, are involved in the resolution of virtually every query, the bulk of both the raw IO and postprocessing is associated with the enormous Fact structures. We saw earlier how the DBMS associates each dimension table record with a surrogate key. It is these integer values, along with the associated measures (e.g., a total sales summation), that are housed within the Fact Structure. Physically, the DBMS stores and indexes data using what is known as a *packed R-tree*[17]. Specifically, the underlying data records are ordered as per the Hilbert space filling curve [5].

Note that for a d-dimensional space of side-length s, the s^d-length curve identifies a unique, strictly increasing order on the s^d point positions. We refer to the numeric representation of a point position as a Hilbert *ordinal*. Figure 4(a) provides an illustration of a simple 2-dimensional space, with each point representing the feature values of two distinct dimensions (e.g., think Product on the vertical axis and Customer on the horizontal). One can clearly see how data points are ordered as per the Hilbert curve (origin at bottom left), with the index itself formed as a series of increasingly broad block-sized bounding boxes. Here, Box *B0* forms the root of the R-tree, B1/B2 are at the second level, and B2-B8 form the leaves. Ultimately, the benefit of Hilbert packing is that it clusters spatially related points into common disk blocks. For OLAP queries that typically identify value ranges on multiple dimensions, this translates into significant IO savings at run-time.

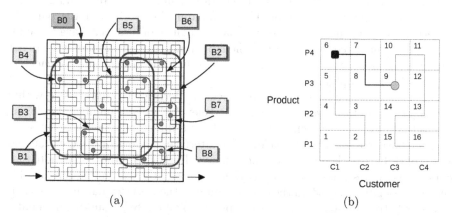

(a)

(b)

Fig. 3. (a) A 2-D Hilbert data set and index (b) Tuple differential compression

In practice, the fact table is stored in a compressed form, using a technique known as *Hilbert tuple differential* compression. Here, data records are first sorted in terms of their relative position in the $\mathcal{H}_k^d$ space. Pairs of *adjacent* points $< i, j >$ along the curve may then be represented in integer form as the difference value $|ordinal_i - ordinal_j|, i < j$. Figure 4(b) provides a simple example in which points are located 6 and 9 steps from the $\mathcal{H}_3^2$ origin. With the first point serving as the *anchor*, the second point — sales of Product 3 for Customer 3 — is stored as $9 - 6 = 3$, or 11 in binary form. This is 62 bits less that the default encoding for a 2-d value (assuming 32-bit integers). When coupled with bit compaction techniques that strip away leading zeros, differential compression can produce storage savings of 20%-80%.

It should be clear that the physical representation of the Fact data, while conceptual encoded as a table of records (i.e., ROLAP), bears little resemblance to a traditional table. In essence, it is a block-based collection of compacted bit strings, each representing a specific point in the multi-dimensional data space. Moreover, its supporting Hilbert R-tree index provide rapid retrieval of points in contiguous ranges. In short, it blurs the line between ROLAP and MOLAP by providing some of the best features of both.

5   Cube Consolidation

In practice, the DBMS allows the administrator to either fully or partially materialize the $O(2^d)$ summary views or *group-bys* in the d-dimensional.This can be done to reflect disk space availability or performance constraints. Each group-by effectively represents a subset of the Fact Structure and is physically represented as a pair of files — one that houses the data in Hilbert sort order and one that defines the R-tree index metadata and bounding boxes. Even for a partially materialized cube, this can represent a large number of files that have to be independently managed by the OS and the DB admin. Moreover, these files are not databases in any sense of the word and lack even basic mechanisms for caching, locking, ACID compliance, etc.

For this reason, we have chosen to embed the Berkeley DB libraries [2] within the larger DBMS framework. While the Berkeley API offers a number of indexing methods (Btree, Hash, Recno, Queue), it has no direct support for R-trees. As such, we have extended the Berkeley C++ interface to allow for the creation and access of Hilbert packed R-trees using standard Berkeley protocols. We note that Berkeley supports the storage of multiple *Database Objects* in one physical file known as an *environment*. In the current context, the Berkeley database contains a *master* B-tree database that, in turn, points to all related group-by meta data,

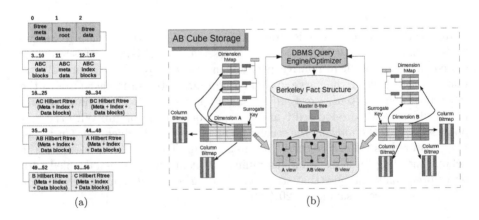

(a)   (b)

Fig. 4. (a) The internal structure of a cube database (b) The full cube storage architecture

indexes, and tuple compressed data. The extended API transparently routes data access requests as required. In Figure 4(a) we see how we store, in one physical file, the seven materialized group-bys for the three-dimensional cube ABC (letters represent dimension names). For each indexed group-by, the following blocks are required: one block to store the metadata, consecutive blocks to store the data blocks in their Hilbert ordered form, and consecutive blocks to store the Hilbert R-tree index. In this case, 56 contiguous blocks are used in total.

Finally, Figure 4(b) provides a more complete illustration of the core components of the storage engine, albeit for a very simple example with just two dimensions. One can see how the dimension tables are accompanied by a (memory-resident) hMap structure to support hierarchical attributes, multiple bitmaps for non-hierarchical attributes, and a system generated surrogate key. During query resolution (discussed below), strings of surrogate keys pass in/through the query engine to the storage backend. At that point records can be matched against Hilbert compressed data, using the Berkeley Master B-tree to locate the required view/block combinations.

5.1   Supporting DBMS Components

While the table storage and indexing components are the focus of the current paper, we note that the DBMS as a whole provides a relatively comprehensive processing stack. Figure 5 illustrates the complete architecture of the DBMS, including the native components as well as the Berkeley extensions. Note that the View Manager is responsible for the identification of the most cost effective group-by — and is initialized by scanning the primary Master B-tree database that contains references to all indexed group-bys — while the Hierarchy Manager builds and maintains the in-memory mapGraph structures. Note as well that OLAP Caching has nothing to do with the Berkeley caching component that stores recently accessed disk blocks, but refers instead to a native, multi-dimensional OLAP query cache.

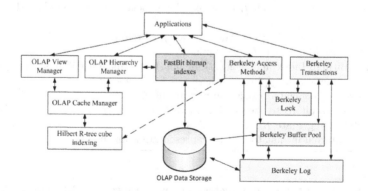

Fig. 5. Single Berkeley environment

6   Query Processing Logic

The DBMS provides full query processing functionality. In particular, it exposes an OLAP-specific algebra (similar to the relational algebra) that allows queries to be transformed into algebraic expression trees and extensively optimized at runtime. While the algebra and optimization methods are the subject of a concurrent submission, it is nonetheless important in the current context to understand how the storage and indexing facilities are integrating into the query engine. In this section, we examine the logic of the query resolution process.

Algorithm 1 is a somewhat simplified representation of the core logic implemented by the query engine. (We note that DBMS is actually a fully parallelized architecture so that the resolution algorithm is executed by each of the backend servers in the cluster federation.) Queries are transmitted to/from the end user in XML format and verified syntactically against an XML Query Grammar. Valid queries are then evaluated for semantic correctness to ensure that they comply with the database schema. If so, an algebraic plan is generated, optimized via a set of *transformation rules*, and then converted into a series of function calls that carry out the plan. It is these execution algorithms that will actually manipulate the indexes and storage structures described in the paper. While the View Manager, Hierarchy Manager, and Bitmap Manager will already have been initialized (on previous queries), they may need to be updated as per the current query parameters. Finally, once the query has been resolved (using buffer pipelining where appropriate), the encoded integer values are converted back into the text-based column values expected by the end user.

Algorithm 1. Query Resolution

1: Receive user's XML-encoded OLAP query Q
2: Perform syntactic and semantic verification on Q
3: IF Q is valid, create parse tree P
4: Generate and optimize algebraic algebraic expression tree E from P
5: **for** each algebraic operation in E **do**
6:     Generate a function call f_i to invoke the associated execution algorithm
7:     Add function f_i to set fC.
8: Update mapGraph Hierarchy Manager (M), as required
9: Update View Manager (M), as required
10: Update Bitmap Index manager (B) for non-hierarchical attributes, as required.
11: **for** each function call f_i in fC, where $i = 1, ..., n - 1$ **do**
12:     Invoke the associated execution algorithm(s)
13:     Pass the result R_i to the pipelined parent function f_{i+1}
14: Re-sort and aggregate R_{n-1} as required.
15: Convert internal format to textual representation for display.

We now turn to the logic implemented by the second FOR loop; that is, the actual data access methods. While each implementation function is associated with a distinct algebraic operator, we will focus here on SELECTION as it is arguably the most important and expensive of the core operations. Given the underlying indexes and storage structures, it is the job of the SELECTION algorithm to map the user's query constraints to the Dimension and Fact structures. This happens in two stages. First, hierarchical and non-hierarchical query attributes are converted as required into the base level attributes found in the Fact table. Algorithm 2 describes this process. Using either the mapGraph Hierarchy Manager or the FastBit bitmap indexes, ranges of contiguous base level IDs are extracted. Logical AND or OR operations are applied as required. The end result is an ordered list of Dimension record IDs (i.e., surrogate keys) that may be passed as input to the Fact Structure search algorithm.

Algorithm 2. SELECTION Transformation

Input: An OLAP selection condition C, a hierarchy manager M, an OP array of
  logical operators, and a bitmap index manager B.
Output: A set S of matching surrogate keys.
1: **for** each dimension condition C_i in C **do**
2:    **for** each expression e_j in C_i **do**
3:      **if** attribute (A) involved in e_j is a hierarchical attribute level **then**
4:        $array_j = M.\textbf{getBaseID}(A, e_j)$
5:      **else**
6:        $array_j = B.\textbf{getBaseID}(A, e_j)$
7:      **if** Logical Operator between e_j and e_{j-1} == AND **then**
8:        $array_j = \textbf{setIntersection}(array_j, array_{j-1})$
9:      **else**
10:       $array_j = \textbf{setUnion}(array_j, array_{j-1})$
11:   Set $R = array_j$, ordered as a set of contiguous ranges
12:   Replace current SELECTION condition C_i with R, the list of record-level IDs
      satisfying condition C_i.

Once the Dimension ID lists have been generated, they are passed to the cube storage engine to be matched against the Hilbert ordinals of the Fact Structure (Note that the View Manager transparently selects the most cost effective group by within the Berkeley database). Given an ordered list of $O(d)$ range sets, the search algorithm traverses the nodes in the selected R-tree based on a *breadth first* traversal strategy, visiting each node in a level-by-level, left-to-right fashion. Queries are answered as follows. For a level i of the tree, the algorithm identifies at level $i-1$ the j nodes (block numbers) that intersect the user query. It places these block numbers into a page list W. Using the block numbers in W, the algorithm traverses the blocks at level $i-1$ and replaces W with a new list $W\prime$. This procedure is repeated until the leaf level has been reached. At this point, the algorithm identifies and returns the d-dimensional records encapsulated by the user query. Figure 6 illustrates the traversal logic of the Fact Structure search. Because the Fact table stores dimension attributes as base level record IDs, and because the input to the search algorithm is a set of base level IDs sorted in ascending order, a breadth first search is able to make a single pass through the table, incrementally adding relevant blocks IDs to the result list (While it is not obvious in the illustration, the levels of the R-tree index are physically ordered on disk in this same root-to-leaf fashion). Moreover, because of the explicit Hilbert ordering of data, target records tend to be clustered into a small number of disk blocks. In fact, even when selectivity is very high, the combination of Hilbert ordering and breadth first search implies that, in the worst case, Fact Structure access can be no worse than a sequential scan (and is typically much better).

7   Experimental Results

We now turn to the effectiveness of the integrated storage engine. To begin, we note that all evaluations, unless otherwise indicated, are conducted on a Linux-based workstation running a standard copy of the 2.6.x kernel, with 8 GB of main memory and a 3.2 GHz CPU. Disks are 160 GB SATA drives operating at 7200 RPM. The Berkeley DB components are taken from version db4.7.25. Data sets are generated using a custom data generator developed specifically for this environment. We first generate a multi-dimensional Fact Table (the dimension count varies with the particular test), with cardinalities arbitrarily chosen in the range 2–10000. Depending on the test involved, row counts typically vary from 100,000 to 10 million records. The primary Fact tables are then used to compute fully materialized data cubes containing hundreds of additional views or cuboids. For example, a 10-dimensional input set of 1,000,000 records produced a data cube of 1024 views and approximately 120 million total records. Once the cubes are materialized, we index the data using the the R-tree and bitmap mechanisms.

Because individual millisecond-scale queries cannot be accurately timed, we use the standard approach of timing queries in batch mode. In the succeeding tests, five batches of queries are generated and the average run-time is computed for each plotted point. Because query benchmarks are not well standardized for OLAP (the OLAP APB benchmark is effectively dead and TPC-H is better

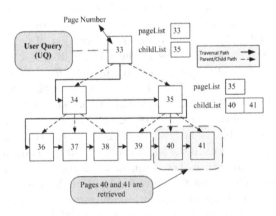

Fig. 6. Breadth First search strategy

suited to long running, ad hoc warehouse queries), we define our own query classes (described below). (We note that space restrictions prevent a full catalogue of the queries. We expect to include the list of queries in the longer version of this paper). The queries themselves are typically written in SQL and then translated to an XML representation as required. Finally, we note that when evaluating query performance, we use the "drop_caches" option available in the newer Linux kernels to delete the OS page cache between runs.

7.1   Non-hierarchical Attributes: FastBit Bitmap Versus Standard B-tree

We begin with a comparison of the FastBit indexing subsystem for non hierarchical attributes versus clusters of standard B-trees (implemented by Berkeley DB). We create a dimension (called `Customer`) with five non-hierarchical attributes (`Age`, `FirstName`, `LastName`, `Balance` and `Nationality`) and 1,000,000 records (i.e., the cardinality of the primary key `CustomerID`). The cardinalities of the non-hierarchical attributes were arbitrarily chosen in the range 100 - 1000.

We constructed 3 sets of queries against the Customer dimension, with each set containing five queries. The SQL format of two sample queries from each category is given in Figure 7(a). We can see, for example, that *Set 1* contains only look-up queries on a single non-hierarchical attribute. *Set 2* includes multi-column constraints, while *Set 3* consists of range queries with one or more attributes.

Figure 7 (b) shows a comparison of the running time using the two indexing implementations. For the first set (simple look-up on one attribute), we can see that the running times are actually quite similar. However, when we move to the more complex queries in *Set 2* and *Set 3*, there is a factor of two to three increase in running time for the B-tree indexing method. The difference is primarily due to the efficient bitwise logical operations (AND and OR) directly supported on

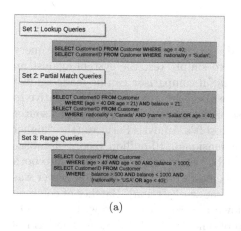 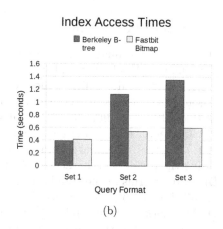

(a)   (b)

Fig. 7. (a) Sample SQL queries (b) Berkeley Btree versus FastBit bitmap

the compressed FastBit bitmap indexes. (For higher numbers of non-hierarchical attributes, the performance of B-trees is quite poor.) Finally, we note that the size of the B-tree indexes in this case is four times greater than the size of the compressed FastBit bitmap indexes — 13.8 MB for the B-trees versus 3.5 MB for the bitmaps.

7.2   Cube Construction

As previously noted, one of the advantages of the use of the Berkeley libraries is that its *environment* construct allows us to encapsulate all views and indexes into a single table space, thereby reducing the burden on the OS (and administrator). We therefore compared cube construction times (indexes and data sets) for the Berkeley environment versus the multi-file approach, as a function of both Fact table size and Dimension count.

In the first test, the full cube (2^d views) was generated from 9-dimensional input sets (i.e., Fact tables) ranging in size from 10,000 records to 1,000,000 records. Figure 8 (a) shows the running time for index cube construction before and after Berkeley DB integration. On average, the integration of Berkeley into our server reduces the index cube construction time by 40% - 60%. The primary reason for this reduction in time is that because the new method uses a single, integrated DB repository, its contiguous block layout allows for very efficient IO, even on larger R-trees. Conversely, use of multiple OS files leads to considerable disk thrashing.

An increase in dimension count has a similar impact in that each additional dimension effectively doubles the number of views to be computed and stored. In Figure 8 (b), we see the results for a data set of one million records and dimension counts of 5, 7, and 9 (common dimension counts in many OLAP environments). Again, we observe that the running time when using Berkeley DB drops by 40% to 60% due to the fact that we are storing the indexed cube in one contiguous physical file.

7.3   Query Performance

Of course, the ultimate purpose of a DBMS server is to provide impressive query performance. While it would be possible to simply test the system against an artificially defined baseline, a more meaningful comparison can be made against existing DBMS servers. Therefore, we have also evaluated the DBMS relative to systems often used in industrial database environments, namely the open source MySQL server (the "lower end") and Microsoft's Analysis Services (the "higher end"). In this case, we generate a 6-dimensional, 10-million record database (using the dimension characteristics described previously) and load it into both DBMS platforms in the standard Star Schema format (the Microsoft server was installed on a Windows partition on the same workstation). Batches of common OLAP queries — slice and dice, drill down and rollup — were written in SQL format, as well as the native XML form of our own OLAP DBMS. The form of these queries is similar to those described above, except that multiple dimensions

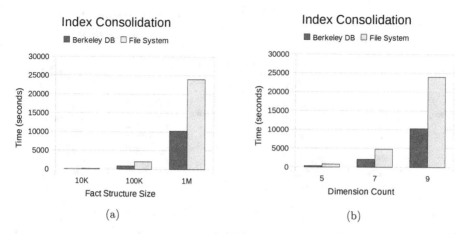

Fig. 8. (a) Index Construction/Fact table size (b) Index Construction/Dimension Count

were used in each query. Figure 9 shows comparative results for both platforms and demonstrates that the MySQL server takes approximately 10-15 times as long to resolve the same queries, while Microsoft's Analysis Services — running in ROLAP mode — is three to six times slower. Note that the term "Sibling Server" refers to a single node of our parallel DBMS.

Of course, one can argue that MOLAP offers superior performance to ROLAP configurations, at least for data sets of this size. So we loaded the same Star Schema data using the MOLAP mode of Microsoft's Analysis Services. Figure 10 (a) shows that MOLAP does indeed outperform our OLAP DBMS by a factor of about 5 to 1. However, we note that in this test, our DBMS was not permitted to materialize any additional data; it was essentially just an efficient Star Schema.

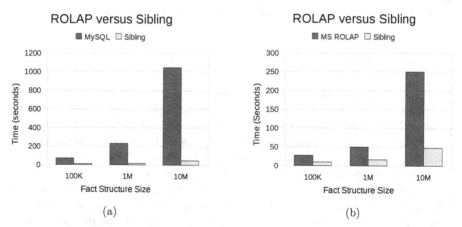

Fig. 9. Single "sibling" server versus (a) MySQL (b) MS Analysis Services (ROLAP)

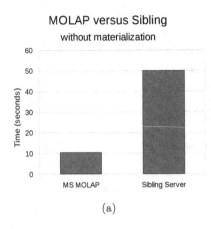
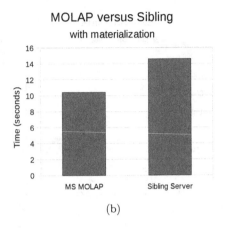

Fig. 10. (a) MOLAP versus non-materialized Sibling (b) MOLAP versus materialized Sibling

In Figure 10 (b), we see the result once aggregate materialization is added to the Fact Structure. (Note that production systems would typically use partial cube materialization consisting of the Fact data and a set of low-dimensional group-bys. In practice, this produces a compressed cube database that is not much bigger than the original Fact and Dimension tables). While Microsoft's MOLAP server still has a slight advantage, we note that (i) the Microsoft DBMS benefits from years of optimization, and (ii) MOLAP is ideally suited to the scale of the current test (i.e., 1-10 million records). Given that our DBMS framework is not constrained by the limits of array-based storage [4], these preliminary results suggest that the current DBMS — and the architecture it represents — does indeed have the potential to provide MOLAP-style performance with ROLAP-style scalability. (We note that a number of legacy components in the code base currently prevent true Terabyte scale testing. However, an ongoing software re-engineering effort is expected to remove these limitations in the coming year).

8   Conclusions

In this paper, we have described the storage and indexing architecture of a high performance OLAP DBMS. Current OLAP DBMS platforms generally take an "either/or" (MOLAP/ROLAP) approach to data representation and query processing, with the result being a very clear tradeoff between the scalability of relational systems and the performance of array-based platforms. The DBMS described in this paper attempts to build on the best features of both. Specifically, it uses a Fact Structure storage model that is constrained primarily by the disk space available, rather than the sparsity of the cube space. At the same time, the use of compressed Hilbert ordered R-tree indexes, mapGraph mapping tables for hierarchical attributes, and bitmap indexes on non-hierarchical attributes, coupled with a linearized Fact Structure search strategy, produces query performance beyond what one would expect with relational systems. In fact, a series of experiments confirmed that not only are the storage structures compact (and easily administered), but that query performance is actually comparable to commercial, and far less scalable, MOLAP servers. Given the enormous size of both existing and projected warehouses, we believe that the principles presented in the current paper offer great potential for the OLAP servers of the future.

References

1. Abouzeid, A., Bajda-Pawlikowski, K., Abadi, D., Silberschatz, A., Rasin, A.: Hadoopdb: an architectural hybrid of mapreduce and dbms technologies for analytical workloads. Proc. VLDB Endow. 2, 922–933 (2009)
2. Berkeley db (2011),
   http://www.oracle.com/technetwork/database/
   berkeleydb/overview/index.html
3. Dean, J., Ghemawat, S.: Mapreduce: a flexible data processing tool. Commununications of the ACM 53, 72–77 (2010)
4. Dehne, F., E.T., Rau-Chaplin, A.: Rcube: Parallel multi-dimensional rolap indexing. Journal of Data Warehousing and Mining 4, 1–14 (2008)
5. Eavis, T., Cueva, D.: The lbf r-tree: Efficient multidimensional indexing with graceful degradation. In: 22nd International Database Engineering and Applications Symposium, IDEAS 2007 (2007)
6. Eavis, T., Taleb, A.: Mapgraph: efficient methods for complex olap hierarchies. In: Conference on Information and Knowledge Management, pp. 465–474 (2007)
7. Zimanyi, E., Malinowski, E.: Hierarchies in a conceptual mode, from conceptual modeling to logical representation. In: Data & KNowledge Engineering (2005)
8. Fastbit (2011), https://sdm.lbl.gov/fastbit/
9. Gray, J., Bosworth, A., Layman, A., Pirahesh, H.: Data cube: A relational aggregation operator generalizing group-by, cross-tab, and sub-total. In: International Conference on Data Engineering (ICDE), pp. 152–159. IEEE Computer Society, Washington, DC (1996)
10. Gupta, H., Harinarayan, V., Rajaraman, A., Ullman, J.D.: Index selection for olap. In: Proceedings of the Thirteenth International Conference on Data Engineering, ICDE 1997, pp. 208–219. IEEE Computer Society, Washington, DC (1997)
11. Lakshmanan, L.V.S., Pei, J., Zhao, Y.: Qc-trees: an efficient summary structure for semantic olap. In: Proceedings of the 2003 ACM SIGMOD International Conference on Management of Data, SIGMOD 2003, pp. 64–75. ACM, New York (2003)
12. Microsoft analysis services (2011),
    http://www.microsoft.com/sqlserver/2008/en/us/analysis-services.aspx
13. Mondrian (2011), http://www.mondrian.pentaho.org
14. Morfonios, K., Ioannidis, Y.: Cure for cubes: cubing using a rolap engine. In: Proceedings of the 32nd International Conference on Very Large Data Bases, VLDB 2006, pp. 379–390. VLDB Endowment (2006)
15. Oracle olap (2011),
    http://www.oracle.com/technology/products/bi/olap/index.html
16. Plattner, H.: A common database approach for oltp and olap using an in-memory column database. In: Proceedings of the 35th SIGMOD International Conference on Management of Data, SIGMOD 2009, pp. 1–2 (2009)

17. Roussopoulos, N., Kotidis, Y., Roussopoulos, M.: Cubetree: organization of and bulk incremental updates on the data cube. In: Proceedings of the 1997 ACM SIGMOD International Conference on Management of Data, SIGMOD 1997, pp. 89–99. ACM, New York (1997)
18. Sismanis, Y., Deligiannakis, A., Roussopoulos, N., Kotidis, Y.: Dwarf: shrinking the PetaCube. In: Proceedings of the 2002 ACM SIGMOD Conference, pp. 464–475 (2002)
19. Stonebraker, M., Abadi, D., DeWitt, D.J., Madden, S., Paulson, E., Pavlo, A., Rasin, A.: Mapreduce and parallel dbmss: friends or foes? Commun. ACM 53, 64–71 (2010)

Highly Scalable Speech Processing on Data Stream Management System

Shunsuke Nishii[1] and Toyotaro Suzumura[1,2]

[1] Tokyo Institute of Technology 2-12-1 Ookayama, Meguro-ku, Tokyo, Japan
[2] IBM Research - Tokyo 1623-14 Shimotsuruma, Yamato-shi, Kanagawa, Japan

Abstract. Today we require sophisticated speech processing technologies that process massive speech data simultaneously. In this paper we describe the implementation and evaluation of a Julius-backended parallel and scalable speech recognition system on the data stream management system "System S" developed by IBM Research. Our experimental result on our parallel and distributed environment with 4 nodes and 16 cores shows that the throughput can be significantly increased by a factor of 13.8 when compared with that on a single core. We also demonstrate that the beam management module in our system can keep throughput and recognition accuracy with varying input data rate.

1 Introduction

Present emphasis on speech communication technologies has triggered the requirement of sophisticated speech processing technologies that process speech simultaneously. For example, in call center of an enterprise there can be simultaneous activities such as checking whether the operator talks the matter that should be told, making sure the operator does not talk the matter that violates compliance, and display information that the operator needs on monitor, etc. Such activities can be done via processing the speech/text communication that happens between customer and the operator. It is important to utilize real-time speech processing technologies, since quality of the call center's service depends up on aforementioned factors. On the other hand, data stream processing [1] is a new computational paradigm for processing large amounts of streaming data in real-time. It is an active area of research that concentrates on processing data in memory without use of an offline storage. Some software systems (e.g. Data Stream Management Systems (DSMS) and Data Stream Processing Systems (DSPS)) and programming models have been proposed, such as IBM's System S [2][3] and Borealis [1] etc.

While there were previous works on real-time speech processing [4], this paper aims at utilizing the generality and high extensibility introduced by DSMS for speech processing. We implemented a highly scalable and distributed parallel speech recognition system on top of System S and evaluated scalability and validity of our approach.

Normally performance of one instance of speech recognition system (e.g. recognition accuracy, throughput, latency) is fixed after it initialize. Recognition accuracy and throughput/latency are in relation of trade-off. Input data rate can

S.-g. Lee et al. (Eds.): DASFAA 2012, Part II, LNCS 7239, pp. 203–212, 2012.

be variable (some times it can be extremely high) in data stream processing systems. If input data rate is low then a high recognition accuracy is required. On the other hand, if the input data rate is high then a high through put is required. Therefore a mechanism that changes priority between recognition accuracy and throughput according to input data rate is needed. In this paper we implemented a mechanism that optimizes "beam width" which is the parameter of a speech recognition engine that largely affects recognition accuracy, throughput and latency according to the current input data rate.

Organization of this paper is as follows. In section 2 we refer to related work. In section 3 we explain data stream processing and DSMS. We briefly explain speech recognition in section 4. Then design and implementation of the system are given in section 5. System evaluation is described in section 6. Finally conclusions and further works are given in section 7.

2   Related Work

Work done by Arakawa *et al* [4] proposes a method to find the appropriate number of servers required for processing assumed services by performance of the servers and scalability evaluation model. Our research is different from this in the perspective of using DSMS and in the point of variability in performance of the servers.

Load shedding is a technique that ignores part of input data to realize real-timeness when input data rate is increased. Many researches about load shedding on DSMS have been performed such as the work by Tatbul *et al* [5]. In this paper, our mechanism that manages beam width resembles a similar load shedding approach.

3   System S

System S [2][6][7][8] is large-scale, distributed data stream processing middleware under development at IBM Research. It processes structured and unstructured data streams and can be scaled to a large numbers of compute nodes. The System S runtime can execute a large number of long-running jobs (queries) in the form of data-flow graphs described in its special stream-application language called SPADE (Stream Processing Application Declarative Engine) [6]. SPADE is a stream-centric and operator-based language for stream processing applications for System S, and also supports all of the basic stream-relational operators with rich windowing semantics. Here are the main built-in operators mentioned in this paper:

- **Functor**: adds attributes, removes attributes, filters tuples, maps output attributes to a function of input attributes
- **Aggregate**: window-based aggregates, with groupings
- **Join**: window-based binary stream join
- **Sort**: window-based approximate sorting

- **Barrier**: synchronizes multiple streams
- **Split**: splits the incoming tuples for different operators
- **Source**: ingests data from outside sources such as network sockets
- **Sink**: publishes data to outside destinations such as network sockets, databases, or file systems

SPADE also allows users to create customized operations with analytic code or legacy code written in C/C++ or Java. Such an operator is a UDOP (User-Defined Operator), and it has a critical role in providing flexibility for System S. Developers can use built-in operators and UDOPs to build data-flow graphs.

After a developer writes a SPADE program, the SPADE compiler creates executables, shell scripts, other configurations, and then assembles the executable files. The compiler optimizes the code with statically available information such as the current status of the CPU utilization or other profiled information. System S also has a runtime optimization system, SODA [2]. For additional details on these techniques, please refer to [2][6][7][8].

SPADE uses code generation to fuse operators with PEs. The PE code generator produces code that (1) fetches tuples from the PE input buffers and relays them to the operators within, (2) receives tuples from operators within and inserts them into the PE output buffers, and (3) for all of the intra-PE connections between the operators, it fuses the outputs of operators with the downstream inputs using function calls. In other words, when going from a SPADE program to the actual deployable distributed program, the logical streams may be implemented as simple function calls (for fused operators) for pointer exchanges (across PEs in the same computational node) to network communications (for PEs sitting on different computational nodes). This code generation approach is extremely powerful because simple recompilation can go from a fully fused application to a fully distributed version, adapting to different ratios of processing to I/O provided by different computational architectures (such as blade centers versus Blue Gene).

4   Speech Recognition with Data Stream Processing

4.1   Outline of Speech Recognition

Speech recognition is the process that converts human's spoken words to text. In speech recognition, speech data is converted to the time series of feature vector: $X = x_1, x_2, \ldots, x_T$, and calculate word sequence W that maximize $P(W|X)$:

$$\tilde{W} = \operatorname*{argmax}_{W} P(W|X)$$

Since calculating $P(W|X)$ directly is difficult, the expression is transformed based on Bayes rule as follows.

$$\tilde{W} = \operatorname*{argmax}_{W} \frac{P(X|W)P(W)}{P(X)}$$

Since $P(X)$ is constant for W:

$$\tilde{W} = \underset{W}{\operatorname{argmax}} P(X|W)P(W)$$

The statistical models that are used to calculate $P(W|X)$ and $P(W)$ are called acoustic model (AM) and language model (LM) respectively, and it is mainstream to use Hidden Markov Model and N-gram model respectively. There are some speech recognition engines that can recognize Japanese such as HTK [9], Julius [10][11], T3 decoder [12] etc. In this paper, we adopted Julius 4.1.4 for the convenience of implementing speech recognition operator as UDOP (User-Defined Operator) on SPADE.

4.2   Beam Search

In Julius, search network is built based on acoustic model and language model, and speech recognition is performed by the algorithm based on tree-trellis search [13]. The algorithm consists of two passes. In 1st pass, Rough search by approximation is conducted to narrow the candidate of the output. Then in 2nd pass, detailed search is done based on the results of first pass to get the final output sentence. Both in 1st and 2nd pass, search is not done to all routes in the network, but calculation is done for the hypotheses with high score, and the other hypotheses are rejected. This rejection method is called "beam search", and the width of calculation range is called "beam width".

The 1st pass takes majority of process time during speech recognition. To perform the accurate speech recognition, it is needed to raise the probability that the correct answer sentence stays in the candidate of the output hypotheses. Therefore, it is needed to enlarge the value of beam width. On the other hand, too large value of beam width makes the recognition process slow. In short, the value of beam width in the 1st pass most greatly affects both recognition accuracy and processing time.

In the following sections, "beam width" simply indicates the beam width in the 1st pass.

5   Design and Implementation of Highly Scalable Speech Recognition System

We implemented a scalable speech recognition system using System S and SPADE. Fig.1 shows rough data flow diagram of the system and fig.2 shows detailed data flow diagram. The idea of system design are described in subsections 5.1 to 5.3. Explanation of each module in the system is given in subsection 5.4. Additionally supplementary explanation of the data flow diagram is described in Section 5.5.

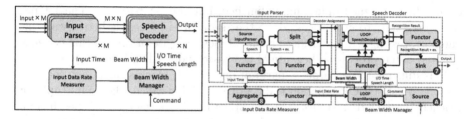

Fig. 1. Rough Data Flow Diagram of the System

Fig. 2. Detailed Data Flow Diagram of the system

5.1 Extensibility and Scalability

The system has been designed considering the scale-out property and extensibility for various speech processing. SPADE source code of this system consists of only 120 lines. Hence it indicates SPADE's ability in describing some parallel and distributed processes over plural nodes easily. Furthermore, to perform scale-out processing, it only has to modify lines 1-2 (number of parallel nodes) and 24-27 (names of using nodes). So it shows that it is easy and straightforward to make the system scale-out. Moreover, if one needs to add extra speech/text processing to the system, that can be done by writing processes after "Decoder@j_Transcripton" stream (lines 71-75).

5.2 Mechanism of Beam Width Management

As described in section 4.2, the 1st pass takes the majority of process time in speech recognition and the process time depends on the value of beam width. It is normal in Julius that the value of beam width is set by users and this value is fixed after the engine is launched. As mentioned previously in speech recognition on DSMS, if input data rate is low then recognition accuracy has the priority. If input data rate is high then throughput has the priority. Therefore in our system the value of beam width is a variable one. We implemented a mechanism to adoptively optimize beam width according to the current input data rate. In this mechanism, user first sets candidates of beam width (e.g. 400, 800 and 1200). After this, user launches the system and inputs the training data to the system. Then the system measures throughput for each candidate of beam width. When the system is in use it monitors input data rate and sets beam width to the max value on condition that throughput is not less than input data rate.

5.3 Batch Processing vs. Sequential Processing

Speech recognition operator (UDOP) is implemented based on libjulius/libsent, the library version of Julius. There are two methods of speech recognition. In one method the recognition starts after input of one speech block has finished (batch processing). In the other method the recognition starts sequentially as parts of one speech block has arrived (sequential processing). The 2nd pass cannot be performed unless input finished, but the 1st pass can. The 1st pass takes long time to get processed, so that sequential process is better way than batch processing when considering the latency. In spite of this, in this paper we adopted batch processing because of the reasons described below.

It is difficult to get/process parts of one speech block sequentially for some reasons. First, UDOP function is data-driven and is called when data tuple has reached the UDOP instance. On the other hand, the method of speech input in libjulius is a callback function, that is, when libjulius needs data to process, libjulius calls user-defined callback function to get speech data. Taking both the data-driven nature of UDOP and dependency of libjulius on user defined call back function, it is needed to make one UDOP instance to multithreading explicitly. In general one UDOP instance runs in single thread. Considering this factor, our system was implemented using batch processing rather than sequential processing.

Batch processing has an advantage compared to sequential processing. In sequential processing, if RTF (real time factor) of the process time is less than 1.0, the CPU needs to idle till the input data appears which reduces the CPU's efficiency. Therefore the capacity of speeches to be processed simultaneously (throughput) cannot exceed 1.0 per one speech recognition operator instance. But it can exceed 1.0 in batch process. Therefore our system with batch process approach has the advantage in terms of throughput when assuming that RFT is less than 1.0.

5.4   The Structure of the System

The system is composed of four modules of "Input Parser", "Speech Decoder", "Input Data Rate Measurer" and "Beam Width Manager" (shown in fig.1). The role of each module is as follows.

"Input Parser" module receives input by socket communications and converts it into tuples that can be processed in System S. This module runs in M-cores parallel (M threads). Each thread of this module has a port of socket communication, so the communication load can be distributed. Additionally, this module adds input time to data tuples for measuring input data rate and throughput.

"Speech Decoder" module decodes input speech data into recognized text. This module runs in N-cores parallel (N threads). The input data of this module are passed from Input Parser and assigned into each module. This assignment method is naive round-robin. Each data input sequence is assigned into 1st, 2nd, ..., N-th core of Speech Decoder module, and the next of N-th is assigned to 1st again. Output data tuples (consist of input time, output time and speech length) are passed into Beam Width Manager module. This data are used for measuring throughput.

"Input Data Rate Measurer" module measures input data rate from input time data given from Input Parser module.

"Beam Width Manager" module runs in 3 modes by external command; (1) set beam width direct, (2) measure throughput for current beam width, (3) optimize beam width for current input data rate. In running mode (3), beam width is chosen according to throughput measured in (2).

5.5   Supplementary Explanation

This section provides a supplementary explanation about data flow diagram of the system shown in fig.2. The system consists of 12 operators. In the fig.2, each operator has the number; (0) to (9), (A) and (B). Among these, (0) to (3) belong to Input Parser, (4) to (7) belong to Speech Decoder, (8) and (9) belong to Input Data Rate Measurer, and (A) and (B) belong to Beam Width Manager. The behavior of each operator is as follows.

(0) Receives input speech data from the outside of the system and converts input speech data into data tuples that are available in System S.
(1) Adds input time and tag, used for decoder assignment, to tuples.
(2) Assigns speech data to each decoder.
(3) Extracts data that are used for measuring input data rate from tuples.
(4) Performs speech recognition. (Beam width can be changed by the operator (B).)
(5) Adds output time to tuples.
(6) Extracts data that are used for measuring throughput from tuples.
(7) Outputs recognition results outside of the system.
(8) Aggregates data for measuring input data rate by a fixed-size window.
(9) Measures input data rate.
(A) Receives commands for Beam Width Manager from the outside of the system.
(B) Integrates data and sets beam width to decoders if needed.

6   Evaluation

We evaluated the scale-out property to number of cores used for Speech Decoder module and effectiveness of Beam Width Manager module experimentally.

6.1   Experimental Environment

Computing environment is as follows. A node that consists of [Opteron 1.6GHz L2 512KB (2 cores), Memory 8GB] was used for I/O from the outside of the system to the inside of the system, a node that consists of [Phenom X4 2.0GHz L2 512KB (4 cores), Memory 3.5GB] was used for Input Parser module and measuring input/output time, four nodes that consist of [Phenom X4 2.5GHz L2 512KB (4 cores), Memory 8GB] respectively were used for Speech Decoder module, and a node that consists of [Phenom X4 2.5GHz L2 512KB (4 cores), Memory 8GB] was used for Input Data Rate Measurer module and Beam Width Manager module. All nodes in computing environment were connected with 1Gpbs Ethernet. As software environment, [CentOS 5.2 2.6.18-92.el5 AMD64, gcc4.1.2, InfoSphere Streams 1.2 (System S)] were used common in all nodes.

6.2 Recognition Models and Julius Parameters

The acoustic model was a tied-state Gaussian mixture triphone HMM that was trained on 52 hours of clean speech data from the Japanese News Article Sentences (JNAS) [14] corpus using the ML (maximum likelihood) methods. The HMM had 3000 states and Gaussian mixture with 16 components per state. Feature vectors had 38 elements comprising 12 MFCCs, their delta, their delta delta, delta of log energy, and delta delta of log energy. The language model was a trigram trained from the newspaper articles of 1991-2002 years on Mainichi Newspapers, and the dictionary size was 60k. Beam width was variable, and the other parameters were set to values that makes RTF less than 1.0 for our test data set.

6.3 Test Data Set

Test data set for evaluation was 20.2 minutes of clean speech data from JNAS (IPA-98-TestSet), which was not overlapped with the one that was used for acoustic model training. We divided it into set0 and set1. Set0 consists of about 20% of total test data set (3.7 minutes), and set1 consists of the rest 80% (16.5 minutes).

6.4 Evaluation of Scale-Out Property

We set the number of cores used for Input Parser (M) to 4, and fixed beam width to 1200. We measured throughputs of Speech Decoder module when the number of cores used for Speech Decoder module (N) was changed from 1 to 16 under this condition. Set0 was used for input data. In detail, each input port received the entire copy of set0. Fig.3 shows the result. The horizontal axis shows the capacity of speeches to be processed simultaneously (throughput). As fig.3 shows, the capacity of speeches to be processed simultaneously when operating in single core was 1.4 and when operating in parallel 16 cores was 19.3. This means that throughput of 16 cores is 13.8 times as much as that of single-core and nodes can be added to the system easily as shown in section 5.1 giving the system a high scale-out property. Word error rate (WER) for set0 in this condition was 5.9%.

6.5 Evaluating Beam Width Manager

We tested the effectiveness of Beam Width Manager module by test input data set with variable input data rate, in running mode (3) (given in section 5.4). We set the number of cores used for Input Parser (M) to 4, and that for Speech Decoder module (N) to 16. We measured throughput for beam width 400, 800, and 1200 by set0 in running mode (2). So the system chose as beam width among 400, 800 and 1200 automatically if needed. Test input data set was set1 with variable input data rate, which is shown in fig.4. In these conditions, we compared WER and response time in variable beam width mode (running mode (3)) to that in fixed beam width mode (running mode (1), values of beam width are 400, 800, and 1200). We call these modes VAR, FIX400, FIX800, and FIX1200 respectively at the following.

Fig.5 shows WERs and fig.6 shows response times for test data set in each running mode. Since beam width varied from 400 to 1200 in VAR, WER in VAR was less than that in FIX400 and more than FIX1200 as fig.5 shows. Response time of VAR did not explode like in FIX1200 when on high input data rate as fig.6 shows. Therefore, it is shown that Beam Width Manager module can keep both throughput and recognition accuracy.

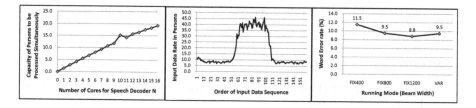

Fig. 3. Scale-Out Property of Speech Decoder Module

Fig. 4. Change in Input Data Rate of Set1

Fig. 5. Word Error Rate for Set 1

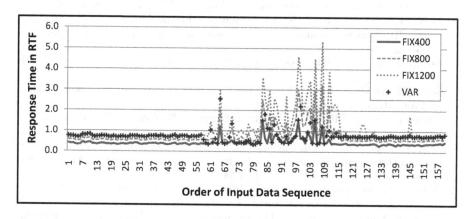

Fig. 6. Response Time in RTF for Set 1 (=Response Time ÷ Speech Length)

7 Concluding Remarks and Future Work

In this work we implemented a highly scalable speech recognition system using System S and SPADE. We developed speech recognition operator used in this system based on Julius 4.1.2. We also implemented the automatic and adaptive mechanism of beam width management that enables to keep both throughput and recognition accuracy. It was experimentally shown that throughput when operating in parallel 16 cores is 13.8 times as much as that when operating in single-core. Moreover it was also experimentally demonstrated that Beam Width Manager module in the system can keep both throughput and recognition accuracy.

In our current implementation, we adopted naive round-robin method for task assignment of Speech Decoder module. Such a simple method causes biases of task loads among cores hence some times response time is increased. This problem can be avoided by making the system to assign tasks evenly based on speech length.

We adopted batch processing method for our speech recognition operator, so high latency cannot be avoided with the current mechanism. To avoid this, it is required to implement a speech recognition operator that runs in sequential process.

References

1. Abadi, D.J., et al.: The Design of the Borealis Stream Processing Engine. In: Proc. CIDR, pp. 277–289 (2005)
2. Wolf, J., Bansal, N., Hildrum, K., Parekh, S., Rajan, D., Wagle, R., Wu, K.-L., Fleischer, L.K.: SODA: An Optimizing Scheduler for Large-Scale Stream-Based Distributed Computer Systems. In: Issarny, V., Schantz, R. (eds.) Middleware 2008. LNCS, vol. 5346, pp. 306–325. Springer, Heidelberg (2008)
3. Gedik, B., et al.: A Code Generation Approach to Optimizing High-Performance Distributed Data Stream Processing. In: Proc. USENIX, pp. 847–856 (2009)
4. Arakawa, Y., et al.: A Study for a Scalability Evaluation Model of Spoken Dialogue System. Transactions of Information Processing Society of Japan 46(9), 2269–2278 (2005) (in Japanese)
5. Tatbul, N., et al.: Load Shedding in a Data Stream Manager. In: Proc. VLDB (2003)
6. Gedik, B., et al.: SPADE: The System S Declarative Stream Processing Engine. In: Proc. SIGMOD, pp. 1123–1134 (2008)
7. Amini, L., et al.: SPC: A Distributed, Scalable Platform for Data Mining. In: DM-SSP, pp. 27–37 (2006)
8. Jain, N., et al.: Design, implementation, and evaluation of the linear road benchmark on the stream processing core. In: International Conference on Management of Data, ACM SIGMOD, Chicago, IL (2006)
9. Young, S., et al.: The HTK book (for HTK Version 3.2) (2002)
10. Lee, A., et al.: Recent Development of Open-Source Speech Recognition Engine Julius. In: Asia-Pacific Signal and Information Processing Association Annual Summit and Conference, APSIPA ASC (2009)
11. Lee, A.: Large Vocabulary Continuous Speech Recognition Engine Julius ver. 4. IEICE technical report. Speech 107(406), pp.307-312 (2007) (in Japanese)
12. Dixon, P.R., et al.: The Titech Large Vocabulary WFST Speech Recognition System. In: IEEE ASRU, pp. 443–448 (2007)
13. Lee, A., et al.: An Efficient Two-pass Search Algorithm using Word Trellis Index. In: Proc. ICSLP, pp. 1831–1834 (1998)
14. Itahashi, S., et al.: Development of ASJ Japanese newspaper article sentences corpus. Annual Meeting of Acoustic Society of Japan 1997(2), 187–188 (1997) (in Japanese)

EVIS: A Fast and Scalable Episode Matching Engine for Massively Parallel Data Streams

Shinichiro Tago, Tatsuya Asai, Takashi Katoh,
Hiroaki Morikawa, and Hiroya Inakoshi

Fujitsu Laboratories Ltd., Kawasaki 211-8588, Japan
{s-tago,asai.tatsuya,kato.takashi_01,h.morikawa,
inakoshi.hiroya}@jp.fujitsu.com

Abstract. We propose a fast episode pattern matching engine *EVIS* that detects all occurrences in massively parallel data streams for an episode pattern, which represents a collection of event types in a given partial order. There should be important applications to be addressed with this technology, such as monitoring stock price movements, and tracking vehicles or merchandise by using GPS or RFID sensors. EVIS employs a variant of non-deterministic finite automata whose states are extended to maintain their activated times and activating streams. This extension allows EVIS's episode pattern to have 1) *interval constraints* that enforce time-bound conditions on every pair of consequent event types in the pattern, and 2) *stream constraints* by which two interested series of events are associated with each other and found in arbitrary pairs of streams. The experimental results show that EVIS performs much faster than a popular CEP engine for both artificial and real world datasets, as well as that EVIS effectively works for over 100,000 streams.

1 Introduction

Recent advances in network and sensor technologies such as GPS and RFID have facilitated their adoption in a growing number of applications, including shoplifting detections [1], stock price movement detections, and car accident detections [2]. Thus, there have been increasing demands for cyber physical systems [11] that enable us to get useful and valuable information for our social issues such as energy, traffic, distribution, finance, or urban problems by integrating both computational and physical processes of our society. There are many complex correlations among tons of natural phenomena and human activities in our society. To understand these correlations and to take measures against them become very important factors for improving our society. Then the successful development of cyber physical systems must address two unique challenges below: 1) Real time processing against massively parallel and heterogeneous event streams that are summaries of various event streams for natural phenomena and human activities in both real and digital worlds. 2) A fast matching algorithm for detecting richer classes of patterns than sequential patterns with more precise temporal constraints than time window constraints.

S.-g. Lee et al. (Eds.): DASFAA 2012, Part II, LNCS 7239, pp. 213–223, 2012.

Fig. 1. An example of our extended episode, which represents that a car reduces its speed quickly and another car immediately changes to the left lane within one hour after the latter car frequently changes lanes in 30 seconds

In this paper, we propose a fast and scalable episode matching engine *EVIS* that detects all occurrences in parallel event streams for an episode pattern. The episode represents a collection of event types in a given partial order [12]. It is a rich class of patterns, and it can represent conjunction patterns. EVIS employs a variant of non-deterministic finite automata whose states are extended to maintain their activated times and activating streams. There are some difficulties for NFAs to detect conjunction patterns against parallel event streams. We resolved these difficulties, moreover, we enable EVIS to detect episodes with the following two constraints for practical use. The first one is an *interval constraint*, which restricts a time bound between a pair of event types. By this constraint, for instance, we can detect a short episode after a long episode. The other is a *stream constraint*, which detects associations in arbitrary pairs of streams, such as an object's episode in a stream causes another object's episode. Our extended episodes can represent various natural phenomena and human activities. Fig. 1 shows an example of our extended episode, which represents that an unsafe driver threatens another driver.

The experimental results show that EVIS performs much faster than a popular Complex Event Processing (CEP) engine for both artificial and real world datasets, and EVIS runs effectively against complex patterns such that the CEP engine cannot finish its execution. EVIS also effectively works on some practical patterns for even over 100,000 streams, and it can be said that EVIS has a potential to solve practically important problems for our society in real time.

This paper is organized as follows: Section 2 introduces related work, Section 3 gives notations and two types of episode matching problems, Section 4 describes an NFA based processing model and algorithms of constructing our NFA for the two problems, Section 5 reports evaluation results, and finally, Section 6 concludes this paper.

2   Related Work

CEP systems appeared as an active database [7,9,5] which detects situation rules. A stream-oriented processing [4] is proposed for CEP systems, rather than no real-time processing for traditional DBMSs. Some CEP engines [1,3] use NFAs to search sequence-like patterns from several types of event streams. There are some difficulties for NFAs in detecting conjunction patterns. An open-source software Esper [8] searches window constrained sequence-like patterns and SQL-like language patterns, that have a capability to detect conjunction patterns within a window by inserting time stamp data for each event set and processing a join operation. However, it takes much time to detect a conjunction pattern by an SQL-like language. Zstream [13] detects conjunction, disjunction, negation, and Kleene Closure patterns by using tree based query plans.

There are various studies for episode mining problems [12,10,14]. Das *et al.* [6] proposed efficient algorithms for the episode matching problem with window constraints. The window constraint is not sufficient for solving many types of practical problems. Another interval constraint [15] between a basis event type and another event type is proposed, while we propose an interval constraint between each pair of event types.

3 Preliminaries

In this section, we define two types of episodes and those detection problems. We denote the set of all natural numbers by $\mathbb{N}$. For a set S, we denote the cardinality of S and the power set of S by $|S|$ and 2^S, respectively. A *digraph* is a graph with directed edges (or, arcs). A digraph is *weakly connected* if there is an undirected path between any pair of nodes. For nodes u and v on a digraph, if there exists an arc directed from u to v, the nodes v and u are called the *head* and *tail* of the arc, respectively. For an arc e, we denote the head and tail of e by $\text{Head}(e)$ and $\text{Tail}(e)$, respectively. For arcs e_1, e_2 such that $\text{Tail}(e_1) = \text{Tail}(e_2)$, we call the tail a *branch tail* and call the arcs *branch arcs*. Additionally, for arcs e_1, e_2 such that $\text{Head}(e_1) = \text{Head}(e_2)$, we call the head a *junction head* and call the arcs *junction arcs*. Let $\Sigma = \{1, \ldots, k\}(k \geq 1)$ be a finite alphabet with a total order $\leq$ over $\mathbb{N}$. Each element $e \in \Sigma$ is called an *event type*.

An *input event stream* (a *stream* for short) S on Σ is a sequence $\langle S_1, S_2, \ldots \rangle \in (2^\Sigma)^*$ of events, where $S_i \subseteq \Sigma$ is called the i-th event set for every $i \geq 1$. For any $i \leq 0$, we define $S_i = \emptyset$. *Parallel input event streams* (*parallel streams* for short) $\mathcal{M}$ on Σ are a sequence of streams $\langle \mathcal{S}_1, \ldots, \mathcal{S}_m \rangle$ of events, where $\mathcal{S}_j$ is called the j-th event stream for every $1 \leq j \leq m$. We denote i-th event set of $\mathcal{S}_j$ by $S_{(i,j)}$.

Definition 1. (Episode: Mannila *et al.* [12]) *A labeled acyclic digraph $X = (V, E, g)$ is an* episode *over Σ where V is a set of nodes, $E \subseteq V \times V$ is a set of arcs and $g : V \to \Sigma$ is a mapping associating each node with an event type.*

An episode is an acyclic digraph in the above definition, while it is defined as a partial order by Mannila et al. [12]. It is not hard to see that the two definitions are essentially the same.

Definition 2. (Interval-Constrained Episode and Matching) *For an episode $X = (V, E, g)$ and a mapping $h : E \to N$, a labeled and weighted acyclic digraph $Y = (V, E, g, h)$ is an interval-constrained episode over Σ. An interval-constrained episode $Y = (V, E, g, h)$ is matched at the i-th event set on a stream S if there exists at least one mapping $f : V \to \mathbb{N}$ such that (i) f preserves the labels of nodes, that is, for all $v \in V, g(v) \in S_{f(v)}$, (ii) f preserves a precedence relation, that is, for all $u, v \in V$ with $u \neq v$, if $(u, v) \in E$, then $f(u) < f(v)$, (iii) the last period of the event sets of nodes is i, that is, $\max_{\forall v \in V} f(v) = i$, and (iv) f satisfies the interval constraints, that is, for all $u, v \in V$ with $u \neq v$*

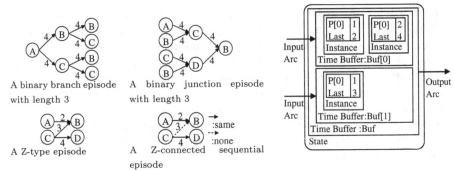

Fig. 2. Examples of stream-constrained episodes   **Fig. 3.** Components of PNFA

if $(u, v) \in E$, then $f(v) - f(u) \leqq h((u, v))$. The mapping f is called a matching of Y at the i-th event set on S, or it is simply called a matching of Y on S.

We note that we define a matching by a many-to-one mapping, while Mannila et al. [12] defines a matching by a one-to-one mapping. Without the loss of generality, we can extend a minimum interval constraint or both the maximum and minimum interval constraints. In this paper, we simply call a maximum interval-constrained episode an interval-constrained episode.

Definition 3. (Stream-Constrained Episode and Matching) *For an interval episode $Y = (V, E, g, h)$ and a mapping associating each arc $z : E \to$ {"same", "differ", "none"}, a double labeled and weighted acyclic digraph $Z = (V, E, g, h, z)$ is a stream-constrained episode over Σ. For a stream-constrained episode $Z = (V, E, g, h, z)$, Z is matched at i-th event sets on parallel streams $\mathcal{M}$ if there exists at least one pair of matchings (1) a period mapping $f_P : V \to \mathbb{N}$ and (2) a track mapping $f_T : V \to \mathbb{N}$ such that (i) f_P and f_T preserves the label of nodes, that is, for all $v \in V, g(v) \in S_{(f_P(v), f_T(v))}$, (ii) f_P preserves the precedence relation, (iii) the last period of the event sets of nodes is i, (iv) f_P satisfies the interval constraints, and (v) f_T satisfies the stream constraints such that for all $u, v \in V$ with $u \neq v$, (a) if $(u, v) \in E$ and $z(u, v) =$ "same", then $f_T(u) = f_T(v)$, and (b) if $(u, v) \in E$ and $z(u, v) =$ "differ", then $f_T(u) \neq f_T(v)$. The pair of mappings f_P and f_T is called a matching of Z at i-th event sets on $\mathcal{M}$, or it is simply called a matching of Z on $\mathcal{M}$.*

In this paper, we treat the "same" and "none" constraints and leave the "differ" constraint for future work. Examples of stream-constrained episodes are in Fig. 2.

4   Processing Model

Our processing model employs a variant of a nondeterministic finite state automaton (NFA). We call our variant of an NFA PNFA. Formally, a PNFA $A = (Q, D, \theta, \mathcal{T}, q_0, F)$ consists of a set of states Q, a set of arcs $D \subset Q \times Q$, a set of *operation sequences* θ for processing PNFA, a mapping $\mathcal{T} : D \to \theta \times \Sigma \times \mathbb{N}^*$, a start state $q_0 \in Q$ and a set of final states $F \subset Q$. Fig. 3 shows components of PNFA and we represent about each of them.

Instances each of them indicates a state of matching and holds a series of event detected periods. Periods $P[0], P[1], \ldots$ of them are used by checking event identifications for branch or junction patterns, and a period Last of them is used by checking interval constraints.

Time Buffers hold instances. Each time buffer is associated with a connected arc. Each time buffer associated with an output arc contains all of the time buffers associated with input arcs.

States have time buffers. Each state represents a state of matching process. The start state q_0 represents the beginning of matching and holds the start instance with Last $= 0$ during the whole time. A final state $q_f \in F$ represents that the pattern is matched completely. A period of matching detection is reported if an instance is created in a time buffer of a final state.

Operation sequence defines a series of actions, such as creation, update, or deletion, against some instances in a defined situation.

Arc represents a transitions of a states and is associated with an operation sequence by $\mathcal{T}$. The basic operation sequence associated with each arc is to check instances in the time buffers of the tail of the arc, and then, according to the result, to create or update an instance in the time buffer of the head of the arc and/or to delete instances in the time buffer of the tail of the arc.

We also represent the following features of PNFA.

Preservative. While each state does not keep a previous state unless it is defined by a transition function in normal NFAs, each instance stays in the previous time buffer in PNFAs. In PNFAs, only after it has occurred a situation defined by an operation sequence, the instances are processed. In other words, each instance in a time buffer of a state is never processed unless it occurs a situation defined by an operation sequences associated with an arcs connected to the state.

Non-deterministic. There may exist several instances on a PNFA because instances may be created without deleting instances. This means that the next state of a state may be any one of several possible states. Then, PNFAs have a non-deterministic character.

Next, we present our method for building and processing a PNFA for an episode pattern detection. We first propose an interval-constrained episode matching detection model, and next, we propose a stream-constrained episode's one.

4.1 Interval-Constrained Episode Matching Detection Model

We present a processing model for interval-constrained episode detection. First, we study the detection of a simple episode, that is, *an interval-constrained sequential episode* for an easy understanding of a PNFA. We can easily solve the pattern detection problem of an interval-constrained sequential episode by using our extended NFA with the event detected period. It is trivial that for each matching state, to just check an event type of a new event and the difference between a detected period of a preceding event type in an instance

```
procedure θ_normal(d, e, c, t)
    input arc d, event type e, interval constraint
    c, time-out t
    if e ∈ S_i then
        for I ∈ Tail(d).Buf do
            if i − I.Last ≤ c then
                if ∃I^x ∈ d.Buf in Head(d) such that
                ∀k, I.P[k] = I^x.P[k]  then
                    I^x.Last ← i
                else
                    Copy I as I' into d.Buf ; I'.Last ← i
                end if
            end if
            if i − I.Last > t then Delete I
        end for
    end if
end procedure
```

```
procedure θ_epsilon(d)
    input arc d
    for I ∈ Tail(d).Buf do
        if I is Updated in this period then
            if ∃I^x ∈ d.Buf in Head(d) such that
            ∀k, I.P[k] = I^x.P[k]  then
                I^x.Last ← i
            else
                Copy I as I' into d.Buf
                I'.Last ← i
            end if
        else
            Delete I
        end if
    end for
end procedure
```

Fig. 4. Algorithms of operation sequences θ_{normal} and θ_{epsilon}

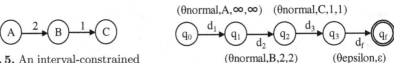

$(\theta\text{normal,A},\infty,\infty)$ $(\theta\text{normal,C},1,1)$

Fig. 5. An interval-constrained sequential episode

$(\theta\text{normal,B},2,2)$ $(\theta\text{epsilon},\varepsilon)$

Fig. 6. An example of PNFA for the episode in Fig. 5

and a detected period of the new event is sufficient to find matching periods for interval-constrained sequential episodes. Moreover, we note that it is sufficient to find matching periods as holding only one instance with the latest period in each time buffer. We use this filtering technique for reducing both memory usage and computation time. Then, we define two operation sequences θ_{normal} for checking the event types and interval constraints and θ_{epsilon} in Fig. 4. Additionally, we prepare a mapping $r : \Sigma \to 2^D$ such that for each $e \in \Sigma, r(e) = \{d|d \in D, \exists\theta_x \in \theta, \mathcal{T}(d) = (\theta_x, e, \ldots)\}$. Because this mapping r gets operation sequences triggered by each detected event e; it enables an event-driven processing.

For example, Fig. 5 and Fig. 6 show a sequential episode and its corresponding PNFA, respectively. Let "AABACBC" be an input event stream. At the beginning, an instance $I_{q_0,0}$ such that $I_{q_0,0}$.Last $= 0$ is created in q_0. At period 1, there is an event $A \in S_1$, $r(A) = \{d_1\}, \mathcal{T}(d_1) = (\theta_{\text{normal}}, A, \infty, \infty)$, and $I_{q_0,0}$ satisfies the interval constraint $i - I_{q_0,0}$.Last $= 1 \leq \infty$. Then $I_{q_1,1}$ is copied from $I_{q_0,0}$ in q_0 to q_1 and $I_{q_1,1}$.Last $= 1$. At period 2, $A \in S_2$ again, $I_{q_0,0}$ satisfies $i - I_{q_0,0}$.Last $= 2 \leq \infty$, and $I_{q_0,0}$ does not have periods $I_{q_0,0}.P$ except for $I_{q_0,0}$.Last. Then $I_{q_1,1}$.Last $= 2$. At period 3, there is $B \in S_3$, $r(B) = \{d_2\}, \mathcal{T}(d_2) = (\theta_{\text{normal}}, B, 2, 2)$, and $I_{q_1,1}$ satisfies $i - I_{q_1,1}$.Last $= 1 \leq 2$, then $I_{q_2,3}$ is copied from $I_{q_1,2}$ in q_1 to q_2 and $I_{q_2,3}$.Last $= 3$. At period 4, $A \in S_4$, $I_{q_0,0}$ satisfies $i - I_{q_0,0}$.Last $= 4 \leq \infty$, then $I_{q_1,1}$.Last $= 4$. At period 5, $C \in S_5$, $r(C) = \{d_3\}, \mathcal{T}(d_3) = (\theta_{\text{normal}}, C, 1, 1)$, and $I_{q_2,3}$ satisfies the delete condition $i - I_{q_2,3}$.Last $= 2 > 1$, then $I_{q_2,3}$ is deleted. At period 6, there is $B \in S_6$, and $I_{q_1,1}$ satisfies $i - I_{q_1,1}$.Last $= 2 \leq 2$, then $I_{q_2,6}$ is copied from $I_{q_1,1}$ in q_1 to q_2 and $I_{q_1,1}$.Last $= 6$. At period 7, $C \in S_7$, and $I_{q_2,6}$ satisfies $i - I_{q_2,6}$.Last $= 1 \leq 1$,

```
procedure θ_save(d, e, c, t, a)            procedure θ_check(d, t, k₁, K)
  input arc d, event type e, interval con-    input arc d, time-out t, save ID for time-out k₁, set
  straint c, time-out t, save ID a           of save ID to check K
  if e ∈ S_i then                            for Updated I_x ∈ Tail(d).Buf[x] do
    for I ∈ Tail(d).Buf do                     for all combination ⟨I_x, ..., I_i^{y_i}, ...⟩ such that
      if i − I.Last ≤ c then                    i ≠ x, I_i^{y_i} ∈ d.Buf[i], ∀k ∈ K, I_x.P[k] = ··· =
        if ∃I^x ∈ d.Buf in Head(d) such        I_i^{y_i}.P[k] = ··· (overlooking I.P[k] = Null) do
    that ∀k, I.P[k] = I^x.P[k] then               Copy I as I′ into d.Buf ; I′.Last ← i
          I^x.Last ← i ; I^x.P[a] ← i            if I′.P[k] = Null then
        else                                       I′.P[k] ← I_y^z.P[k] such that I_y^z.P[k] ≠ Null
          Copy I as I′ into d.Buf               end if
          I′.Last ← i                         end for
        end if                              end for
      end if                              for I ∈ Tail(d).Buf do
      if i − I.Last > t then Delete I       if i − I.P[k₁] > t then Delete I
    end for                               end for
  end if                                 end procedure
end procedure
```

Fig. 7. Algorithms of operation sequences θ_{save} and θ_{check}

then $I_{q_3,7}$ is copied from $I_{q_2,6}$ in q_2 to q_3 and $I_{q_3,7}.\text{Last} = 7$, moreover, $I_{q_f,7}$ is copied from $I_{q_3,7}$ in q_3 to q_f, the period number of S_7, that is, 7 is reported as the period of matching, and $I_{q_f,7}$ is deleted.

Next, we show a PNFA of an interval-constrained episode $Y = (V, E, g, h)$. We consider an episode as a concatenation of some sequential episodes to treat an interval-constrained episode. Our strategy is to detect each divided sequential episode pattern and to check the identification of detected events associated with each branch tail or junction head. To check the identification, we introduce *save states* and *check states*. It holds event detected periods in the corresponding instance at the save state and finds all the combinations of instances such that the held periods are same at the check state. If each instance corresponding to the matching of each divided sequential episode has the same detected periods for each branch tail and junction head, it is guaranteed that there exists a matching for the episode. Even though each instance actually held a detected period of a different event of the same event type, if the different events have been detected as the event type of the branch tail or junction head at the same period, there exists another instance associated with the same event. Because two events with the same event type detected at the same period, if one of them satisfies interval constraints, the other also satisfies the interval constraints. Thus, to just check a detected period and an event type is sufficient to find the periods of matchings for an interval-constrained episode.

We define two operation sequences θ_{save} and θ_{check} in Fig. 7. Then we propose an algorithm for constructing PNFAs from interval-constrained episodes in Fig. 8. For instance, Fig. 9 shows a PNFA for the Z-type episode in Fig. 2.

A PNFA A for Y is processed as follows:

1. when an event $e \in \Sigma$ is detected, for each $d \in r(e)$, operate $\theta_d(d, e, \ldots)$ such that $\mathcal{T}(d) = (\theta_d, e, \ldots)$.

2. when an instance is created or updated in the time buffer of the tail of $d \in D$, operate $\theta_d(d, \ldots)$ such that $\mathcal{T}(d) = (\theta_d, \varepsilon, \ldots)$.

```
Input Y = (V, E, g, h)                              T(d) ← (θ_check, ε, t, v_s.k, K)
Output A = (Q, D, θ, T, q_0, F)                   end if
θ ← {θ_normal, θ_epsilon, θ_save, θ_check}        if v is source node then
Create q_0, q_sf, q_f ∈ Q ; F ← {q_f} ; l = 0       Create d = (q_0, v.f_I) ; T(d) ← (θ_epsilon, ε)
for v ∈ V do                                         Create d = (v.f_I, v.f_N) ∈ D
   Create q ∈ Q ; v.f_N ← q                          T(d) ← (θ_normal, g(v), ∞, ∞)
   if v is branch tail then                       else if v is sink node then
      Create q ∈ Q ; v.f_T ← q ; v.k ← l ; l++       Create d = (v.f_N, q_sf) ∈ D ; T(d) ← (θ_epsilon, ε)
   else if v is junction head then                end if
      Create q ∈ Q ; v.f_H ← q ; v.k ← l ; l++   end for
   end if                                         for e ∈ E do
   if v is source node then                          if e is branch arc then
      Create q ∈ Q ; v.f_I ← q                         Create d = (Tail(e).f_N, e.f_B) ; T(d) ← (θ_eplison, ε)
   end if                                               Create d = (e.f_B, Head(e).f_N) ∈ D
end for                                                  T(d) ← (θ_save, g(Head(e)), h(e), e.k)
for e ∈ E do                                         else if e is junction arc then
   if e is branch arc then                              Create d = (Tail(e).f_N, e.f_J)
      Create q ∈ Q ; e.f_B ← q                          T(d) ← (θ_save, g(Head(e)), h(e), e.k)
   else if e is junction arc then                       Create d = (e.f_J, Head(e).f_H) ∈ D ; T(d) ← (θ_eplison, ε)
      Create q ∈ Q ; e.f_J ← q                       else
   end if                                                Create d = (Tail(e).f_N, Head(e).f_N) ∈ D
end for                                                  T(d) ← (θ_normal, g(Head(e)), h(e))
for v ∈ V do                                         end if
   if v is junction head then                      end for
      Create d = (v.f_H, v.f_N) ∈ D                Create d = (q_sf, q_f) ∈ D
      t ← max_{p∈P} Σ h(v_x) (for v_s ∈ V_Bran. ∪  t ← max_{p∈P} Σ h(v_x) (for v_s ∈ V_Bran., v_e ∈ V_Sink, P = {p|p
      V_Junc., P = {p|p is path (q_0,                 is path (v_s, v_e)})
      v.f_N) and v.k = u.k}                         K ← {v.k|v ∈ V, v is branch tail }
      K ← {v.k|v, u on two different paths          T(d) ← (θ_check, ε, t, v_s.k, K)
```

Fig. 8. An algorithm for constructing a PNFA from an interval-constrained episode

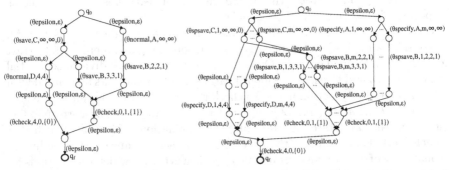

Fig. 9. An example of PNFA for the Z-type episode in Fig. 2

Fig. 10. An example of PNFA for the Z-connected sequential episode in Fig. 2

4.2   Stream-Constrained Episode Matching Detection Model

In this section, we present a processing model for stream-constrained episode detections extended from the interval-constrained one. The basic idea is quite simple. We first divide a stream-constrained episode into sub-episodes with same stream constraints. We prepare m-duplicated PNFAs for each sub-episode with the same stream constraints against m event streams, while we prepare one PNFA for each sub-episode without stream constraints. Each of the m PNFAs detects the matchings of the sub-episode against the corresponding event stream. Then, detecting the matchings on one of the PNFAs means finding the matchings of the sub-episode. Thus, we construct a PNFA for the original episode by connecting the sub-episode's PNFAs by arcs with $\theta_{epsilon}$. We can eliminate redundant ε-transitions from PNFAs like ordinary NFAs.

We define two operation sequences $\theta_{specify}$ and θ_{spsave} for stream-constrained episode matching detections.

```
Input Z = (V, E, g, h, z) , m = # of streams              for each q_j ∈ Q_i^j such that q_j is sink
Output PNFA A = (Q, D, θ, T, q_0, F)                      node on G_i^j or q_j.O = v.f_N, (v, u) ∈
Create PNFA A_s = (Q_s, D_s, θ_s, T_s, q_0, F) for Y = (V, E, g, h)    E, (v, u) ∉ E_i do
Q ← Q_s; D ← D_s                                             Create q_c ∈ Q
θ ← {θ_normal, θ_epsilon, θ_save, θ_check, θ_specify, θ_spsave}    for q such that q.O = q_j do
for each weakly connected subgraph G_i = (V_i, E_i) of G_s = (V, E_s)    Create d = (q, q_c) ∈ D
such that |V_i| ≥ 2, E_s = {e|e ∈ E, z(e) = same} do               T(d) ← (θ_epsilon, ε)
    Q_i^1 ← {v.f_N, v.f_T, v.f_H, e.f_B, e.f_J|v ∈ V_i, e ∈ E_i} ⊆ Q_s    end for
    D_i^1 ← {d|d ∈ D_s, Head(d), Tail(d) ∈ Q_i^1} ⊆ D_s           for d ∈ D such that Tail(d) = q_j do
    Duplicate G_i^1 = (Q_i^1, D_i^1)                               Reconnect arc as Tail(d) ← q_c
into G_i^2 = (Q_i^2, D_i^2), ..., G_i^m = (Q_i^m, D_i^m)          end for
    q_j.O ← original q ∈ Q_i^1 ; d_j.O ← original d ∈ D_i^1    end for
    Q = Q ∪ Q_i^2 ∪ ··· ∪ Q_i^m ; D = D ∪ D_i^2 ∪ ··· ∪ D_i^m    end for
    For d ∈ D_i^j do CHANGEOPERATION(d) end for               function CHANGEOPERATION(d)
    for each q_j ∈ Q_i^j such that q_j is source node on G_i^j or q_j.O =    if T_s(d.O) = (θ_normal, e, c, t) then
v.f_N, (u, v) ∈ E_i (u, v) ∉ E_i do                             T(d) ← (θ_specify, e, j, c, t)
        Duplicate each (q, q_j.O) ∈ D_s into d = (q, q_j) ∈ D    else if T_s(d.O) = (θ_save, e, c, t, a)
    CHANGEOPERATION(d)                                        then
    end for                                                      T(d) ← (θ_spsave, e, j, c, t, a)
                                                             else    T(d) ← T_s(d.O)
                                                             end if
                                                         end function
```

Fig. 11. An algorithm for constructing a PNFA from a stream-constrained episode

$\theta_{\text{specify}}(d, e, l, c, t)$ is defined by as well as $\theta_{\text{normal}}(d, e, c, t)$, but it is triggered by only $e \in S_{(i,l)}$ instead of S_i.

$\theta_{\text{spsave}}(d, e, l, c, t, a)$ is defined by as well as $\theta_{\text{save}}(d, e, c, t, a)$, but it is triggered by only $e \in S_{(i,l)}$ instead of S_i.

Then we propose an algorithm for constructing PNFAs from stream-constrained episodes in Fig. 11. For instance, Fig. 10 shows a PNFA for the Z-connected sequential episode in Fig. 2.

5   Experimental Results

We test EVIS on the Linear Road Benchmark Data [2] and real world taxi probe data, which is emitted by over 3,000 taxis belong to several taxi companies in Tokyo. We transformed the numerical data into parallel event streams in advance. The experiments were run on a PC (Intel Core i7 3.07 GHz quad core, Ubuntu 11.04 64 bit OS) with 6 GB of main memory.

We compared the performance of EVIS with a popular open source CEP engine Esper [8] on sequential (with length 2-4), binary branch (with length 2-4), binary junction (with length 2-4), Z-type, and Z-connected sequential episodes such as Fig. 2 against 1,000 cars of the benchmark data. Esper treats a SQL-like query language, while EVIS treats an episode pattern. Then we transformed episodes into SQL-like queries by using join operations against time-stamps and car IDs. Fig. 12 shows the throughputs of these patterns. The results show that EVIS performs over seventy times faster than Esper on simple patterns, and Esper could not finish its execution on some complex patterns because of out-of-memory (signed with X in Fig. 12). The throughput of EVIS becomes lower for longer branch and junction patterns, because we cannot apply the filtering technique for sequential patterns to these patterns.

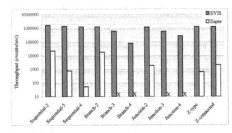 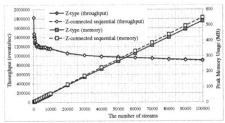

Fig. 12. Throughput of EVIS and Esper for various patterns on 1,000 streams

Fig. 13. Throughput and memory usage with varying the number of streams

Table 1. Throughput for a Z-connected sequential episode on taxi probe data

	EVIS	Esper
Throughput (events/sec)	958095	2854

Next, we examined the scalability of EVIS. Fig. 13 shows the throughputs and memory usages for two episodes in Fig. 2 with varying the number of streams (cars) from 300 to 100,000. The throughputs for the two episodes are about the same. From the results, we can conclude that EVIS runs fast and scales against the number of streams.

Finally, we verified usefulness about the throughput on real world data. Table 1 shows the throughput for the Z-connected sequential episode in Fig. 2 on the taxi probe data. This shows that EVIS runs fast on real world data as well as on artificial data. Then we can suppose that EVIS has a potential to detect a pattern on probes emitted from even 100,000 cars per 1 second.

6   Conclusion

This paper presented EVIS a fast and scalable episode matching engine for massively parallel data streams. EVIS efficiently detects an episode with interval constraints and stream constraints against parallel data streams. EVIS employs a variant of non-deterministic finite automata whose states are extended to maintain their activated times and activating streams. The experimental results showed that EVIS performs much faster than Esper for both artificial and real world datasets, and EVIS also effectively runs against complex patterns such that Esper cannot finish its execution. Furthermore, EVIS could process about nine hundreds of thousands of events per second on even over 100,000 streams. EVIS has the potential to lead our society in the coming age of cyber physical systems by detecting a lot of valuable correlations among tons of natural phenomena and human activities.

References

1. Agrawal, J., Diao, Y., Gyllstrom, D., Immerman, N.: Efficient pattern matching over event streams. In: SIGMOD, pp. 147–160 (2008)
2. Arasu, A., Cherniack, M., Galvez, E., Maier, D., Maskey, A.S., Ryvkina, E., Stonebraker, M., Tibbetts, R.: Linear road: a stream data management benchmark. In: VLDB, pp. 480–491 (2004)
3. Brenna, L., Demers, A., Gehrke, J., Hong, M., Ossher, J., Panda, B., Riedewald, M., Thatte, M., White, W.: Cayuga: a high-performance event processing engine. In: SIGMOD, pp. 1100–1102 (2007)
4. Carney, D., Çetintemel, U., Cherniack, M., Convey, C., Lee, S., Seidman, G., Stonebraker, M., Tatbul, N., Zdonik, S.: Monitoring streams: a new class of data management applications. In: VLDB, pp. 215–226 (2002)
5. Chakravarthy, S., Krishnaprasad, V., Anwar, E., Kim, S.K.: Composite events for active databases: Semantics, contexts and detection. In: VLDB, pp. 606–617 (1994)
6. Das, G., Fleischer, R., Gasieniec, L., Gunopulos, D., Kärkkäinen, J.: Episode Matching. In: Hein, J., Apostolico, A. (eds.) CPM 1997. LNCS, vol. 1264, pp. 12–27. Springer, Heidelberg (1997)
7. Dayal, U., Blaustein, B., Buchmann, A., Chakravarthy, U., Hsu, M., Ledin, R., McCarthy, D., Rosenthal, A., Sarin, S., Carey, M.J., Livny, M., Jauhari, R.: The hipac project: combining active databases and timing constraints. SIGMOD Rec. 17, 51–70 (1998)
8. Espertech, http://www.espertech.com/
9. Gehani, N.H., Jagadish, H.V.: Ode as an active database: Constraints and triggers. In: VLDB, pp. 327–336 (1991)
10. Katoh, T., Arimura, H., Hirata, K.: Mining Frequent k-Partite Episodes from Event Sequences. In: Nakakoji, K., Murakami, Y., McCready, E. (eds.) JSAI-isAI 2009. LNCS (LNAI), vol. 6284, pp. 331–344. Springer, Heidelberg (2010)
11. Lee, E.A.: Cyber physical systems: Design challenges. In: ISORC, pp. 363–369 (2008)
12. Mannila, H., Toivonen, H., Verkamo, A.I.: Discovery of frequent episodes in event sequences. Data Mining and Knowledge Discovery 1(3), 259–289 (1997)
13. Mei, Y., Madden, S.: Zstream: a cost-based query processor for adaptively detecting composite events. In: SIGMOD, pp. 193–206 (2009)
14. Tatti, N., Cule, B.: Mining closed episodes with simultaneous events. In: SIGKDD, pp. 1172–1180 (2011)
15. White, W., Riedewald, M., Gehrke, J., Demers, A.: What is "next" in event processing? In: PODS, pp. 263–272 (2007)

Real-Time Analysis of ECG Data Using Mobile Data Stream Management System

Seokjin Hong, Rana Prasad Sahu, M.R. Srikanth, Supriya Mandal,
Kyoung-Gu Woo, and Il-Pyung Park

Samsung Advanced Institute of Technology, South Korea
{s.jin.hong,rana.prasad,srikanth.mr,
mandal.s,kg.woo,ilpyung.park}@samsung.com
http://www.sait.samsung.co.kr

Abstract. Monitoring and analyzing electrocardiogram(ECG) signals
for the purpose of detecting cardiac arrhythmia is a challenging task,
and often requires a Complex Event Processing (CEP) system to analyze real-time streamed data. Various server-based CEP engines exist today. However, they have practical limitations to be used in environments
where network connectivity is poor and yet continuous real-time monitoring and analysis is critical. In this paper, we introduce a lightweight
mobile-based CEP engine called Mobile Data Stream Management System (MDSMS) that runs on the smart phone. MDSMS is built on an
extensible architecture with concepts such as lightweight scheduling and
efficient tuple representation. MDSMS enables developers to easily incorporate domain specific functionalities with User Defined Operator
(UDO) and User Defined Function (UDF). MDSMS also has other useful features, such as mechanisms for archiving streamed data in local
or remote data stores. We also show effectiveness of our MDSMS by
implementing a portable, continuous, and real-time cardiac arrhythmia
detection system based on the MDSMS. The system consists of ECG
sensor and a smart phone connected to each other via a wireless connec-
tion. MDSMS can detect and classify various arrhythmia conditions from
ECG streams by executing arrhythmia detection algorithms written in
Continuous Query Language.

Keywords: Stream Data Processing, Complex Event Processing, ECG,
Arrhythmia detection, Telemedicine, Mobile device.

1 Introduction

Real-time electronic detection systems of impending cardiac attacks could prove
life saver for mid-risk patients by giving warnings during the golden hour. A system of this nature, when put in place, presents physicians with an opportunity to
get details of the cardiac episodes as it happens which significantly improves the
accuracy of decision making while selecting appropriate treatment procedures.
With the advent of new technology in biomedical device design and a flurry
of publications seen in the recent times associated with efficient algorithms for

S.-g. Lee et al. (Eds.): DASFAA 2012, Part II, LNCS 7239, pp. 224–233, 2012.

cardiac signal processing, it is becoming increasingly essential to develop powerful systems that could aid in accurate real-time monitoring of cardiovascular diseases. Detection of such cardiovascular abnormal conditions is commonly enabled by analysis of electrocardiogram(ECG). ECG signal is an across the chest wall interpretation of electrical activity of the human heart over a period of time. The electrical activity is detected by electrodes attached to the surface of skin and recorded by a device external to the body. The conditions that can be detected by analyzing ECG signals range from minor, chronic, to life threatening in nature. A large and heterogeneous group of heart rate abnormalities are called as arrhythmia which can be broadly classified as un-sustained with a single irregular heartbeat or as sustained with an irregular group of heart beats. While many arrhythmias are less serious in nature, some are to be treated as life-threatening medical emergencies that may lead to stroke or sudden cardiac death. Various algorithmic methods have been discussed in literature for classification of arrhythmia based on analysis of ECG signals[1,2,3]. Recently we have witnessed a paradigm shift in the healthcare domain where the disease management strategy is moving from a reactive treatment approach to a predictive and preventive mode. This often necessitates continuous monitoring of bio signals without affecting the normal life of a patient. Ubiquitous nature of a smart phone with computing capabilities and an array of connectivity features proves to be a promising candidate for implementing such a system.

In this paper, we describe a software system that we have developed for real-time monitoring and analysis of ECG signals specifically aimed for a smart phone. A system of this nature should deal with, (a) Multiple data streams originating from varying data sources and their correlation, (b) Synchronizing event streams especially if they are delayed or at times arrive out of order and (c) Handling large volumes of data from input sources yet ensure low latency for real-time analysis. Considering all these points, a Complex Event Processing (CEP) system stands out to be a right choice for realizing an alert system based on real-time analysis of ECG signals. A CEP system typically deployed on a server is capable of handling streams from multiple sources, performing real-time analysis on large volumes of data with low latency and allowing application developers to express most of the complex logic using queries based on Continuous Query Language (CQL). Unlike the existing CEP systems that run on a server, the system that we have realized is a mobile based CEP engine called Mobile Data Stream Management System henceforth referred to as MDSMS.

The remainder of this paper is organized as follows. We provide background on healthcare monitoring systems and continuous query processing models in section 2. We describe architecture of portable arrhythmia detection system and its components in section 3. Architecture and key features of MDSMS are also described in section 4. Subsequently we discuss the experiment results for the arrhythmia detection application in section 5, and conclude the paper with future scope of our work in section 6.

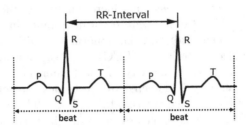

Fig. 1. ECG Signal

2   Preliminaries and Related Work

A typical ECG pattern represents a cyclic signal composed of a series of beats as shown in Fig. 1. Each beat is an outcome of polarization and depolarization of the heart muscles initiated by the sinus node. ECG signal has several kinds of prominent features. Some features are intra-beat features such as P, Q, R, S, and T which are medically meaningful points in a beat of the signal. There are also inter-beat features that are derived from intra-beat features such as RR-interval.

The process of detecting abnormal heart beat conditions known as arrhythmia begins with the extraction and annotation of several kinds of features from each beat of the ECG signal streams. Arrhythmic beat or episode can be classified using these ECG features. Various algorithms have been developed to detect arrhythmias from ECG features such as RR Interval based[1], time domain analysis[2] and frequency domain analysis[3].

There have been systems proposed in literature for real-time monitoring of ECG data[4], but these systems rely on transmitting the data to a networked server and perform analysis remotely. Such systems are prone to disruption of services in situations where network connectivity is poor or nonexistent. Applications performing such analysis locally face various challenges of handling high volumes of data, yet guaranteeing low latency for providing the results. CEP systems are most appropriate to analyze continuous streams of data. Such CEP systems[5][6] have been designed to run on server machines. In contrast, our system is designed with the goal to run on smart phones possible through a light-weight architecture and a small foot print.

3   Portable Arrhythmia Detection System

3.1   System Architecture

The system components consist of a single channel ECG sensor, preprocessor, MDSMS execution engine and a monitoring application as shown in Fig. 2. The ECG sensor is enabled by bluetooth connectivity that continuously monitors ECG signals while the subject goes about with his daily routine. Internally the

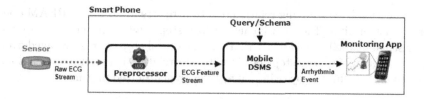

Fig. 2. System Architecture

preprocessor is responsible for noise removal from the sensor data and apply feature extraction techniques for accurate R-peak detection[7]. It then transforms the detected R-peak data into feature streams consisting of tuples with an associated timestamp called RR-tuples. Arrhythmia detection algorithms have been implemented as continuous queries and registered in MDSMS. MDSMS continuously executes these registered queries over ECG feature stream received from the preprocessor. Whenever an event satisfied with the query is detected, MDSMS notifies the event to the monitoring application.

3.2 Continuous Query Language for Arrhythmia Detection

Continuous queries are queries for data streams written using CQL, which are issued once and run continually over the database[8]. The CQL language of STREAM[9] which is a powerful SQL based approach for CEP systems has been adopted. Our system implements the semantics of CQL with added features like pattern matching, UDF and UDO.

Continuous queries act on either a stream or a relation. A stream is an unbounded collection of timestamped tuples associated with a schema similar to that of a schema for relational database. A relation is a finite set of tuples derived from streams but constrained by time or number of tuples within a given window, such relations are obtained by applying window operators on streams.

Life cycle of a typical stream begins by getting processed into a finite subset of tuples called as a relation using window operator which is a stream-to-relation operation. These relations are then subjected to relation-to-relation operators for further processing such as join, aggregate etc. Output from relational operators are then passed through other relational operators as input or as output stream to the applications.

We have implemented continuous queries for various arrhythmia conditions such as Bradycardia (BC), Sinus Tachycardia (STC), Ventricular Tachycardia (VTC) and Premature Ventricular Complexes (PVC) as a variant of algorithm developed by Tsipouras et.al.[1]. Given below are examples of continuous queries for Sinus Tachycardia and Premature Ventricular Complexes. *EcgStream* is the input data stream consisting of time varied tuples which consist of three RR-interval values (RR1, RR2, RR3) computed over four successive beats. They are passed through ROW window operator to split the stream into one tuple per relation and predicates are evaluated on the relation. The output is propagated to the application by converting relational output to stream using ISTREAM operator. Query for Sinus Tachycardia also utilizes step count from the accelerometer sensor stream called *AccelStream* that contains total step counts averaged every 5 seconds extrapolated to steps per minute. Correlation between *EcgStream* and *AccelStream* is achieved by using binary join synchronized on time.

Sinus Tachycardia

```
ISTREAM ( SELECT E.RR1, E.RR2, E.RR3
          FROM EcgStream [ROWS 1] AS E, AccelStream [ROWS 1] AS A
          WHERE (60.0/E.RR1 > 100.0) AND (60.0/E.RR1 <= 120.0)
          AND (A.StepsPerMinute < 250)
          AND (E.timestamp >= A.timestamp ))
```

Premature Ventricular Complexes

```
IStream ( SELECT * FROM EcgStream [ROWS 1]
          WHERE ( (1.15*RR2 < RR1) AND (1.15*RR2 < RR3) )
          OR ((((RR1-RR2 > 0)
            AND (RR1-RR2 < 0.3)) OR ((RR2-RR1 >= 0)
            AND (RR2-RR1 < 0.3))) AND ((RR1<0.8 )
            AND (RR2<0.8)) AND (RR3 > 1.2*((RR1+RR2)/2)))
          OR ((((RR2-RR3 > 0) AND (RR2-RR3 < 0.3))
                OR ((RR3-RR2 >= 0) AND (RR3-RR2 < 0.3)))
            AND ((RR2<0.8) AND (RR3<0.8))
            AND (RR1 > 1.2*((RR2+RR3)/2)))))
```

4   Mobile Data Stream Management System

In this section we present the architecture of MDSMS. It was designed primarily to be deployed on a mobile device. Such a system should be aware of the limited resources and provide capabilities of existing CEP systems. At the same time design of such a system should focus on providing low latency, high throughput stream processing while ensuring lower battery consumption and limited memory usage. Architecture of such a system should be lightweight and extensible in providing flexible feature set to the applications.

4.1   MDSMS Architecture

General architecture of the system is shown in Fig. 3. MDSMS is a 3-layered architecture which consists of MDSMS API, MDSMS core, and platform abstraction layers. Functionality of the core is exposed to the applications through an API layer. MDSMS exposes API to perform operations such as registration of queries, push input tuples and retrieve query results. Applications have the flexibility of expressing the logic by composing continuous queries. The core of

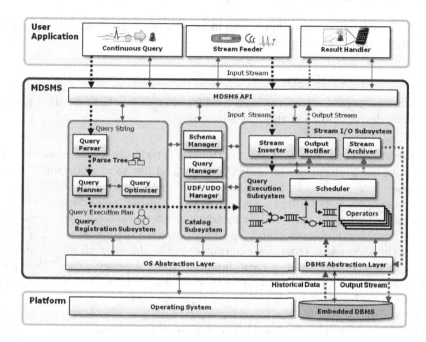

Fig. 3. MDSMS Architecture

MDSMS is comprised of query registration, catalog, stream IO, and query execution subsystems. A platform abstraction layer ensures portability of the system across various OS and DBMS.

When the application registers a query, it is parsed into a parse tree with syntactic and semantic verification, and the parse tree is then transformed into an execution plan graph. Each vertex of this plan graph is an executable operator which is scheduled independently by the scheduler. Every operator is associated with one or more input as well as output queues. Cataloging subsystem maintains the registered queries as forms of execution plan graph and meta-data related to all the schema in the system. A registered query can be used as a view in other queries that is, an output stream of a query can be used as an input stream to subsequent queries.

MDSMS provides extensibility features for applications to express domain specific functionality via User Defined Operators (UDO) and User Defined Functions (UDF) to be discussed in section 4.2. The catalog subsystem is also responsible for maintaining UDO, UDF meta data information. Query registration and execution subsystems uses UDO, UDF meta data from the repository subsystem. The Query execution subsystem that includes the scheduler is discussed in the subsequent sections. The IO subsystem is in charge of stream input, output, and archive. The stream inserter module receives stream input data and inserts it into MDSMS. Incoming stream data at times suffers from transmission anomalies causing the data to arrive out of order or at irregular intervals or data being delivered in bursts. The system provides heartbeat mechanism which guarantees that the stream will contain tuples with increasing timestamp and to ensure that the system generates timely output.

4.2   Features of MDSMS

MDSMS incorporates various system level features during the execution phase. We briefly describe some of these as following.

Optimization techniques. Several optimization techniques have been adopted to improve the performance of MDSMS. We would like to mention some of the techniques that resulted in performance boost in the streaming data context. Query plan optimization plays a major role in enhancing the performance of the system by reducing the number of operators and interactions between them. The planner optimizes the query plan by pushing down the predicates which reduces load on the successors. Operator's output queue is shared with all the successors input queues to minimize data duplication and memory usage. We follow tuple sharing across multiple operators to avoid duplication. This is attained by reference counting mechanism within the tuple structure.

Fig. 4. depicts plan for the following query with join over two input stream ranges of 30 seconds each. Selection and projection are combined along with binary join operator without extra selection and projection operators. In addition the output queue of the range window is used as the shared queue for the input queue of the binary join operator. It can reduce the number of operators and query execution latency.

```
SELECT S1.A, S2.D
FROM Stream1[RANGE 30 SECONDS] as S1,
     Stream2[RANGE 30 SECONDS] as S2
WHERE S1.B > 0 AND S1.A > S2.E
```

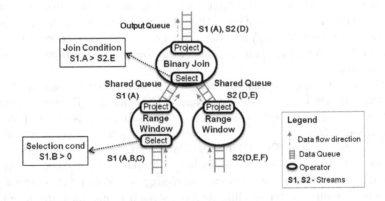

Fig. 4. Queue sharing mechanism

Lightweight Scheduler. Based on the lightweight architecture, MDSMS implements a single threaded execution engine which comprises query execution subsystem and result handler. The scheduler is based on round robin algorithm, where number of tuples to be processed for each job is computed via statistics based adaptive load aware scheduling strategy. If an operator has backlog of tuples to be processed, it pushes itself to the ready queue after each run. As an optimization for the scheduling strategy, each operator after its execution places its successor to the ready queue to improve query response times.

User Defined Components. MDSMS provides an extensible framework for applications by adding the functionality of the system through UDO and UDF. A UDO is a user defined stream-to-stream operator that consumes a stream and produces a stream. A UDF is a user defined function that takes scalar values across one or multiple tuples and returns a scalar value. These components are compiled into native libraries and linked with MDSMS which are seamlessly invoked at runtime from continuous queries. In our application we chose to implement feature extraction algorithms as UDO which accepts raw ECG data streams and produce RR-interval feature stream.

Local/Remote Persistent Store. MDSMS supports a feature to store and retrieve stream data using a persistent store. This is required to perform historical analysis as well as saving subset of the actual stream for future references. In our analysis with arrhythmia monitoring application, we have realized that we need to store arrhythmic episode data to be reviewed by a physician at any point of time. We have exposed this feature through CQL to support local and remote data stores to tackle limited memory capabilities on mobile phones.

5   Experimental Result

The system that we have developed has been extensively tested with ECG data from the MIT-BIH database[10]. It provides a total of 48 data sets, lasting for 30 minutes each. All the data sets contain full information of beat types reviewed and manually annotated by cardiac experts. From these we have chosen 30 data sets, which do not contain overlapping regions among various arrhythmias with the main purpose of accurately validating the existence of a particular arrhythmia. Performance of system was measured based on the doctor's annotation provided in the database. The system was demonstrated on a smart phone with a processor clock of 800 MHz and memory of 256 MB RAM with the embedded operating system as Microsoft Windows Mobile 6.5 Professional. Table 1 provides the accuracy of the system for various arrhythmia conditions.

ECG sensors operate on various sampling frequencies, hence the data was categorized into three sampling zones consisting of 60, 300 and 600 readings per second. Fig. 5. shows time delays associated with identification of arrhythmia conditions once the preprocessing of raw tuples is carried out. Arrhythmia

Table 1. Accuracy Measurements

Accuracy	No. of Data set	True Positive (%)	False Positive (%)	False Negative (%)
BC	30	98.8	0	1.20
STC	30	98.20	0	1.80
VTC	30	97.50	0	2.50
PVC	30	95.20	14.60	4.80

detection latency was measured as a difference between the output tuple emit time and the tuple's input timestamp. Fig. 6. presents memory usage of the system. Profiling data shows that overhead of running MDSMS in the context of real-time ECG detection system is acceptable to run in mobile devices.

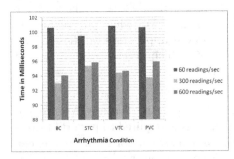
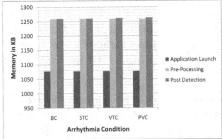

Fig. 5. Arrhythmia Detection Latency   **Fig. 6.** Memory footprint

6   Conclusion and Future Work

In this paper, we have presented mobile data stream management system that executes continuous queries on stream data in mobile devices such as smart phones. The key features of this MDSMS system are a lightweight architecture, extensible user defined modules and mechanisms for efficiently storing the streams. Based on MDSMS we have successfully demonstrated a portable arrhythmia detection system that is capable of real-time detection and analysis of four different arrhythmia conditions. We are also working towards enhancing the present system to a signal based MDSMS that provides specialized operators capable of handling raw sensor data streams there by extending its application to any bio-sensor. We believe that such as system when put in place in an ambulatory medical scenario will be an useful tool in the hands of physicians for the treatment of patients with cardiac and neurological disorders.

Acknowledgment. The authors would like to thank Prof. S Sudarshan from Dept. of Computer Science and Engineering, IIT Bombay for the discussion and valuable reviews of this work.

References

1. Tsipouras, M., Fotiadis, D., Sideris, D.: Arrhythmia classification using the RR-interval duration signal. Computers in Cardiology, 485–488 (2002)
2. Throne, R., Jenkins, J., DiCarlo, L.: A comparison of four new time-domain techniques for discriminating monomorphic ventricular tachycardia from sinus rhythm using ventricular waveform morphology. IEEE Transactions on Biomedical Engineering 38(6), 561–570 (1991)
3. Afonso, V., Tompkins, W., Nguyen, T., Luo, S.: ECG beat detection using filter banks. IEEE Transactions on Biomedical Engineering 46(2), 192–202 (1991)
4. Goni, A., Rodriguez, J., Burgos, A., Illarramendi, A., Dranca, L.: Real-time monitoring of mobile biological sensor data-streams: Architecture and cost model. In: Ninth International Conference on Mobile Data Management Workshops, pp. 97–105 (2008)
5. Abadi, D., Carney, D., Çetintemel, U., Cherniack, M., Convey, C., Lee, S., Stonebraker, M., Tatbul, N., Zdonik, S.: Aurora: A new model and architecture for data stream management. The VLDB Journal 12(2), 120–139 (2003)
6. Arasu, A., et al.: Stream: The stanford stream data manager. IEEE Data Engineering Bulletin 26(1), 19–26 (2003)
7. Bera, D., Bopardikar, A.S., Narayanan, R.: A robust algorithm for R-peak detection in an ECG waveform using local threshold computed over a sliding window. In: Third International Conference on Bioinformatics and Biomedical Technology, pp. 278–283 (2011)
8. Terry, D., Goldberg, D., Nichols, D., Oki, B.: Continuous queries over append-only databases. In: ACM SIGMOD International Conference on Management of Data, vol. 21(2), pp. 321–330 (1992)
9. Arasu, A., Babu, S., Widom, J.: The CQL continuous query language: semantic foundations and query execution. The VLDB Journal 15(2), 121–142 (2006)
10. MIT-BIH database distribution (1998),
http://www.physionet.org/physiobank/database/mitdb

A Main Memory Based Spatial DBMS: Kairos

Hyeok Han and Seong-il Jin

Realtimetech Inc., Suite 207 IT Venture Town,
694 Tamnip-Dong, Daejeon, Korea
{hhan,sijin}@realtimetech.co.kr

Abstract. The spatial database management system that supports the real-time processing for spatial data is emerging with the Location-based Services(LBS) that uses the location information as the key information in ubiquitous environment. Kairos Spatial is a spatial database management system which is based on the main-memory database technology. It supports both relational data model and spatial data model in the forms of the engine-level integration and supports an efficient spatial data management and real-time spatial data processing in GIS applications handling large amounts of spatial data. Using Kairos Spatial as the DBMS for the spatial data management will be able to get effect reducing the development time and improving the performance of spatial data processing.

In this paper, we present the introduction to spatial DBMS technology, Spatial DBMS features and functionality, short introduction to a Main memory-based Spatial DBMS and its case study.

Keywords: Main Memory DBMS, Spatial DBMS.

1 Introduction

As Location-based Services(LBS) for providing optimal resources and information services based on the geographical location is emerging with the development of mobile infrastructure and the growth of the customers demand for leisure. Many companies such as web portal company, mobile handset manufacturers and communication carriers have recognized the LBS as a key strategic element of the business and have entered into the LBS market.

The map search service for Navigation and POI search is already generalized on the web site such as Google and Yahoo. Its applicable range is gradually expanding into the mobile market like the Google's Map/Earth service using Apple's iPhone.

The essential technical components to enable LBS services are the location positioning technology and the GIS(Geographical Information System) related technology. The former is technology to identify the location of moving objects such as cars or mobile users by GPS, RFID and Wireless network. The latter is the set of information technology to store, manage and analyze the spatial data and attributes of entities(topography, streets, buildings, etc.) representing the geometric characteristics of the real-world.

S.-g. Lee et al. (Eds.): DASFAA 2012, Part II, LNCS 7239, pp. 234–242, 2012.

In the 1970s, GIS system was used as a utility for managing communication lines, water lines, transformers, and other public facilities. Spatial data and geographic information systems (GIS) had a professional systematic nature primarily used by a limited user and tailored to the special purpose application such as land information systems, urban information systems and traffic information systems.

As the increase of internet users and the development of web technologies have create a new environment which is easy to provide geographic information services to an unspecified number of users, GIS system deviated from the concept of professional GIS system is recently changing into the concept of integrated information system in order to accommodate the needs in a variety of user-layer.

In addition to traditional applications, GIS is used in a broad range of application domains such as Education, military, weather forecasts, sales analysis, demographic and land planning. As the GIS is changing into the service-oriented paradigm through convergence with communication technologies, the Spatial DBMS available for storing, managing and analyzing spatial data has been the subject of interest by the many GIS companies.

2   The Emergence Background of Spatial DBMS

In this paper, before the introduction to the Spatial DBMS, the historical transition process of the GIS application architecture will be discussed to understand the emergence background of Spatial DBMS

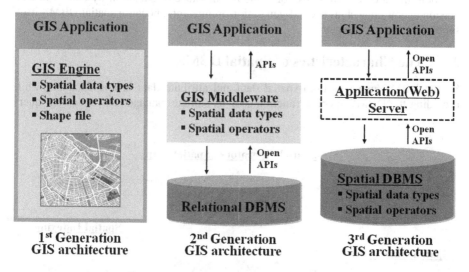

Fig. 1. Architecture evolution of GIS system

The architecture of first-generation GIS system stores spatial data in the file system and supports spatial data types, spatial indexes and spatial operators for the processing of spatial data in GIS engine. In this architecture, GIS application using non-standard API offered exclusively by the GIS engine must implement spatial data processing capabilities and store spatial data in the file system. So, it has the disadvantage that the sharing of the application data is difficult.

In the early 1990s, major GIS vendors such as IBM, ESRI and MapInfo developed second generation GIS system architecture in which stored spatial data in the Relational DBMS by using conventional database data types (BLOB, Integer) and served spatial data operations by a GIS middleware. Second-generation architecture with the introduction of RDBMS technology in the GIS system takes advantage of DBMS-specific functionalities such as concurrency control, data backup and recovery. But, the interface to develop GIS application provided by each middleware vendor can still not support an open architecture which is the pursuit of the recent enterprise solutions.

In addition, this architecture is inefficient in the case of the web GIS service due to the duality of this architecture, in which spatial operation is handled by GIS middleware and spatial data is stored in the RDBMS as a converted form of BLOB type.

As increasing needs of customers who need GIS applications not only taking full advantage of the DBMS benefits such as Data sharing, management, and security but also handling the high-speed spatial operations, interoperability between GIS solutions, and web-based geographic information services, the third-generation GIS system architecture managing spatial data and attribute data in one database and providing spatial indexes, spatial operators, and access of spatial data through standard SQL was emerged. The Spatial DBMS supports the storage and operation for spatial data in this third-generation GIS system architecture. In third generation of architecture, GIS applications using the Open API such as C, JDBC, OLEDB, and .NET Data Provider can use the features provided by Web Application Server and Spatial DBMS. In addition, this architecture does not cause performance degradation by data type conversion operation such as second-generation architecture, because spatial DBMS itself supports spatial data types

3   The Characteristics of Spatial DBMS

Spatial DBMS manages both spatial data and attribute data in one database, while providing the DBMS-specific function such as storage management, indexing, query

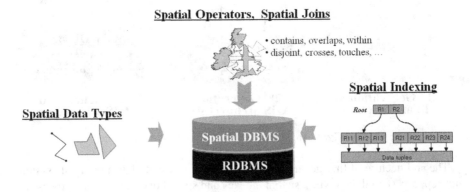

Fig. 2. Concept of Spatial DBMS

optimization, and user management capabilities and GIS engine data processing function such as the spatial data types, spatial operators, spatial indexing techniques and the spatial join operations. GIS Applications can achieve increasing performance and reducing system complexity and overhead of spatial data management by taking advantage of the convenient features of spatial data management such as SQL and ODBC databases via the standard programming interface provided by Spatial DBMS instead of middleware.

Spatial DBMS can be used for the following many applications.

Table 1. Application area of Spatial DBMS

Application area	Example
Location-based Living information	Interest information service associated with the user's current location, Address search, finding travel routes, POI search.
Telematics	Vehicle Information and management services, Navigation, weather/traffic information service.
Map-based web site	Travel maps service, geo-coding service, etc.
Geographic Information system	Land Use Management and flood control, population/geographic data management for disaster and rescue.
Public facilities management	Public Facilities network management like Utility poles, telephone lines, water supplying line
Personal information management	Increasing work efficiency from combining PIM(Personal Information Management) system to mange conferences, meetings, business trip and daily task with location information.

The Classification of Spatial DBMS. Spatial DBMS can be divided into Disk-based Spatial DBMS (Spatial DDBMS) and the Main Memory-based Spatial DBMS (Spatial MMDBMS) by the physical storage which loads the database. Spatial DDBMS has mainly released by major DBMS vendors on the market and MMDBMS companies such as Realtimetech Inc.[6], comes with Spatial MMDBMS packages reflecting change of service-oriented GIS systems technology flow. MMDBMS improves the response time and processing ratio for transaction by optimizing data and index structures, query processing algorithms and managing the entire database in memory.

The Main Memory Based Spatial DBMS. The benefits of Main Memory DBMS can be applied in Spatial DBMS. Disk-based DBMS has the fundamental problems such as slowing down of data processing speed due to the frequent disk I/O. Typically, the complexity of spatial queries on Spatial DBMS may be faced far more serious performance problems. In contrast, Spatial MMDBMS has the advantage of being able to provide performance of spatial query processing faster than Spatial DDBMS because the entire database is loaded into the system memory. Particularly, it shows the excellent spatial data processing performance in GIS applications using search oriented operations. But, in terms of the database size which can be loaded, Spatial DDBMS can be said to dominate rather than Spatial MMDBMS. However, as this restriction on the amount of memory has been improved by drop of main memory price and the generalization of 64bit computing, large database can be handled in MMDBMS[1].

4   Kairos Spatial : A Main Memory Based Spatial DBMS

Kairos Spatial was initially released in 2004 as the world's first commercial memory-resident Spatial DBMS with 2D spatial model implementation. It is based on main-memory database technology to provide high-speed transaction processing performance. It delivers core components of GIS specific functions as the DBMS engine functionalities. Recently, its functionalities were extended to the 3D spatial model and network data model in order to meet the requirement of various GIS markets. The spatial data in Kairos Spatial is based on 'Simple Feature Access' specification which is defined by OGC, an international GIS standard group[5].

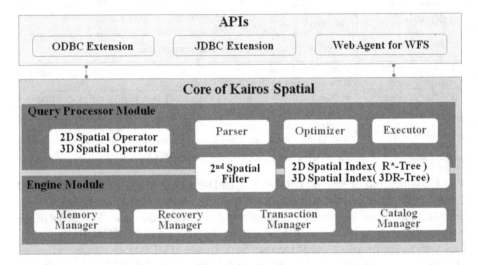

Fig. 3. Architecture of Kairos Spatial

As you can see in Fig.3, Kairos Spatial has the architecture of the form that is integrated with one DBMS in order to provide spatial data management facilities on underlying database engine modules. Thus, Kairos Spatial takes the advantage of being able to manage the spatial data and relational data in a single integrated database. Due to the use of main-memory database techniques and the support of the integrated service with the spatial data and relational data in one DBMS, Kairos Spatial is suitable not only for the conventional GIS application, but also for the ubiquitous application which needs the high speed spatial data process. Also, we can expect productivity improvement through integrating the spatial data model into the RDBMS components and reusing both the environment and experience which are used in RDBMS development.

From now, Let us observe Kairos Spatial functions and key features.

Spatial Data Model. As shown in Fig.4, Kairos Spatial provides spatial data model including 3- dimensional spatial objects that conform to OGC Simple Feature specification. This specification supports data formats such as Well Known Binary (WKB) and Well Known Text (WKT) which is commonly used in industry in order to represent real-world spatial objects having geometric features such as rivers, roads and buildings. Also, it can be used for modeling spatial objects.

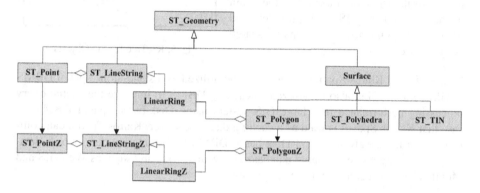

Fig. 4. Data model of Kairos Spatial

Spatial Operators. Kairos Spatial can evaluate the correlation between the various spatial objects(including 3D spatial Objects[4]) by providing spatial relational functions based on the OGC 9-Intersection model such as Contains, Intersects and Overlaps etc. In addition, it provides various spatial analytical functions conforming OGC standard such as Distance, Area, Length, Convex-hull and Volume etc. Table 2 shows Kairos spatial operators based on the conformance of standard specifications.

Table 2. Spatial Operators in Kairos Spatial

OGC standatd functions			Non-standard extended spatial functions		
AREA	GEOMFROMTEXT	NUMGEOMETRIES		ST_ADDPOINT	ST_ADDPOINTTOLS
ASBINARY	GEOMFROMWKB	NUMINTERIORRING		ST_ADDPOINTTOMP	ST_ADDPOINTTOPG
ASTEXT	INTERIORRINGN	NUMPOINTS		ST_DUPPOINT	ST_MOVE
BOUNDARY	INTERSECTION	OVERLAPS	Edit	ST_REPLACEPOINT	ST_REVERSE
BUFFER	INTERSECTS	POINTFROMTEXT		ST_RMDUPPOINTS	ST_RMPOINT
CENTROID	ISCLOSED	POINTFROMWKB		ST_RMSAMESLOPE	ST_SELFINTERSECTION
CONTAINS	ISEMPTY	POINTN		ST_TOPOINT	ST_MERGE
CONVEXHULL	ISRING	POINTONSURFACE		ST_IN	ST_INGEOM
CROSSES	ISSIMPLE	POLYFROMTEXT		ST_ISCLOCKWISE	ST_LARGEPOLYGON
DIFFERENCE	LENGTH	POLYFROMWKB		ST_MINX	ST_MINY
DIMENSION	LINEFROMTEXT	RELATE	Analyze	ST_MAXX	ST_MAXY
DISJOINT	LINEFROMWKB	SRID		ST_MBRINTERSECTS	ST_MBRWITHIN
DISTANCE	MLINEFROMTEXT	STARTPOINT		ST_MBRCONTAINS	ST_OUT
ENDPOINT	MLINEFROMWKB	SYMDIFFERENCE		ST_PERIMETER	ST_RECTFROMTEXT
ENVELOPE	MOVE	TOUCHES		ST_RECTFROMWKB	
EQUALS	MPOINTFROMTEXT	UNION	LBS	ST_LEFTBUFFER	ST_RIGHTBUFFER
EXTERIORRING	MPOINTFROMWKB	WITHIN		ST_WITHINDISTANCE	ST_NEAREST
GEOMETRYN	MPOLYFROMTEXT	X		ST_ASGML	ST_GEOCODE
GEOMETRYTYPE	MPOLYFROMWKB	Y	WEB/Open API	ST_GEOCODEASGEOM	ST_GEOSTRTOTEXT
				ST_GETFULLADDRESS	ST_REVERSEGEOCODE

Spatial Data Access. As shown in Fig.5, Kairos Spatial provides R*Tree spatial index[2] using the information of the minimum bounding rectangles (MBR) of spatial objects in order to optimize the spatial query performance. In addition, it provides spatial join algorithms[3] optimized for the main memory based DBMS and provides 2nd filters for the enhancement of spatial search performance.

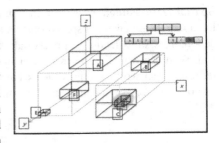

Fig. 5. R*-Tree for 3D Objects

Usage Model. Kairos Spatial can not only be utilized as a stand-alone spatial DBMS in GIS solutions, but also be utilized as caching DBMS to improve the spatial query processing performance in a large GIS system using Disk-based spatial DBMS. In case that Kairos Spatial is used as a caching database server, Kairos Spatial can compensate for the disadvantages of Disk-based DBMS by loading hot data into its database and having a role of spatial database server of service area such as Web GIS and Mobile GIS that requires real-time processing

5   Case Study : u-Statistical System of Statistics Korea

In this chapter, the applied case of Kairos Spatial in u-Statistical System of Statistics Korea will be introduced.[7]

System Overview. In April 2007, Statistics Korea launched the infrastructure project of u-Statistics service to accommodate the need for spatial statistic service to promote national policies and for the production of sub-regional statistics in order to help policy and decision-making with highlighting the importance of the spatial statistical services.

The u-Statistical system software is divided into the service area and data management area. Data management area is a part of u-Statistics system creating and managing u-Statistics database. It is implemented by Oracle 10g RDBMS and ESRI's ArcGIS solution in order to edit and manage the GIS data.

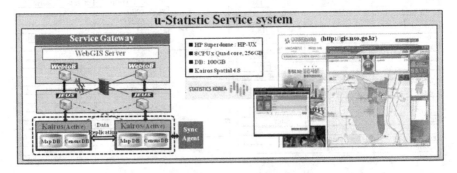

Fig. 6. u-Statistic service system architecture

Service area provides the portal service of real-time statistics to public and end users through the Web using u-Statistic database. As shown in Fig.6, spatial MMDBMS(Kairos Spatial) and web server technology is applied in order to implement this area. The spatial data and Census data of u-Statistics database in Oracle is exported as shape file and then imported to Kairos Spatial using the utility which is provided by Kairos Spatial.

The Description of Problem and Solution. Performance problem arose while processing the editing transaction on the large u-Statistics database and providing web portal service of high-quality real-time statistics with only single Disk-based DBMS simultaneously.

Kairos Spatial was adopted as the solution to this problem. They utilized Kairos Spatial as the cashing server of Oracle in order to be dedicated to process the real-time statistical search from web users. As a result of this approach, the workload was distributed by separating the spatial data and census data from the entire u-Statistics database.

The Effect of Introduction. As an example, the response time of the statistical theme query shortened to less than 2 seconds while the conventional service took an hour or more in the worst case. This approach enabled to provide the portal service for real-time statistics. Furthermore, it contributed to improving Citizen Service of Statistical Office and to obtaining economic benefits by improving the ROI.

6   Conclusion

In this paper, we presented the introduction to spatial DBMS technology, Spatial DBMS features and functionality, short introduction to a Main memory-based Spatial DBMS and its case study.

LBS that provide a variety of services with location and geo-information breathe new life into the business around the world. Also, Spatial DBMS, especially main memory based spatial DBMS that provide high-speed processing for the spatial data, is expected to increase the utilization in LBS and GIS market.

References

1. Devitt, D.J., et al.: Implementation techniques for main memory data systems. In: Proc. ACM SIGMOD Conf. (1984)
2. Beckmann, N., Kriegel, H.-P., Schneider, R., Seeger, B.: The R*-tree: an efficient and robust access method for points and rectangles. In: ACM SIGMOD Conf., pp. 322–331 (1990)

3. Brinkhoff, T., Kriegel, H.P., Seeger, B.: Efficient Processing of Spatial Joins Using R-Trees. In: Proc. ACM SIGMOD Conf., pp. 237–246 (1993)
4. Zlatanova, S.: 3D geometries in DBMS. In: Proc.: Innovations in 3D GeoInformation Systems, Kualla Lumpur, Malaysia, pp. 1–14 (2006)
5. OpenGIS Consortium, OpenGIS Implementation Specification for Geographic information – Simple feature access – Part 2: SQL option, Revision 1.2.1 (2010)
6. Realtimetech Inc., http://www.realtimetech.co.kr
7. SGIS of Statistics Korea, http://sgis.kostat.go.kr

Study on the International Standardization for the Semantic Metadata Mapping Procedure

Sungjoon Lim[1], Taesul Seo[2], Changhan Lee[1], and Soungsoo Shin[1]

[1] Korea Database Agency, 9th Floor, KNTO Bldg.,
10 Da-dong, Jung-gu, Seoul 100-180, Korea
{joon,leech,kolatree}@kdb.or.kr
[2] Korea Institute of Science and Technology Information,
66 Hoegi-ro, Dondaemun-gu, Seoul, 130-741 Korea
tsseo@kisti.re.kr

Abstract. This paper introduces a semantic metadata mapping procedure which is able to maximize the interoperability among metadata. The methodology consists of three processes such as identifying metadata element sets, grouping data elements, and mapping semantically. And also this paper shows applicable example for the e-book cataloging domain. The methodology in this paper is recommended for use in a specific subject domain because the procedure can be more meaningful when a specific information object is concerned.

Keywords: metadata, metadata registry, semantic interoperability, international standard, ISO.

1 Introduction

Usually, two systems do not share the same model, and that is because the categories represented in the models were not factored in the same way. This situation inhibits interoperability. For example, there may be two or more metadata element sets applicable to an information object. For example, metadata such as DC (Dublin Core), MARC (MAchine Readable Cataloguing), and MODS (Metadata Object Description Schema) can be used to describe a book. Thus, a data element for an information object may be differently named due to the preferences of individual database developers. Consequently, data exchange among databases becomes difficult or almost impossible.

ISO/IEC 11179[1] provides a framework for achieving semantic interoperability of metadata between systems. A metadata registry based on ISO/IEC 11179 offers a good way to secure interoperability among databases. However, there are many metadata sets which are not following ISO/IEC 11179. In order to mediate among plural data elements already developed or used, other measures are necessary. In general, interoperability may be achieved through conformity to some set of provisions. The basic concept of ISO/IEC 11179 can still be applicable to the

---

[1] An international standard for representing metadata for an organization in a metadata registry.

S.-g. Lee et al. (Eds.): DASFAA 2012, Part II, LNCS 7239, pp. 243–249, 2012.

improvement of semantic metadata crosswalk because it addresses the semantics of metadata and naming principles for data elements.

This paper describes a semantic metadata mapping procedure (SMMP) based on ISO/IEC 11179, which can maximize the interoperability among data elements. The procedure consists of three main processes: identifying metadata element sets, grouping data elements, and semantic mapping.

2   Previous Studies for Metadata Interoperability

Metadata crosswalk [1] is the most commonly used way to map a data element to another data element. However, the metadata crosswalk has poor semantics; it provides a simple one-to-one mapping table among data elements without any explanation about the semantic relationship. Therefore, the metadata crosswalk needs to be elaborated in order to give semantics and to cover cases other than one-to-one mapping.

Some other approaches have been tried to provide guidelines or a model for harmonization of metadata and data in especially transport industry. Piprani, B. [2] suggests a model for semantic mapping of master data harmonizing. This master data harmonizing is including rationale for the need of semantic metadata mapping in general.

3   Semantic Metadata Mapping Procedure

3.1   General

In this paper, the procedure for data element mapping consists of three main processes as shown in Fig. 1.

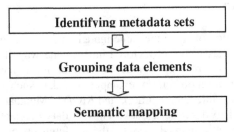

Fig. 1. Procedure for data element mapping

The first process is to identify metadata element sets required to be mapped. It is necessary to survey available metadata element sets (in a specific domain).

The second process is to group data elements obtained from the identified metadata element sets, including four consecutive sub-processes namely, finding objects, grouping all data elements by object, finding their properties, and grouping all data elements by property.

The last process involves mapping data elements semantically. In this process, it is necessary to arrange all data elements into a table. Notes on the accuracy of matching are included in every slot of the table. A recommended set of metadata can also be provided in the process for guiding future standardization.

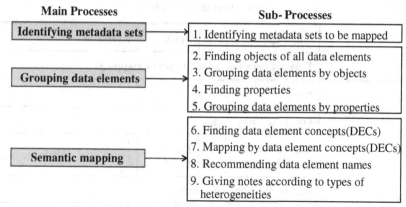

Fig. 2. Main and sub-processes for data element mapping

Fig. 2. shows all sub-processes related to corresponding main processes.

3.2   Identifying Metadata Element Sets

First, it is necessary to collect available metadata element sets and to identify candidate metadata element sets to be mapped. Then, what the domain or service database is should be checked, how many numbers of fields should be counted, and whether sample data exists or not should be checked. Who or which organization has the authority over each metadata element set should also be checked.

If the metadata element set or target object is not suitable for mapping, it may not be chosen.

3.3   Grouping Data Elements

The next process is to group data elements by object class, and then to find properties involved in the objects and sub-grouped data elements by properties. For convenience, it is helpful to select a primary data element among the collected data elements and aggregate data elements by the primary data element. The simplest or the highest-level data element is recommended to be the primary data element.

All data elements included in the candidate metadata element sets should be aggregated by property. Data elements relatively less important may be eliminated. Some data elements, which cannot be grouped, are supposed to be set aside.

In this process, metadata experts should perform the work along with domain experts.

3.4   Semantic Mapping

After identifying object classes and properties hidden in and related to all the primary data elements, we can create common Data Element Concepts (DECs) according to ISO/IEC 11179-1:2004[2] and ISO/IEC 11179-3:2003[3].

---

[2] Information technology -- Metadata registries (MDR) -- Part 1: Framework.

[3] Information technology -- Metadata registries (MDR) -- Part 3: Registry metamodel and basic attributes.

Table 1. Types of heterogeneity

Type	Sub-Type	Mark	Examples
	Ways of harmonization (Types of mapping)		
Same	Same		
	One-to-one mapping		
Hierarchical	Generalization Specialization	H/gen H/spe	Price Retail price, Wholesale price, …
	One-to-one mapping (dumb down)		
	Composition Decomposition	H/com H/dec	Name Family name, given name, …
	One-to-many or many-to-one mapping (if required)		
Domain	Domain	D	Summary : Synopsis
	One-to-one mapping (if required)		
Lexical	Synonyms	L/syn	First name : Given name
	Abbreviation	L/abb	Address : Addr.
	Acronyms	L/acr	Serial Number :SN
	Case sensitivity	L/cas	Address : ADDRESS
	Language	L/lan	Name : 이름
	Variation	L/var	Color : Colour
	One-to-one mapping		
Syntactic	Ordering	S/ord	Family name : Name (family)
	Delimiters	S/del	Family-name : Family_name
	Missing	S/mis	Author name : Author
	One-to-one mapping		
Complicated	Complicated	C	
	Mapping is impossible.		

The third process starts from finding common data element concepts in each group of data elements based on objects and properties found in the second process. If the domain ontology or taxonomy is known, it will be very helpful to construct common DECs.

Finally, all candidate data elements are arranged into a table by the common DECs. Types of heterogeneity can be described in the table. The types consist of six categories. Complicated difference is the type of heterogeneity that cannot be solved without human intervention. A recommended data element can be provided for the future standardization work.

4   Example for SMMP

This example shows mapping procedure using SMMP for the e-book cataloging domain. In the e-book cataloging domain, three available metadata element set are identified, such as OEBPS [3], MODS [4] and TEI [5].

- Domain: e-book cataloging
- Available metadata element sets: OEBPS, MODS and TEI

Table 2. Analysing available metadata element sets

Metadata element set name	OEBPS	MODS	TEI header
Domain or service database	Description of electronic book	Description of library resources	Encoding methods for machine-readable texts
Number of fields	15	About 60 (top level: 20)	Over 20
Sample data	Yes	No	Yes
Authority	Open eBook Forum	Library of Congress	TEI Consortium

In the sample object class, the e-Book has plural properties as shown below.

- Object class: e-Book
- Properties: title, author, subject, ..., edition

Similar properties of MODS and TEI are grouped according to those of the primary data elements from OEBPS. In the table, the italicized parts refer to properties considered less important in the target application domain.

The data element concepts found can be shown as follow:

- DECs: ebookTitle, ebookAuthor, ebookSubject, ..., ebookEdition

Table 3. Example of grouping data elements by property

OEBPS (primary data elements)	MODS	TEI
Title	Title subTitle *partNumber* *partName* *nonSort*	title seriesStmt:title *seriesStmt:idno*
Creator(role) Creator(file-as)	name:role name:namePart *name:displayForm* *name:affiliation* *name:discription*	author
Subject	Topic classification *catographics* *occupation*	keyword classCode *catRef*
...	...	...
(no data element)	Edition	*fileDesc_editionStmt_date* fileDesc_editionStmt_edition *fileDesc_editionStmt_respStmt* *fileDesc_editionStmt_respStmt_name* *fileDesc_editionStmt_respStmt_resp*

Table 4. Semantic mapping of metadata

Common DEC	OEBPS		MODS		TEI		Recommended DE[4]
ebookTitle	Title		Title	H/dec	Title	H/dec	ebookTitle
			Subtitle	H/dec	seriesStmt:title	H/dec S/del	
ebookAuthor	Creator(role)	C	name:role	C	Author	S/mis	ebookAuthorName
	Creator(file-as)	D	name:namePart	D			
ebookSubject	Subject		Topic	L/syn	Keyword	L/syn	ebookSubject
			Classification	L/syn	Class	L/syn	
ebookEdition			Edition	L/cas S/mis	Edition	L/cas S/mis	ebookEditionNumber

---

[4] Recommended DE (Data Element) is provided according to the DE naming rule from ISO/IEC 11179-5:2005 Information technology -- Metadata registries (MDR) -- Part 5: Naming and identification principles.

Finally, we can create DECs according to ISO/IEC 11179-1. New DECs should be also created for data elements that could not be grouped during the second process.

Table 4 shows the final result obtained through the procedure. The common DECs are described in the first column while the recommended data elements are done in the right end column. Between them are data elements from the candidate metadata element sets.

5   Conclusions

Data elements having different names even though they have the same meanings may cause a data discrepancy problem when such data are shared or interchanged. Thus, semantic metadata mapping is required to mediate among these data elements to allow sharing or interoperable use. A metadata crosswalk is the most commonly used method to map a data element to another data element. However, it has poor semantics because it is meaningful only for simple one-to-one mapping. Therefore, the metadata crosswalk needs to be elaborated in order to have semantics and to cover cases other than one-to-one mapping.

This paper describes a semantic metadata mapping procedure (SMMP), which can maximize the interoperability among data elements. The procedure of data element mapping consists of three main processes, which are divided into nine sub-processes. The main processes are identifying metadata element sets, grouping data elements, and semantic mapping. This paper also includes a simple example to explain each process.

This paper is recommended for use in a specific subject domain because the procedure can be more meaningful when a specific information object is concerned.

References

1. Zeng, M.L., Chan, L.M.: Metadata Interoperability and Standardization – A Study of Methodology Part I. D-Lib Magazine 12(6) (2006)
2. Piprani, B.: A Model for Semantic Equivalence Discovery for Harmonizing Master Data. In: Meersman, R., Herrero, P., Dillon, T. (eds.) OTM 2009 Workshops. LNCS, vol. 5872, pp. 649–658. Springer, Heidelberg (2009)
3. Open eBook Publication Structure, http://old.idpf.org/oebps/oebps1.2/index.html
4. Metadata Object Description Schema, http://www.loc.gov/standards/mods
5. Text Encoding Initiative, http://www.tei-c.org/index.xml

Semantics and Usage Statistics
for Multi-dimensional Query Expansion

Raphaël Thollot[1], Nicolas Kuchmann-Beauger[1], and Marie-Aude Aufaure[2]

[1] SAP Research - BI Practice, 157 rue Anatole France, 92309 Levallois-Perret, France
raphael.thollot@sap.com, nicolas.kuchmann-beauger@sap.com
[2] Ecole Centrale Paris, MAS laboratory, 92290 Chatenay-Malabry, France
marie-aude.aufaure@ecp.fr

Abstract. As the amount and complexity of data keep increasing in data warehouses, their exploration for analytical purposes may be hindered. Recommender systems have grown very popular on the Web with sites like Amazon, Netflix, etc. These systems proved successful to help users explore available content related to what they are currently looking at. Recent systems consider the use of recommendation techniques to suggest data warehouse queries and help an analyst pursue its exploration. In this paper, we present a personalized query expansion component which suggests measures and dimensions to iteratively build consistent queries over a data warehouse. Our approach leverages (a) semantics defined in multi-dimensional domain models, (b) collaborative usage statistics derived from existing repositories of Business Intelligence documents like dashboards and reports and (c) preferences defined in a user profile. We finally present results obtained with a prototype implementation of an interactive query designer.

Keywords: query, expansion, recommendation, multi-dimensional, OLAP.

1 Introduction

Data warehouses (DW) are designed to integrate and prepare data from production systems to be analyzed with Business Intelligence (BI) tools. End-users can navigate through and analyze large amounts of data thanks to a significant effort from IT and domain experts to first model domains of interests. However, exploiting these multi-dimensional models may become challenging in important deployments of production systems. Indeed, they can grow extremely complex with thousands of BI entities, measures and dimensions used, e.g., to build OLAP cubes.

In common reporting and analysis tools, end-users can design data queries using some kind of query panel. For instance, a user may drag and drop measures and dimensions she wants to use to create a new visualization, e.g., showing the Sales revenue aggregated by City. Given the number of available measures and dimensions, it is crucial to help the user iteratively build her query. We

S.-g. Lee et al. (Eds.): DASFAA 2012, Part II, LNCS 7239, pp. 250–260, 2012.

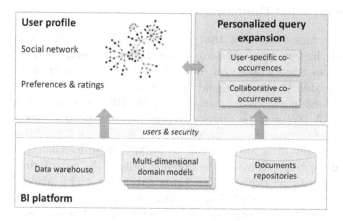

Fig. 1. Architecture overview of the proposed personalized query expansion system for multi-dimensional models

define the iterative query expansion problem as a function QE taking as input a user u, the current query q and additional parameters $params$. This function returns a collection of scored queries (q_i, r_i) such that, for all i from 1 to n, $|q_i| = |q| + 1$ and $q \subset q_i$:

$$QE \colon (u, q, params) \mapsto \{(q_1, r_1), \ldots, (q_n, r_n)\}$$

The problem is thus to find candidate measures and dimensions that are best associated with q. In response to this, this paper presents our solution which was experimented by building an interactive and personalized query designer. Our method leverages semantics of multi-dimensional models, collaborative usage statistics derived from repositories of BI documents and user preferences to iteratively suggest relevant measures and dimensions. Figure 1 illustrates the main components involved in the architecture of our system.

In the rest of this article, Section 2 introduces multi-dimensional domain models and their semantics. Section 3 presents a collaborative measure of co-occurrence between entities of these models. Then, Section 4 introduces users' preferences and our personalized query expansion component. Section 5 describes the system architecture and presents results obtained with the implementation of an interactive query designer. Finally, we review related work in Section 6.

2   Semantics of Multi-dimensional Domain Models

Multi-dimensional models for DWs define concepts of the business domain with key indicators (*measures*) and axis of analysis (*dimensions*).

2.1   Measures and Dimensions

Measures are numerical facts that can be aggregated against various dimensions [3]. For instance, the measure `Sales revenue` could be aggregated (e.g., from unit sales transactions) on dimensions `Country` and `Product` to get the revenue generated by products sold in different countries.

Business domain models are used to query the DW for actual data and perform calculations. A DW may be materialized as a relational database, and queries thus have to be expressed accordingly, for instance as SQL. From a calculation point of view, it is also possible to build multi-dimensional OLAP cubes. Measures are aggregated inside the different cells of the cube formed by dimensions. Queries can be expressed on these cubes, e.g., with Multi-Dimensional eXpressions (MDX). Beyond this, modern DWs provide SQL/MDX generation algorithms to enable non-expert users to formulate ad-hoc queries.

In the next section, we present hierarchies and functional dependencies (between measures and dimensions) that multi-dimensional domain models may also define.

2.2   Functional Dependencies and Hierarchies

Two objects (measures or dimensions) are functionally dependant if one *determines* the other. For instance, knowing the `City` determines the related `State`. Another example that involves a measure and a dimension is to say that the `Sales revenue` is determined by a `Customer` (e.g., aggregated from unit sales in a fact table). Functional dependencies are transitive: if `A` determines `B` which determines `C`, then `A` determines `C`. In the most simple scenario, all measures are determined by all dimensions. This is the case when using a basic dataset, for instance reduced to one fact table with dimensions in a *star schema*.

Functional dependencies are important to compose meaningful queries. For instance, they can be used to ensure suggested queries do not contain incompatible objects which would prevent their execution. However, business domain models do not necessarily capture and expose this information. Hierarchies of dimensions are more common though, usually exploited in reporting and analysis tools to enable the *drill-down* operation. For instance, if a `Year` - `Quarter` hierarchy is defined, the result of a user drilling down on `Year` 2010 is a more fine-grained query with the `Quarter` dimension, filtered on `Year` = 2010. If hierarchies of dimensions can be used to determine minimal dependency chains, techniques are required to help with automatic detection of functional dependencies. In particular, the approach presented by [12] is to create *DL-Lite* domain ontologies from conceptual schemas and use inferencing capabilities.

3   Usage Statistics in BI Documents

Functional dependencies and hierarchies previously presented provide very structural knowledge regarding associations between BI entities. Beyond this, some

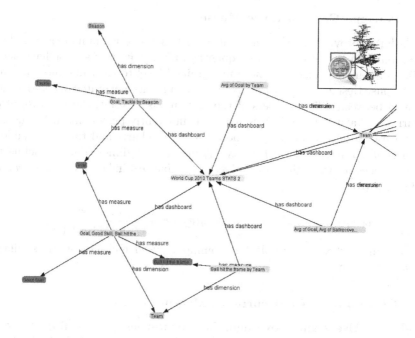

Fig. 2. Graph describing a dashboard (orange) and associated charts (blue), with referenced measures (purple) and dimensions (yellow)

BI platforms propose repositories of documents like reports or dashboards which can be used to compute actual usage statistics for measures and dimensions. This kind of information is extremely valuable in our use case, since query expansion implies to find the best candidate to associate to a given set of measures and dimensions.

3.1 Structure of BI Documents and Co-occurrence

We use the structure of BI documents to define co-occurrences between measures and dimensions. For instance, BI reports are roughly composed of sections which may contain charts, tables, text areas for comments, etc. Charts and tables define important units of sense. Measures and dimensions associated in a same table/chart are likely to be strongly related and represent an analysis of specific interest to the user. Similarly, dashboards can be composed of different pages or views which contain charts and tables. Figure 2 illustrates the graph representation of the dashboard *World Cup Team STATS 2* and its associated charts. More generally, any BI document referencing measures and dimensions could be used to derive consolidated co-occurrences or usage statistics.

3.2   Personal Co-occurrence Measure

BI platforms provide access control rules for business domain models and documents built on top of them. Consequently, different users may not have access to the same models and at a more fine-grained level to the same measures and dimensions. Besides, repositories contain documents generated by and shared (or not) between different users of the system. As a result, the measure of co-occurrence that we define in this section is inherently personalized. Let us consider a user u and let $occ_u(e_1)$ denote the set of charts and tables – visible to the user u – referencing a BI entity e_1, measure or dimension. We define the co-occurrence of two entities e_1 and e_2 as the Jaccard index of the sets $occ_u(e_1)$ and $occ_u(e_2)$:

$$cooc_u(e_1, e_2) = J(occ_u(e_1), occ_u(e_2)) = \frac{|occ_u(e_1) \cap occ_u(e_2)|}{|occ_u(e_1) \cup occ_u(e_2)|} \qquad (1)$$

The Jaccard index is a simple but commonly used measure of the similarity between two sample sets.

3.3   Collaborative Co-occurrence Measure

Cold-start Users and Coverage. In recommender systems (RS), the *coverage* is the percentage of items that can actually be recommended, similar to the recall in information retrieval. Formula 1 presents a problem for *cold-start* users, i.e. those new to the system. Indeed, these users do not have stored documents from which co-occurrences can be computed. Collaborative RS introduce the contribution of other users in the item scoring function to improve the system's coverage and enable the exploration of resources previously unknown (or unused) by the user. A simple approach consists in using a linear combination of the user-specific value and the average over the set of all users.

Using the Social/Trust Network. The simple approach previously described broadens the collaborative contribution to "the whole world" and all users have the same weight. Trust-based RS have illustrated the importance of considering the user's social network and, e.g., favoring users close to the current user [7]. Narrowing the collaborative contribution down to close users presents benefits at two levels: (a) results are more precisely personalized and (b) potential pre-computation is reduced.

Let us note $SN(u)$ the set of users in u's social network which can be filtered, e.g., to keep only users up to a certain maximum distance. We propose the following refined co-occurrence measure, were α and β are positive coefficients to be adjusted experimentally such that $\alpha + \beta = 1$:

$$\begin{aligned} cooc(u, e_1, e_2) = {} & \alpha \cdot cooc_u(e_1, e_2) \\ & + \frac{\beta}{|SN(u)|} \cdot \sum_{u' \in SN(u)} \frac{1}{d(u, u')} cooc_{u'}(e_1, e_2) \end{aligned} \qquad (2)$$

This measure $cooc(u, e_1, e_2)$ is defined for entities e_1 and e_2 exposed to the user u by access control rules. The contribution of each user u' is weighted by the inverse of the distance $d(u, u')$.

Relations between users can be obtained from a variety of sources, including popular social networks on the Web. However, this does not necessarily match corporate requirements since users of the system are actual employees of a same company. In this context, enterprise directories can be used to extract, e.g., hierarchical relations between employees. Clearly, other types of relations may be considered but the actual construction of the social network is beyond the scope of this paper.

4 Personalized Query Expansion

In this section, we describe our approach to design a personalized query expansion component leveraging models semantics, co-occurrences and user preferences.

4.1 User Preferences

We distinguish *explicit* and *implicit* preferences, respectively noted $pref_{u,expl}$ and $pref_{u,impl}$. For a given entity e, we define the user's preference function $pref_u$ as a linear combination of both preferences, for instance simply:

$$pref_u(e) = \frac{1}{2}\left(pref_{u,impl}(e) + pref_{u,expl}(e)\right) \tag{3}$$

Explicit preferences are feedback received from the user, e.g., in the form of ratings (in $[0, 1]$) assigned to measures and dimensions. Let us note $r_{u,e}$ the rating given by u to e and $\overline{r_u}$ the average rating given by u. We define $pref_{u,expl}(e) = r_{u,e}$ if u has already rated e, and $pref_{u,expl}(e) = \overline{r_u}$ otherwise.

Implicit preferences can be derived from a variety of sources, for instance by analyzing logs of queries executed in users' sessions [4]. In our case, we consider occurrences of BI entities in documents manipulated by the user as a simple indicator of such preferences:

$$pref_{u,impl}(e) = \frac{|occ_u(e)|}{\max_{e'} |occ_u(e')|} \tag{4}$$

4.2 Query Expansion

The aim of our system is to assist the user in the query design phase by offering suggestions of measures and dimensions she could use to explore data. When she selects a measure or a dimension, it is added to the query being built and suggestions are refreshed to form new consistently augmented queries.

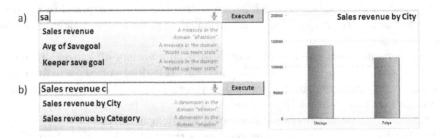

Fig. 3. Screenshot of auto-completion used in an interactive query designer. (a) First suggestions after characters "sa" and (b) suggestions following the selection of measure Sales revenue and character "c". On the right is a sample visualization that can be built with the query Sales revenue by City.

Ranking. To complete a given query $q = \{e_1, \ldots, e_n\}$ with an additional measure or dimension, we need to find candidate entities and rank them. Candidate entities, $c_j, j = 1..p$, are those defined in the same domain and compatible with every e_i, determined using functional dependencies (see Section 2.2). We then use the following personalized function to rank each candidate c_j:

$$rank_u(c_j, q) = \begin{cases} pref_u(c_j) & \text{if } q = \emptyset \\ pref_u(c_j) \cdot \frac{1}{n} \sum_{i=1}^{n} cooc(u, c_j, e_i) & \text{otherwise} \end{cases} \quad (5)$$

To conclude with the notation of the query expansion problem introduced in Section 1, we define our component QE as:

$$QE: (u, q, params) \mapsto \{(q_1, rank_u(c_1, q)), \ldots, (q_p, rank_u(c_p, q))\}$$

Parameters. Beyond ranking, suggestions of the query expansion component can be fine-tuned using various parameters:

- The maximum number of results.
- The type of suggested entities can be limited to measures and/or dimensions.
- The domain can be restricted to a list of accepted models.
- Suggested dimensions can be grouped by and limited to certain hierarchies. This may be used to reduce the number of suggestions and encourage the user explore varied axis of analysis.

5   Experimentation: Auto-Completion in a Query Designer

In previous sections we presented tools used to implement an interactive query designer. In this section, we illustrate results obtained with a prototype implementation. We developed a query designer which simply presents a search text box to the user. As she types, candidate measures and dimensions are proposed as auto-completion suggestions.

Table 1. Top-5 dimensions that most co-occur (in a collection of dashboards) with the Sales Revenue measure

Measure	Dimension	Co-occurrence
Sales Revenue	Quarter	0,38
	State	0,25
	Year	0,25
	Category	0,25
	Lines	0,22

Figure 3.a) shows measures (from distinct domain models) suggested when the user starts typing "sa": Sales revenue, Avg of savegoal and Keeper save goal. In Figure 3.b), the user has selected the first suggestion Sales revenue and keeps typing "c". The system suggests the two dimensions City and Category. The auto-completion initialization requires that the user roughly knows the names of objects she wants to manipulate, which may be a barrier to adoption. To help her get started and explore available data, suggestions can be surfaced to the user before she even starts typing. For instance, the most commonly used measures and dimensions of various domain models could be suggested to start with.

In our implementation of the architecture presented in Figure 1, we relied on the BI platform *SAP BI 4*. Documents wise, this platform proposes reporting and dashboarding solutions respectively named *WebIntelligence* and *SAP BusinessObjects Explorer* (or *Exploration Views*). We experimented the computation of co-occurrences using dashboards accessible through the demonstration account of *SAP Exploration Views* on-demand[1]. This account exposes 9 dashboards which contain 31 charts. The 7 underlying domain models define 54 dimensions and 86 measures. Table 1 presents the 5 dimensions that most co-occur with a given measure named Sales Revenue. From the social network point of view, we build on the prototype *Social Network Analyzer*[2]. In particular, APIs of this prototype expose the user's social graph at a depth of 2.

6 Related Work

In this section we briefly review previous work related to personalization and recommendation in multi-dimensional DWs. Various types of OLAP recommendations can be considered with *interactive assistance for query design*, *anticipatory recommendations* and *alternative results* [8]. The work presented in this paper best corresponds to the first type which, to the best of our knowledge, has not been investigated much by previous research.

Techniques employed for query recommendations in DWs have been thoroughly reviewed by Marcel et al. [10]. In particular, the authors provide a formal framework to express query recommendations and divide them in methods (a)

---

[1] http://exploration-views.ondemand.com
[2] http://sna-demo.ondemand.com/

based on user profiles, (b) using query logs, (c) based on expectations and (d) hybrid ones. The first ones include user profiles (e.g., preferences) in the recommendation process to maximize the user's interest for suggested queries [6,9]. Interestingly, recommendations may integrate visualization-related constraints [2]. Second, methods based on query logs mainly address predictive recommendations of forthcoming queries and position analysis sessions as first-class citizens [4]. One approach is to model query logs using a probabilistic Markov model. Another is to use a distance metric between sessions to extract recommended queries from past sessions similar to the current one. Methods based on expectations aim at determining and guiding the user toward zones of a cube that present unexpected data, for instance by maximizing the entropy [13]. Finally, these approaches may be combined in various ways with hybrid methods [5].

Recommendations of multi-dimensional queries is a fairly recent topic. However, RS have become a popular research area thanks to successful commercial applications, e.g., on e-commerce Web sites. RS are commonly categorized in content-based (CB) and collaborative filtering (CF) approaches [1]. CF methods usually rely on *user* × *item* matrices to compute similarities between items and users based on ratings, preferences, etc. [14]. The main assumption behind such techniques is that users with similar rating schemes will react similarly to other items. Preferences can be either explicit (ratings) or implicit (e.g., derived from click-through data). Metrics used in CF techniques build for instance on vector-based cosine similarity and *Pearson* correlation. The collaborative contribution can be refined by considering the use of a trust network between users [7]. CF techniques present a certain number of issues dealing with matrix sparsity and cold-start users. Besides, they remain superficial since they lack a representation of the actual item content. On the other hand, CB methods use descriptions of items' features (like weighted keywords in a vector space model) and assume that the user will like items similar to those he liked in the past. Finally, hybrid methods often combine CB and CF approaches to overcome their respective drawbacks [11].

7   Conclusion and Future Work

In this paper we presented a personalized query expansion system that leverages (a) semantics of multi-dimensional domain models, (b) usage statistics derived from (co-)occurrences of *measures* and *dimensions* in repositories of BI documents and (c) user preferences. The system was experimented with a prototype of interactive query designer, assisting the user with auto-completion suggestions. This experimentation showed encouraging usability results.

However, we did not manage to obtain a joint dataset between deployments of *SAP Exploration Views* and *Social Network Analyzer*. Therefore, we would like to focus in future work on the generation of such a dataset to conduct user-satisfaction tests. In particular, it would be interesting to highlight and measure the benefits of the collaborative contribution introduced in formula 2.

We illustrated our approach with an interactive query designer. Beyond this, we are currently investigating other promising applications of the concepts presented in this paper. For instance, the search-like interface of Figure 3 could be extended. In particular, we are considering the retrieval of charts – from existing reports and dashboards – with similar data. Also, user preferences and (co-)occurrences may be used in a *question answering* system to help, e.g., with personalized query reformulation.

More generally, we reckon that recommendations in the context of DWs and BI platforms could benefit much further from techniques developed in the RS area. However, taking into account the specific semantics of multi-dimensional models is also key to provide relevant structured analytics.

References

1. Adomavicius, G., Tuzhilin, A.: Toward the next generation of recommender systems: A survey of the state-of-the-art and possible extensions. IEEE Trans. Knowl. Data Eng. 17(6), 734–749 (2005)
2. Bellatreche, L., Giacometti, A., Marcel, P., Mouloudi, H., Laurent, D.: A personalization framework for olap queries. In: Proceedings of the 8th ACM International Workshop on Data Warehousing and OLAP, DOLAP 2005, pp. 9–18. ACM, New York (2005)
3. Bhide, M., Chakravarthy, V., Gupta, A., Gupta, H., Mohania, M.K., Puniyani, K., Roy, P., Roy, S., Sengar, V.S.: Enhanced business intelligence using erocs. In: ICDE, pp. 1616–1619. IEEE (2008)
4. Giacometti, A., Marcel, P., Negre, E.: A framework for recommending olap queries. In: Proceeding of the ACM 11th International Workshop on Data Warehousing and OLAP, DOLAP 2008, pp. 73–80. ACM, New York (2008)
5. Giacometti, A., Marcel, P., Negre, E., Soulet, A.: Query recommendations for olap discovery driven analysis. In: Proceeding of the 12th ACM Workshop on Data Warehousing and OLAP, DOLAP 2009, pp. 81–88. ACM, New York (2009)
6. Golfarelli, M., Rizzi, S., Biondi, P.: myolap: An approach to express and evaluate olap preferences. IEEE Transactions on Knowledge and Data Engineering 23, 1050–1064 (2011)
7. Jamali, M., Ester, M.: TrustWalker: a random walk model for combining trust-based and item-based recommendation. In: Elder IV, J.F., Fogelman-Soulié, F., Flach, P.A., Zaki, M.J. (eds.) KDD, pp. 397–406. ACM (2009)
8. Jerbi, H., Ravat, F., Teste, O., Zurfluh, G.: Preference-Based Recommendations for OLAP Analysis. In: Pedersen, T.B., Mohania, M.K., Tjoa, A.M. (eds.) DaWaK 2009. LNCS, vol. 5691, pp. 467–478. Springer, Heidelberg (2009)
9. Kozmina, N., Niedrite, L.: OLAP Personalization with User-Describing Profiles. In: Forbrig, P., Günther, H. (eds.) BIR 2010. LNBIP, vol. 64, pp. 188–202. Springer, Heidelberg (2010)
10. Marcel, P., Negre, E.: A survey of query recommendation techniques for datawarehouse exploration. In: Proceedings of 7th Conference on Data Warehousing and On-Line Analysis (Entrepôts de Données et Analyse), EDA 2011 (2011)

11. Melville, P., Mooney, R.J., Nagarajan, R.: Content-boosted collaborative filtering for improved recommendations. In: Eighteenth National Conference on Artificial Intelligence, pp. 187–192 (2002)
12. Romero, O., Calvanese, D., Abelló, A., Rodríguez-Muro, M.: Discovering functional dependencies for multidimensional design. In: Proceeding of the 12th ACM Workshop on Data Warehousing and OLAP, DOLAP 2009, pp. 1–8. ACM, New York (2009)
13. Sarawagi, S.: User-adaptive exploration of multidimensional data. In: Proceedings of the 26th Conference on Very Large DataBases (VLDB), Cairo, Egypt (2000)
14. Su, X., Khoshgoftaar, T.M.: A survey of collaborative filtering techniques. Advances in Artificial Intelligence, 4:2–4:2 (January 2009)

Hierarchy-Based Update Propagation in Decision Support Systems⋆

Haitang Feng[1,2], Nicolas Lumineau[1],
Mohand-Saíd Hacid[1], and Richard Domps[2]

[1] Université de Lyon, CNRS
Université Lyon 1, LIRIS, UMR5205, F-69622, France
`firstname.lastname@liris.cnrs.fr`
[2] Anticipeo, 4 bis impasse Courteline, 94800, Villejuif, France
`firstname.lastname@anticipeo.com`

Abstract. Sales forecasting systems are used by enterprise managers and executives to better understand the market trends and prepare appropriate business plans. These decision support systems usually use a data warehouse to store data and OLAP tools to visualize query results. A specific feature of sales forecasting systems regarding future predictions modification is backward propagation of updates, which is the computation of the impact of modifications on summaries over base data. In Data warehouses, some methods propagate updates in hierarchies when data sources are subject to modifications. However, very few works have been devoted so far, regarding update propagation from summaries to data sources. This paper proposes an algorithm called PAM (Propagation of Aggregate Modification), to efficiently propagate modifications on summaries over base data. Experiments on an operational application (Anticipeo[1]) have been conducted.

Keywords: Aggregate update propagation, Data warehouses, Decision support systems, Dimension-Hierarchy.

1 Introduction

A forecasting system is a set of techniques or tools that are mainly used for analysis of historical data, selection of most appropriate modeling structure, model validation, development of forecasts, and monitoring and adjustment of forecasts[2]. The most frequently used forecasting systems relate to domains like weather, transportation or sales.

A sales forecasting system (SFS), also called a business forecasting system is a kind of forecasting system allowing achievable sales revenue, based on historical

---

⋆ Research partially supported by the French Agency ANRT (www.anrt.asso.fr), Anticipeo (www.anticipeo.com) and the Rhône-Alpes region (projet Web Intelligence, http://www.web-intelligence-rhone-alpes.org).
[1] See http://www.anticipeo.com
[2] See http://www.businessdictionary.com/definition/forecasting-system.html.

S.-g. Lee et al. (Eds.): DASFAA 2012, Part II, LNCS 7239, pp. 261–271, 2012.

sales data, analysis of market surveys and trends, and salespersons' estimates[3]. To be effective, a SFS must adhere to some fundamental principles such as the use of a suite of time-series techniques [11].

The basic functionalities a SFS supports are: computation, visualization and modification. The first functionality, computation of forecasts, uses specific methods (typically statistical models) to derive sales forecasts. Several predictive methods have been introduced in the domain of statistics (see, e.g., [9][10]). The second functionality, visualization of computed forecasts, uses OLAP (online analytical processing) tools to visualize data stored in a DW (data warehouse). The visualization methods are investigated, especially by resorting to "roll-up" and "drill-down" operators of OLAP or reporting tools of BI (Business Intelligence) (see, e.g., [7][12][13]). However, the third functionality, modification of computed forecasts during visualization, is a specific problem. In SFSs, source data are composed of historical data and predictive data while predictive data are not as stable as traditional source data that we employ in a traditional DW. Sales managers could make some modifications to adjust computed forecasts to some specific situations, e.g., occasional offers. These adjustments occur on summarized data and should be propagated to base data (facts and forecasts) and then to other summarized data. Some approaches dealing with view maintenance in OLAP were proposed. Some of them focus on the evolution of multidimensional structure (see, e.g., [1][8]) and others focus on the optimization of OLAP operators such as pivot and unpivot (see, e.g., [3]). Approaches to view maintenance in DWs were also investigated. They are concerned with the combination of updates before propagation (see, e.g., [14]), multi-view consistency over distributed data sources (e.g., [2]) and many others[4]. The main context of these approaches is the propagation of updates over sources on materialized views. To the best of our knowledge, the problem of updating summaries and computing the effect on row data has not been investigated so far. The motivating case study we consider is a real sales forecasting system called Anticipeo. In this application, we are facing the optimization problem regarding update propagation from summarized data to base data and to other summarized data.

The rest of this paper is organized as follows : Section 2 defines the sales forecasts modification problem. Section 3 discusses the current and ad-hoc solutions to support this issue. In Section 4, we describe a novel algorithm based on intelligent exploitation of dimension-hierarchies. Section 5 describes the evaluation and the experimental results. We conclude in Section 6.

2   Problem Statement and Motivations

To clearly define our problem, we first review the use of dimensions (and hierarchies) and the basic data schema by visualization tools of OLAP systems [4]. OLAP systems employ multidimensional data models to structure "raw" data into multidimensional structures in which each data entry is shaped into a fact

---

[3] See http://www.businessdictionary.com/definition/sales-forecast.html

[4] See http://www.cise.ufl.edu/~jhammer/online-bib.htm

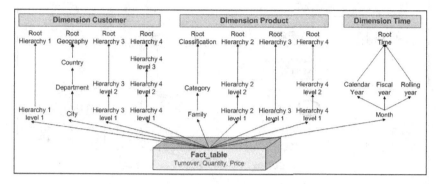

Fig. 1. Example of dimension-hierarchies and fact tables

with associated measure(s) and descriptive dimensions that characterize the fact. The values within a dimension can be further organized in a containment type hierarchy to support multiple granularities. In our case, we are interested in quantitative measures, not qualitative measures.

In the example shown Fig. 1, we present the data model in a sales forecasting system. This data model is based on one fact table with three measures: *turnover*, *quantity* and *price*; and three different dimensions: *customer*, *product* and *time*. Each dimension has its hierarchies to describe the organization of data. Customer dimension has 4 hierarchies, product dimension has 4 hierarchies and time dimension has 3 hierarchies. For instance, the second hierarchy of customer dimension is a geographical hierarchy. Customers are grouped by city for level 1, by department for level 2 and by country for level 3. Base sales are aggregated on each level according to this geographical organization when one analyzes the sales through this hierarchy.

Regarding the visualization, the calculation of aggregated information needs to be performed on the fly. OLAP systems employ materialized views or fictive information to avoid extra response time. In this example, fictive customers and fictive products are added to represent elements in superior hierarchy levels, such as a fictive customer for the city of Lyon, another one for the department Rhône and a third one for the country France. Thus, the system has three new entries in the customer dimension and accordingly some aggregated sales in the fact table regarding these newly created customers. Finally, all elements of every hierarchy level from every dimension are aggregated and added to the dimension and fact tables. This pre-calculation guarantees an immediate access to any direct aggregated information while users perform visualization demands.

Regarding forecasting systems, the visualization is not the last operation as in other OLAP systems. The sales forecasts produced by a system are a first version which are reviewed by experienced salespersons. Salespersons check these automatically generated sales forecasts, take into consideration some issues not considered by the system (e.g. promotional offers) and perform some necessary

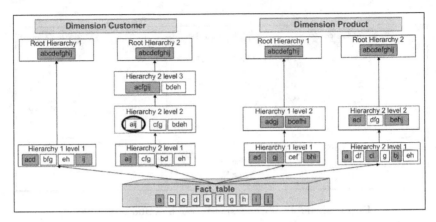

Fig. 2. Example of data modification on an aggregated level

modifications. In other cases, salespersons can also perform some modifications on summaries in order to simulate a new marketing target. This update taking place on an aggregated level constitutes the major feature of sales forecasting systems. Compared to traditional OLAP systems in which source data are considered to be static, data in sales forecasting systems can be modified many times to obtain a final result.

Hence, sales forecasting systems need to have the ability to quickly react to data modification on an aggregated level. Fig. 2 is an example to show how an aggregation-level modification impacts all the data.

In this example and for the sake of simplicity, we consider only the first two hierarchies respectively from customer dimension and product dimension of the previous data model (see Fig. 1). In the fact table, we group all the tuples into 10 sets: named from a to j (In the following, these sets are named "base tuple"). Aggregates at superior hierarchical levels are presented by rectangles including the composing base tuples and are formalized by the function α with the composing base tuples in parameters: $\alpha(x,y,...)$. For instance, the circled rectangle, or the aggregate $\alpha(a,i,j)$ on level 2 of hierarchy 2 of the customer dimension containing aij means that the result of this aggregate is generated from the base tuples a, i and j. In this specific case, the result of the aggregate $\alpha(a,i,j)$ is the sum of sales: a, i and j. Other aggregates are generated and presented in the same manner. The root rectangle of every hierarchy represents all the sales.

Figure 2 shows the underlying data structure when the system presents the prediction result to sales managers. Sales managers analyze the sales and then decide to modify the sales of the aggregate $\alpha(a,i,j)$ (i.e. to evaluate beforehand the impact of a strategical or tactical move). Let us see the impact of this modification. As the aggregate $\alpha(a,i,j)$ is generated from a, i and j, once its result is modified, the results of the three tuples should be afterwards updated. Meanwhile, these three tuples are also the base tuples which are involved in the calculation of other aggregates in all dimension-hierarchies. All aggregates containing any of these three tuples in its composition should be updated as well. The aggregates impacted by the modification on the aggregate α(a,i,j) are darkened in Fig. 2.

Sales managers may perform modifications many times to obtain a satisfying result. SFSs should provide a short delay between the modification and the visualization of modified data. Hence, the problem we need to deal with is how to efficiently update aggregated data through a dimension-hierarchy structure.

3   Current Solution

A current solution consists in identifying approaches to similar problems and building on the implemented solutions. In this system, methods to calculate the aggregates are well defined. The steps of the current solution which recomputes everything are as follows:

1. calculate the base tuples wrt the modification and the decomposition rules,
2. recompute all the aggregates.

To illustrate this process, consider the example shown Fig. 2. The actual result of the aggregate α(a,i,j) is 500 000 euros, but the sales manager has a new marketing plan and (s)he estimates the sales can achieve 600 000 euros. Then (s)he does this modification which is then processed as follows:

Step 1: Calculation of Modified Base Tuples
The example in Fig. 2 indicates that the result of the aggregate α(a,i,j) is composed of three tuples: a, i and j. Assume that the distribution of sales is respectively 100 000 euros, 200 000 euros and 200 000 euros for the three tuples. When the sales of the aggregate α(a,i,j) raise to 600 000 euros, the modification is spread over these tuples with respect to the weight of every tuple according to the company's previously defined rules; the weight in this case is the actual sales. For a, i and j, their weights are respectively 100 000, 200 000 and 200 000 (i.e. 1:2:2). The formula to calculate the new result for a tuple t is

$$eval_{\alpha,T}(t) = val(t) + (val'(\alpha, T) - val(\alpha, T)) * \frac{weight(t)}{\sum_{t' \in T} weight(t')}$$

where α is an aggregation operator, T is a set of tuples, val returns the current value of a tuple or an aggregate and val' returns the new value of a tuple or an aggregate.

The new values for T={a,i,j} are then:
$$eval_{\alpha,T}(a) = 100000 + (600000 - 500000) * \tfrac{1}{1+2+2} = 120000$$

$$eval_{\alpha,T}(i) = 200000 + (600000 - 500000) * \tfrac{2}{1+2+2} = 240000$$
$$eval_{\alpha,T}(j) = 200000 + (600000 - 500000) * \tfrac{2}{1+2+2} = 240000$$

Step 2: Recalculation of Aggregated Information
The second step consists in recomputing the aggregates on higher levels. We follow the same process as when the aggregates are created using the hierarchy dependencies. For instance, the aggregate α(a,c,d) is an aggregate of the base

tuples a, c and d; so its new result is calculated by summing the sales of a, c and d.

Following this straightforward solution, we can regenerate all the hierarchies of the whole schema with updated data.

4   Update Propagation Algorithm

The current solution advocates the calculation of all the aggregates of all the hierarchies. However, this solution performs some useless work. If we closely look at the recomputed aggregates in Fig. 2, only the dark ones are concerned with the modification and need to be updated, that is, **18** aggregates out of **32**. Hence, the current solution leads to the calculation of **14** aggregates in vain. The key idea is thus to be able to identify and recompute only the concerned elements. By considering the dependencies between aggregates and base tuples, we can identify the exact aggregates to modify and hence avoid useless work.

Another drawback with the current solution is its heavy recomputing procedure. Operations of removing and adding aggregates demand heavy maintenance of index tables and physical storage. Nevertheless, our approach can keep the aggregates at their logical and physical location and avoid extra effort.

To summarize, our approach follows the following steps:

1. retrieval of participating base tuples of the modified aggregate
2. creation of a temporary table for the base tuples to be updated and calculation of the differences resulting from the old values and the new ones
3. update of impacted base tuples
4. identification of impacted aggregates
5. update of impacted aggregates with the previously calculated differences of base tuples

The algorithm for the update propagation through dimension-hierarchies is shown Table 1. Line 1 to line 4 identify the base tuples involved in the modification and calculate their differences. Line 5 allows to update these base tuples. Line 6 to line 9 identify impacted aggregates and perform the update.

Let us take the previous example to illustrate the approach. A sales manager changes the sales of the aggregate $\alpha(a,i,j)$ from 500 000 euros to 600 000 euros. Once (s)he confirms the modification, the system will proceed using the algorithm in Table 1.

Step 1: Retrieval of the Participating Tuples to the Aggregate
Retrieve the composition of the aggregate $\alpha(a,i,j)$: sales of the aggregate $\alpha(a,i,j)$ is the sum of a, i and j. Hence, the composing tuples are a, i and j.

Step 2: Creation of a Temporary Table and Calculation of Differences
Create a temporary table ΔX for the base tuples identified in step 1.
Calculate the δ for the aggregate $\alpha(a,i,j)$: $\delta = 600000 - 500000 = 100000$
Considering $W = \sum_{t \in \{a,i,j\}} weight(t)$, calculate the difference for every tuple using the weight coefficient.

Table 1. Algorithm for the update propagation

Algorithm PAM (Propagation of Aggregate Modification)

Input: Schema S, aggregate X, its actual result AR and its objective result OR
Output: An updated schema S' of all hierarchies
Algorithm:

1: Calculate the modification of the aggregate X: δ = OR - AR
2: Retrieve participating base tuples of X : CX = $\{x_1, x_2, ..., x_n\}$
3: Create a temporary table: ΔX = CX
4: Calculate the differences for every base tuple: $\forall x_i \in \Delta X$: $\delta_i = \delta$ * weight_coeff$_i$;
   Add the result to ΔX
5: Update all the base tuples impacted: $\forall bt_i \in \Delta X$: $val'(bt_i) = val(bt_i) + \delta_{bt_i}$
6: Identify aggregates to update:
   A = {results after filtering by dependencies to ΔX}
7: For A_i in A
8:    Retrieve its composition: $CA_i = \{a_1, a_2, ..., a_m\}$
      Identify base tuples which are also in ΔX: U = $\Delta X \cap CA_i$
      Calculate new result of A_i: $\forall u_i \in U$ $val'(A_i) = val(A_i) + \delta_{u_i}$
9: End for

$$\delta_a = \delta * \frac{weight(a)}{W} = 100000 * \frac{1}{1+2+2} = 20000$$
$$\delta_i = \delta * \frac{weight(i)}{W} = 100000 * \frac{2}{1+2+2} = 40000$$
$$\delta_j = \delta * \frac{weight(j)}{W} = 100000 * \frac{2}{1+2+2} = 40000$$

The resulting differences of base tuples are added to the temporary table. This table also contains the dependency information to higher hierarchical levels (shown Table 2).

Step 3: Update of Base Tuples

Update the base tuples impacted by the aggregate modification (same procedure as in the current solution). The new values of these base tuples are computed by their actual values and the differences calculated in step 2. In this case, a is updated to 120 000, i to 240 000 and j to 240 000.

Table 2. Temporary table ΔX created to store influenced base tuples

element_identifier	customer_dim_link	product_dim_link	δ_x
a	customer_link$_a$	product_link$_a$	20000
i	customer_link$_i$	product_link$_i$	40000
j	customer_link$_j$	product_link$_j$	40000

Step 4: Identification of Impacted Aggregates
Identify all the aggregates concerned with the modification of the sales of the aggregate α(a,i,j) by using the links between aggregates and registered base tuples in the temporary table ΔX. In this case, we identify all the dark rectangles.

Step 5: Update of Impacted Aggregates
Propagate the changes to every concerned aggregate. Let us illustrate this issue with the customer dimension hierarchy 1. We loop for every level of the hierarchy. For level 1, two aggregates need to be updated: α(a,c,d) and α(i,j). The aggregate α(a,c,d) is composed of a, c and d and among these base tuples, only one is registered in the table ΔX, namely, the base tuple a. Hence, the value of α(a,c,d) is changed only by adding the δ_a (here 20 000).

$$val'(\alpha, \{a, c, d\}) = val(\alpha, \{a, c, d\}) + \delta_a = val(\alpha, \{a, c, d\}) + 20\ 000$$

Another element ij of level 1 and the root element $abcdefghij$ on level 2 can be calculated in a similar way: $val'(\alpha, \{i, j\}) = val(\alpha, \{i, j\}) + 40\ 000 + 40\ 000$; $val'(\alpha, \{a, b, c, d, e, f, g, h, i, j\}) = val(\alpha, \{a, b, c, d, e, f, g, h, i, j\}) + 20\ 000 + 40\ 000 + 40\ 000$.

Doing this way, we update only the aggregates impacted by the modification for hierarchy 1 of the customer dimension. The propagation in other hierarchies are processed in the same manner. Finally, we obtain updated data over the entire schema.

5   Experiments

The main technical characteristics of the server on which we run the evaluation are: two Intel Quad core Xeon-based 2.4GHz, 16GB RAM and one SAS disk of 500GB. The operating system is a 64-bit Linux Debian system using EXT3 file system. Our evaluation has been performed on real data (copy of Anticipeo database) implemented on MySQL.

The total size of the database is 50 GB of which 50% is used in the computation engine, 45% for result visualization and 5% for the application. Our test only focuses on the data used by the update: one fact table and two dimension tables: *customer* and *product*. The fact table has about 300 MB with 257.8MB of data and 40.1 MB of index. The customer dimension table contains 5240 real customers and 1319 fictive customers (6559 in total) and the product dimension table contains 8256 real products and 404 fictive products (8660 in total). Each of these dimension tables is composed of 4 hierarchies. It presents a similar structure to Fig. 1 with different number of levels in each hierarchy (from 2 to 4 levels). Note that the time dimension is investigated within the fact table for some performance issues [6][5]. Hence, only two explicit dimensions are materialized in dimension tables.

The objective of the evaluation is to compare the time of the whole schema update using the current solution and our algorithm. The tests are performed on 3 hierarchies which have 2, 3 and 4 levels, respectively. In our evaluation,

Table 3. Evaluation time of an aggregate modification using the current solution

	Hierarchy H1		Hierarchy H2			Hierarchy H3			
	Level 1	Level 2	Level 1	Level 2	Level 3	Level 1	Level 2	Level 3	Level 4
Step 1*	0.9	7.9	0.9	1.0	7.5	0.08	0.8	2.9	7.8
Step 2*	179.5	182.1	185.7	181.4	188.4	181.1	179.6	179.9	176.6
Total	180.4	190.0	186.6	182.4	195.9	181.2	180.4	182.8	184.4

* Step 1: update base tuples; Step 2: reconstruct all aggregates

Table 4. Evaluation time of an aggregate modification using our algorithm

	Hierarchy H1		Hierarchy H2			Hierarchy H3			
	Level 1	Level 2	Level 1	Level 2	Level 3	Level 1	Level 2	Level 3	Level 4
Stage 1*	0.3	2.9	0.3	0.3	2.9	0.04	0.3	1.2	3.0
Stage 2*	0.9	7.8	0.9	2.0	7.2	0.1	0.8	2.8	7.7
Stage 3*	5.4	53.5	5.2	5.4	56.1	0.6	4.3	21.0	54.4
Total	6.6	64.2	6.4	6.7	66.2	0.7	5.4	25.0	65.1

* Stage 1: create a temporary table; Stage 2: update base tuples; Stage 3: propagate

we modify one aggregate from each level of every hierarchy to compare the evaluation time resulting from the current solution and from our approach.

We first perform tests with the current solution. The result is shown Table 3. Step 2 stays almost the same for different hierarchies because it concerns the recomputation of the whole schema.

The same tests are performed with our algorithm. During the implementation, we group some logical steps introduced in Section 4 to optimize the execution time: Step 1 and Step 2 are merged to form Stage 1; Step 4 and Step 5 are merged to form Stage 3. The result is shown Table 4.

We then compare the total evaluation time using the two solutions in one chart shown Fig. 3. In some cases, e.g., the level 1 of hierarchy H3, the propagation time is only 0.7 seconds. Compared to 181.2 seconds consumed by the current

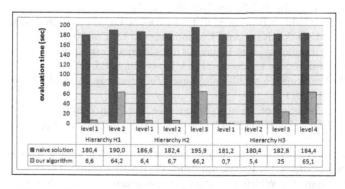

Fig. 3. Comparison of evaluation time using two solutions

solution, the gain of performance reaches 25786%. Even in the worst case where the root aggregate (top level of every hierarchy) is subject to modifications, we get a nearly 200% better performance. In these practical cases, our algorithm presents more than 3000% better performance in average. The result confirms that, instead of recalculating all the aggregates as the current solution does, our solution is more efficient by identifying and updating the **exact** set of aggregates impacted by the modification.

6   Conclusion

In this paper, we discussed the problem of efficiently propagating an aggregate modification through a dimension-hierarchy structure. A current solution naively recomputes all the aggregates of all the hierarchies, which is time-consuming and does not fulfill the performance needs. We proposed the algorithm PAM to reduce the modification cost. Our algorithm is based on the dependencies that may exist between aggregates and base data. It identifies the exact sets of aggregates to be updated and calculates the delta for each aggregate. We conducted experiments to show that with our approach, the update propagation time can be considerably reduced compared to the current solution implemented in a real application.

For further work, we will take into consideration more factors that could affect the evaluation time, such as complexity of dimensions, complexity of hierarchies, complexity of queries, etc. We will also investigate the update propagation of an aggregate which results from multiple hierarchies.

References

1. Body, M., Miquel, M., Bedard, Y., Tchounikine, A.: Handling Evolutions in Multidimensional Structures. In: ICDE, pp. 581–591 (2003)
2. Chen, S., Liu, B., Rundensteiner, E.A.: Multiversion-based view maintenance over distributed data sources. ACM Trans. Database Syst. 29(4), 675–709 (2004)
3. Chen, S., Rundensteiner, E.A.: Gpivot: Efficient incremental maintenance of complex rolap views. In: ICDE, pp. 552–563 (2005)
4. Codd, E.F., Codd, S.B., Salley, C.T.: Providing OLAP (On-Line Analytical Processing) to User-Analysis: An IT Mandate, vol. 32. Codd & Date, Inc. (1993)
5. Feng, H.: Data management in forecasting systems: Case study - performance problems and preliminary results. In: BDA Proceedings (2011)
6. Feng, H.: Performance problems of forecasting systems. In: ADBIS Proceedings II, pp. 254–261 (2011)
7. Gupta, H.: Selection of Views to Materialize in a Data Warehouse. In: Afrati, F.N., Kolaitis, P.G. (eds.) ICDT 1997. LNCS, vol. 1186, pp. 98–112. Springer, Heidelberg (1997)
8. Hurtado, C.A., Mendelzon, A.O., Vaisman, A.A.: Maintaining data cubes under dimension updates. In: ICDE, pp. 346–355 (1999)
9. Jain, Chaman, L.: Which forecasting model should we use? Journal of Business Forecasting Methods & Systems 19(3), 2 (2000)

10. McKeefry, Lynne, H.: Adding more science to the art of forecasting. In: EBN, vol. 1252, p. 46 (2001)
11. Mentzer, J.T., Bienstock, C.C.: The seven principles of sales-forecasting systems. Supply Chain Management Review (Fall 1998)
12. Ross, K.A., Srivastava, D., Chatziantoniou, D.: Complex Aggregation at Multiple Granularities. In: Schek, H.-J., Saltor, F., Ramos, I., Alonso, G. (eds.) EDBT 1998. LNCS, vol. 1377, pp. 263–277. Springer, Heidelberg (1998)
13. Theodoratos, D.: Exploiting hierarchical clustering in evaluating multidimensional aggregation queries. In: DOLAP, pp. 63–70 (2003)
14. Zhou, J., Larson, P.Å., Elmongui, H.G.: Lazy maintenance of materialized views. In: VLDB, pp. 231–242 (2007)

An Experiment with Asymmetric Algorithm: CPU Vs. GPU

Sujatha R. Upadhyaya[1] and David Toth[2,*]

[1] Infosys Limited, Bangalore, India
[2] Imperial College, London, UK
sujatha_upadhyaya@infosys.com, david.toth10@imperial.ac.uk

Abstract. Discovery of sequential patterns in large transaction databases for personalized services is gaining importance in several industries. Although a huge amount of mobile location data of consumers is available with the service providers, it is hardly put to use owing its complexity and size. To facilitate this, an approach that represents the entire area by a location grid and records the movements across the cells as sequences has been proposed. A new algorithm for mining sequential data is devised to find frequent travel patterns from location data and analyze user travel patterns. The algorithm is asymmetric in nature and is parallelized on the GPGPU processor and tested for performance. Our experiments assert that asymmetric nature of the algorithm doesn't allow the performance to elevate despite parallelization, especially with large data.

Keywords: Sequence Mining, Parallelization, Location Data, Personalization.

1 Introduction

Mobile towers record the movement of the devices as a response to the signals sent by these devices. The frequency with which the towers respond, the phase of the towers that acknowledges the signal and other parameters can point to the exact location of the device. Given that the process of signal search and acknowledgement is a continuous one; this data can be used to study the travel patterns of individuals. Such analysis of consumer data can be eventually used for personalization of services, where consumers receive the right information at the right time. However, analysis of device location data is not simple owing to its complexity and size. A method has been proposed to convert this data into sequences of locations [1]. This method suggests that geographic area under consideration is divided into a grid like structure where each cell is uniquely identified by a cell ID as shown in Figure 1. Here, movements of individuals are represented as sequences formed by the cell IDs in the order of travel.

---

* David worked as an intern with Infosys Limited. This work was carried out during his internship.

S.-g. Lee et al. (Eds.): DASFAA 2012, Part II, LNCS 7239, pp. 272–281, 2012.

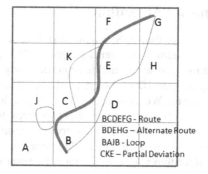

Fig. 1. A Sketch Showing the Location Grid

A sequence that represents movements in a day would start and end with the same location ID as in BCDEFGHEDB. Analysis of such sequences collected over the years reveals travel patterns. An algorithm called Frequent Sequence Algorithm (FS algorithm), particularly used for the purpose of identifying the most frequent and frequent routes used by consumers is proposed in this paper. This algorithm processes the data of people who live in same location.

Processing terabytes of data calls for optimization. Our effort for performance enhancement is brought about by making an effort to parallelize the FS algorithm and running it on a parallel architecture. The modified algorithm is called P-FS algorithm here on. We chose the NVIDIA's GPGPU as the hardware platform, with CUDA as the coding language.

In this paper, a background to the research has been discussed in Section 2. The FS algorithm has been described Section 3. Section 4 elaborates on P-FS Algorithm, the parallelization of FS Algorithm. Section 5 describes the experiments we ran through and discusses the results. The last section presents the conclusions from the experience and future research directions.

2   Background and Literature

A very short background has been presented here. From the beginning of GPU era, we have witnessed several tasks being subjected to parallelization on GPUs. In the initial years, many efforts [1] [2] [6] have taken up tasks such as matrix multiplication, sorting, sparse matrix solving, etc. for parallelization on GPUs. Years later, the focus moved on to parallelizing machine learning algorithms. There have been numerous efforts [3] [4] [5] [7] [8] [9] [10] [11] in this direction staring from itemset mining [10] to the recent text mining efforts [11]. One of the efforts [3] speaks about stream mining of quantiles and frequencies that exploits the computational abilities of GPU to build a new rasterization scheme. Another interesting work [4] refers to a new sorting algorithm that uses blending and mapping functionalities of GPU. Two of the efforts [7] [9] represent parallelization of SVM approach. As against these approaches, our effort describes an effort to parallelize an asymmetric algorithm and studies the effect of parallelization on GPU.

3   Frequent Sequence Algorithm

The Frequent Sequence algorithm (FS Algorithm) is devised especially for the sequences generated from ordering the events generated. Here, the event refers to the movement of the device into a new cell and is represented by its cell ID. A sequence is generated by stringing the cell ID in the order of even occurrence. One sequence corresponds to a day's travel. We collect the data of people living at a particular location, for a couple of years and analyze the sequences to determine the travel patterns of individuals. The data is huge and can contains multitude of travel patterns, depending on the number of people living in that area and directions they take to work. Objective of this algorithm is to find the most frequent and frequent sequences with specific minimum support in the database.

3.1   Data Representation

An example sequence data shown Figure 2 is used for illustrating data representation. The typical sequences have the following properties:

1. They have a common starting and ending event, which refers to the base point or 'home'. In the example set 'A' is the base point.
2. It can also be observed that a certain length of typical sequence represents the forward path and the rest of the sequence represents the reverse path. For example, in sequence 'ABCDCBA', 'ABCD' represents the forward path and 'CBA' represents the reverse path for that sequence. Beginning of the reverse path may be identified by the re-occurrence of an event. While 'forward journeys' and 'return journeys' exist in real sense, it is difficult to identify the respective forward and return paths unless the destination is known.

Example Sequence		A	B	C	D	E	F
A B C D C B A							
A B E D C B A	A	0	6	0	0	0	0
A B C D C B A	B	5	0	3	0	3	1
A B C D C B F A	C	0	5	0	3	0	0
A B E D E B A	D	0	0	5	0	1	0
A B E D C B A	E	0	1	0	3	0	0
	F	1	0	0	0	0	0

Example Sequence   **Matrix Illustration**

Fig. 2. Matrix Representation of Sequence Data

A set of sequences with 'n' unique events is represented with a matrix of size n×n, where 'n' is the number of unique events in the sequence set. The one path frequencies are stored in a matrix as shown above. A careful examination shows that most of the cells in the matrix are empty. This is because, movements occur across the adjacent cells and every cell is adjacent to few other cells. As the length of the sequences increases more entries in the matrix will be zero as the number of cells that cannot be reached from a specific cell also increases. It may also be noted that if the frequency of sequence 'AB' is represented in the upper triangle of the matrix; then the frequency of sequence 'BA' is represented in the lower triangle. If sequence 'AB' represents a forward path, then 'BA' represents the reverse path. In an ideal situation, one must be able to   represent the entire forward paths in the upper triangle and the reverse paths in the lower triangle. But, in reality, this can never happen as it is extremely difficult to find the right order of representation of events in the matrix, given that there could be multiple paths, partial deviations, loops etc. In the illustration example, the representation of subsequence BF appears in the upper triangle of the matrix, although it is a part of the backward path.

3.2   Algorithm

The skeleton of the algorithm has been shown in Figure 2. Other than the frequent sequence matrix, we have two lists L_1, L_2 containing sequences of lengths 1 to n-1, whose frequencies are greater than the minimum support. A sequence with a frequency greater than minimum support is a 'frequent sequence'.

```
Frequent Sequence Algorithm:

Input:
        Sequential Data , SD
        User Defined Support, MinSupport
        User Defined Base Point, BasePoint

Output:
        Set of Candidate Sequences, CS

Variables:
Fr_Matrix:   Frequent Sequence Algorithm
Max_Len:     Maximum Length of the Sequence
Min_Len:     Minimum Length of the sequence

1.  Scan SD
2.  Build Frequent Sequence Matrix, Fr_Matrix
3.  Build L1
4.  Build  L2
5.  Call Candidate_Gen(L1, L2)
6.  Return CS;
```

Fig. 3. FS Algorithm

Procedure for Building L_1, L_2: Our database contains sequence corresponding to the entire community of people who live at a location; however, it would contain equal number of sequences corresponding to each user. The value of minimum support is decided as a percentage of the total number sequences corresponding to one user. We build these lists as preparation to generate the candidate sequences and at this stage we ignore all 1-path sequences with frequencies less than the minimum support.

To build L_1, we start with the 1-path sequences with frequency greater than the minimum support and in each step; grow them by one length by appending suitable 1-length sequences. For illustration example, assuming a minimum support of 2,

Step 1: L_1={ ab, bc, be, cd}
Step 2: L_1={ ab, bc, be, cd, abc, abe, bcd }
Finally, Step 3: L_1={ ab, bc, be, cd, abc, abe, bcd, abcd} – Note that the sequences do not grow further.
Similarly,
Step 1: L_2={ed, dc, cb, ba,}
Step 2: L_2={ed, dc, cb, ba, edc, dcb, cba}
Step 3: L_2={ed, dc, cb, ba, edc, dcb, cba, edcb, dcba}
And, finally, Step 4: L_2={ed, dc, cb, ba, edc, dcb, cba, edcb, dcba, edcba}

Generation of Candidate Set:
In an ideal situation, all paths in L_1 would represent forward paths; all paths in L_2 represent reverse paths. Although in a real situation, this cannot be completely true; we follow this notion and grow sequences by appending sequences in L_1 with sequences in L_2 and then, resultant sequences with sequences in L_1 again and then with sequences in L_2 and so on. The result is similar to growing of sequences in Apriori algorithm; the only difference being; in Apriori procedure, the sequences grew by one length in each stage; One the other hand, sequences of unequal lengths grow by unequal lengths in this context..

In an ideal situation, most of the frequent sequences would have been generated in initial steps of combining L_1 and L_2. (Longest forward sequences would combine with longest reverse sequences to make the full sequences, in step 1).

```
Candidate Gen (L₁, L₂)
Input: L₁, L₂, Fr_Matrix
Output: CS
Variables: New_Seq – newly generated sequence
           To_Grow – List of sequences to be grown further

1.  Append (L₁, L₂)
2.  Append (L₁, L₂)
3.  For all sequences Starting with BasePoint in L₁,
4.  Grow_Seq (L₁, L₂)
       •  Append (L₁, L₂)
       •  If length(New_Seq)==Max_Len .and. New_Seq terminates with BasePoint
          •  Add to CS;
       •  If length(New_Seq)==Max_Len .and. New_Seq does not terminate with BasePoint
          •  Delete;
       •  If (Min_len<=len(New_Seq)>Max_Len) .and. New_Seq terminates with BasePoint
          •  Add to CS – if terminates with BasePoint
          •  Add to To_Grow;
       •  If len(New_Seq)<Min_Len
          •  Add to To_Grow;
5.  Repeat Until To_Grow is empty;
       •  Grow_Seq(To_Grow, L1);
       •  Grow_Seq(To_Grow, L2);
6.  Repeat 3,4,5 with L2
7.  Output CS
```

Fig. 4. Algorithm for Candidate Generation

The algorithm for candidate generation has been depicted in Figure 4. The candidate generation is similar to the procedure in Apriori algorithm. However, each step will not produce frequent sequences of the same length. Candidate sequences will be produced in n-1 trials of combining L_1 with L_2. As may be observed, the sequence generation can be parallelized easily, however, it is not possible to ensure that each thread carries equal load. This makes the algorithm highly asymmetric.

4   Parallelization on GPU

The FS algorithm gives into parallelization very well; in the sense that the algorithm can be split into processes that can run independent of the other in every stage. However, one or two of the threads end up generating useful sequences while the rest complete their part and wait for longer threads to complete. We implemented the parallelized version of the same on NVidia Tesla S870. The parallelization scheme is as shown in Figure 5.

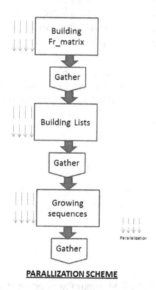

PARALLIZATION SCHEME

Fig. 5. Parallelization Scheme

Parallelization Component 1 - Building Fr_Matrix: The first step in the algorithm builds the Fr_Matrix that counts the frequency of one length sequences in the data. When the size of Fr_Matrix is 'n × n', n^2 threads are introduced. Each thread counts one type of sequence. This portion gives considerable gain, depending on the actual number of 1-length sequences present in the data. However, we notice that most of the threads wait for others to finish, especially if the 1-path sequences are unequally distributed. In such cases, the speed up may not be proportional to the number of threads introduced.

Parallelization Component 2 - Building L_1 and L_2: One thread gets initiated for each 1-length sequence in top/bottom triangle and L_1 and L_2 are built from the 1-length sequences of top and bottom triangles respectively.

Parallelization Component 3 – Generation of Candidate Sequences: We grow sequences from the base point. In the first step, sequences starting with the base point in L_1 / L_2 are appended with appropriate sequences in L_2 / L_1. Number of threads initiated in this step depends on number of sequences that start with the base point. This is one of the stages, where several threads are introduced. During experimentation we observed that the asymmetry of the algorithm causes some of the threads to run much longer than the rest.

5   Experiments with Sequential and Parallel Algorithm

Here, we must mention the nature of the algorithm we just discussed and the architecture of GPGPU. The algorithm is symmetric in the sense that the threads take unequal time to complete. It might so happen that the one or two threads take the entire load while the others finish quickly. However, the available memory is shared equally amongst the threads. As a result, while much of memory does not get used up; a few threads run out of memory soon.

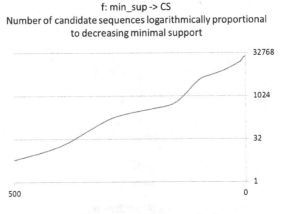

Fig. 6. Effect of Minimum Support on Performance

Figure 6. shows the number of candidate sequences generated for different values of minimum support. These experiments were run on artificially generated data, where the deviation, (the probability that a deviation is introduced while generating data sequences), number of records, length of sequences are some of parameters are controlled. In the above experiment is deviation remains constant.

Figure 7. shows that parallelization does not improve performance as the number of records increase. Length of sequence affects performance. The performance improves a little with the increase in length of sequences. This is possibly because the number of threads that take up load increases with the length of the sequences.

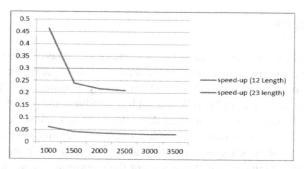

Fig. 7. Speed Up vs. Number of Records for a Fixed Value of Minimal Support

Figure 8. shows the deterioration of speed up as the number of candidates generated increases. There's every chance that the thread uses up the memory allocated quickly.

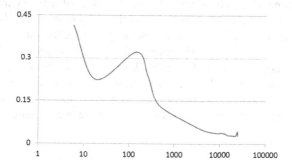

Fig. 8. Speed up Vs. the Number of Candidate Generated

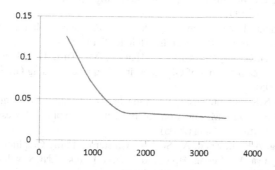

Fig. 9. Speed up Vs. Number of records

Larger the number of records worse is the speed-up. Figure 7. points out the same. Figure 9. emphasizes the same, but also shows that the performance dips sharply for sequences of greater length. We also observed performance for a slightly higher minimum support does not deteriorate as bad for the same number of records.

6   Conclusions

We have implemented the parallelization exercise of an asymmetric algorithm on GPU and have recorded our experience in this paper. As against the many efforts in the past where parallelization is brought about by dividing the data among the processors, we have achieved parallelization by identifying the part of the process that can be run independently on complete data. We have arrived at the conclusion that parallelization of an asymmetric algorithm results in deterioration of performance, despite the algorithm being well parallelizable. The reason for the deterioration is mainly attributed to the asymmetric nature of the algorithm. There are other reasons that contribute to this. One of them is that memory overflow that can be caused easily because available memory gets distributed among the threads. This results in inefficient memory usage. Then, there is the effect of memory transfers that can cause further delay.

Special APIs are now available to deal with asymmetricity. Use of such APIs and the use of streaming mechanisms for reducing the effect of memory transfers would help in bringing about greater speed-up. Another important factor is to devise an efficient threading scheme. The present scheme generates the threads depending on the initial number of sequences. However, it is possible to recursively allocate the new threads to carry on the task, while closing the old threads and make them available for fresh allocation. This scheme in a way could ensure usage of maximum threads. Since a portion of the threads get generated half way through, memory requirement at that stage would be lower as there are many more threads sharing the task.

References

1. Scott Larsen, E., McAllister, D.: Fast Matrix Multiplies Using Graphics Hardware. In: Super Computing (2001)
2. Bolz, J., Farmer, I., Grinspun, E., Schrooder, P.: Sparse Matrix solver on the GPU: Conjugate Gradients and Multigrid. In: SIGGRAPH (2003)
3. Govindaraju, N.K., Raghuvanshi, N., Henson, M., Tuft, D., Manocha, D.: Fast and Approximate Stream Mining of Quantiles and Frequencies Using Graphics Processors. In: SIGMOD (2005)
4. Govindaraju, N.K., Raghuvanshi, N., Manocha, D.: A Cache-Efficient Sorting Algorithm for Database and Data Mining Computations using Graphics Processors, Tech. Report, University of North Caroloina (2005)
5. Cao, F., Tung, A.K.H., Zhou, A.: Scalable Clustering Using Graphics Processors. In: Yu, J.X., Kitsuregawa, M., Leong, H.-V. (eds.) WAIM 2006. LNCS, vol. 4016, pp. 372–384. Springer, Heidelberg (2006), doi:10.1007/11775300_32
6. Govindaraju, N.K., Gray, J., Kumar, R., Manocha, D.: GPUTeraSort: High Performance Graphics Coprocessor Sorting for Large Database Management. In: SIGMOD 2006 (2006)
7. Catanzaro, B., Sundaram, N., Keutzer, K.: Fast Support Vector Machine Training and Classication on Graphics Processors. In: ICML 2008 (2008)

8. Fang, W., Keung Lau, K., Lu, M., Xiao, X., Lam, C.K., Yang Yang, P., He, B., Luo, Q., Sander, P.V., Yang, K.: Parallel Data Mining on Graphics Processors, GPUComputing.net (October 2008)
9. Carpenter, A.: CUSVM: A CUDA Implementation of Support Vector Classification and Regression
10. Fang, W., Lu, M., Xiao, X., He, B., Luo, Q.: Frequent Itemset Mining on Graphics Processors. In: DaMon 2009 (2009)

Tag Association Based Graphical Password
Using Image Feature Matching

Kyoji Kawagoe, Shinichi Sakaguchi, Yuki Sakon, and Hung-Hsuan Huang

College of Information Science and Engineering
Ritsumeikan University, Japan
{kawagoe@is,sakaguchi@coms.is.,sakon@coms.ics,huang@fc.}
ritsumei.ac.jp

Abstract. Much work on graphical password has been proposed in order to realize easier and more secure authentication with use of images as a password rather than text based passwords. We have proposed a *Tag* Association *Based* graphical password called TAB. In TAB, a set of images which are presented to a user is determined among a large collection in image search or sharing web services using user pre-registered pass-terms, while the typical graphical password presents a user a set of images including one of the user pre-registered images. In our demo, we present the novel prototype system with an extension of TAB. The extended TAB is incorporated a well-known image recognition algorithm, such as SIFT (Scale Invariant Feature Transform) or SURF (Speeded Up Robust Feature) in order to increase both Shoulder Surfing Unsuccess and Authentication Success Ratios.

Keywords: graphical password, image search, authentication, image features, tags.

1 Introduction

There have been many research activities on graphical password [1–9], which is a new type of passwords by using images instead of common textual password. The main advantage of graphical password over textual password is its ease to be remembered and recognized [7]. Moreover, a system in which graphical password is adopted is less expensive than other types of authentication methods such as fingerprints and face recognition. In a mobile multimedia application, it is important to provide easy and quick authentication. Such an application can benefit from user friendly direct manipulation interface such as multi-touch input feature provided by many smartphone devices. One of the most important problems in the current graphical passwords using an image selection based method is shoulder surfing attack vulnerability. The shoulder surfing is an attack using direct observation to get users' critical information. As it is easier to make shoulder surfing on mobile multimedia applications, than PC-based ones, the vulnerability is a more critical issue should be addressed.

In order to prevent this shoulder surfing problem, we proposed a new graphical password method, called TAB which stands for *Tag* Association *Based* graphical

S.-g. Lee et al. (Eds.): DASFAA 2012, Part II, LNCS 7239, pp. 282–286, 2012.

password [10]. This method can avoid shoulder surfing by using the association between an image and the tags attached to it. In our demo, we present a prototype implementation of extended TAB. The extension made over the original TAB [10] is on matching images which are related to one of the user pre-registered pass-terms, in order to obtain better Shoulder Surfing Unsuccess Ratio as well as Authentication Success Ratio.

2   Tag Association Based Graphical Password (TAB)

Fig.1 shows a brief process of the typical graphical password (*left*) and that of our proposed original TAB (*right*). In the case of the typical graphical password, a user pre-registers some pass-images among a set of stored images. On the contrary, a user pre-registers some pass-terms for graphical password images in TAB. During the authentication process, the user selects a correct answer image which is associated with one of the pre-registered pass-terms, not from one of an identical predefined set. In TAB, it is important to present images which contain one correct answer image and the other incorrect images which are never associated with any pass-term pre-registered by the user.

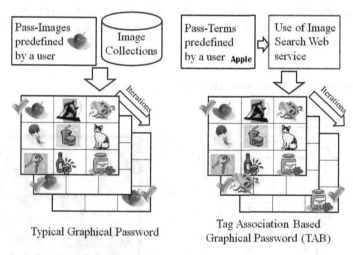

Typical Graphical Password

Tag Association Based
Graphical Password (TAB)

Fig. 1. Typical Graphical Password (*left*) and Tag Association Based Graphical Password (*right*)

The main advantage of TAB is the possibility to provide a set of images that is different each time. Therefore, shoulder surfing is more difficult to be made on TAB based applications. Moreover, TAB is suitable for authentication in mobile applications due to simple layout and easy usage.

It is important to present a user a set of images so as for an attacker not to be able to guess a user pass-term. If the attacker views only apple images including no other objects as the correct answer images, he/she can easily guess the user pass-term is Apple. Therefore, all the images which are presented to a user should include several sub-images.

3   TAB Using Image Feature Matching

In the demo, we present the prototype implementation of extended TAB. It incorporates feature detection algorithms such as SIFT [11] or SURF [12].The similarity between an image obtained by searching images on the Internet and an image associated with a set of user pre-registered pass-terms can be calculated using the features of the two images.

There are not only a set of pass-terms determined in an authentication system beforehand but also a collection of image set appropriately representing each pass-term, which we call *Imagetag*. The feature vectors are built from the feature points found by feature detection algorithms such as SIFT or SURF. During authentication, the system acquires a set of correct answer image candidates which matches the user pre-registered pass-terms as well as a set of incorrect images from image search web services. Then, the feature points of these images are detected with the same algorithm as in the case of *Imagetags*. The similarity between an image and an *Imagetag* is defined as the number of matched pairs of the feature points over the smaller value of the numbers of feature points of two images. Matched pairs are obtained by extracting a pair of feature points whose cosine similarity measure is less than a threshold. Therefore, we can select one correct answer image by getting the image with the highest similarity value between an image candidate and an *Imagetag* associated with one of the user pre-registered pass-terms. Also, we can obtain an incorrect answer set where all images are not similar to any *Imagetag* of user pass-terms.

4   Demonstration System and Overview

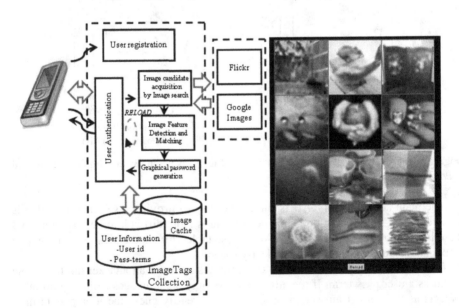

Fig. 2. Prototype system structure (*Left*) and a screenshot (*Right*)

Fig. 2 shows the structure of the prototype system on the left, and a snapshot of the user interface on the right. As shown in the figure, the system has two main components: user registration function and user authentication function. A set of images is presented to a user during authentication, after generating one correct answer image and the other incorrect answer images by calculating the similarity value between an image obtained from image search web services and an *Imagetag* stored in the system. For the sake of avoidance of miss-selection by a user, we provide the RELOAD function which can change a set of images presented to a user. The system keeps the images in a cache after the system obtains a collection of images with use of image search web services, in order to improve the performance on image presentation to a user. The prototype system is mainly implemented in JAVA.

5   Demo Scenario

We demonstrate our prototype system with the following scenario:

Step-1: A guest registers his/her pass-terms by selecting from a list of pass-terms.
Step-2: Our demo system displays N images to the guest. In the demo, N=12 is used.
Step-3: The guest select one image from N images, which is the most related image to one of his/her pass--terms. If the guest cannot select it because of no related images or multiple related images, RELOAD function can be selected.
Step-4: The guest and the system iterate Step-3, K times. In the demo, K is set to 2. The number of RELOAD usage is limited to K.
Step-5: When the guest can select correct images K times, then the system displays "Authentication Success" message to the guest. Otherwise, the system displays "Authentication Failed".

6   Conclusion

In this demo, we present a prototype system implementing a novel graphical password method, called TAB which stands for *T*ag *A*ssociation *B*ased graphical password. The method was designed and developed so as to apply it for mobile multimedia applications in an easy authentication way, with higher tolerance toward shoulder surfing. The prototype system enables a user to make his/her authentication easier and more reliable by using TAB.

References

1. Bicakci, K., Yuceel, M., Erdeniz, B., Gurbaslar, H., Atalay, N.B.: Graphical Passwords as Browser Extension: Implementation and Usability Study. In: Ferrari, E., Li, N., Bertino, E., Karabulut, Y. (eds.) IFIPTM 2009. IFIP AICT, vol. 300, pp. 15–29. Springer, Heidelberg (2009)

2. Gao, H., Guo, X., Chen, X., Wang, L., Liu, X.: Yagp: Yet another graphical password strategy. In: ACSAC 2008, pp. 121–129 (2008)
3. Kumar, M., Garfinkel, T., Boneh, D., Winograd, T.: Reducing shoulder-surfing by using gaze based password entry. In: 3rd Symposium on Usable Privacy and Security, pp. 13–19 (2007)
4. Li, Z., Sun, Q., Lian, Y., Giusto, D.: An association-based graphical password design resistant to shoulder-surfing attack. In: Int. Conf. on Multimedia and Expo., pp. 245–248 (2005)
5. Perra, C., Giusto, D.D.: A framework for image based authentication. In: ICASSP 2005, pp. II-521–II-524 (2005)
6. Abari, A.S., Thorpe, J., van Oorschot, P.C.: On purely automated attacks and click-based graphical passwords. In: 24th ACSAC, pp. 111–120 (2008)
7. Suo, X., Zhu, Y., Owen, G.S.: Graphical passwords: A survey. In: 21st ACSAC, pp. 463–472 (2005)
8. Takada, T., Koike, H.: Awase-e: Image-based authentication for mobile phones using user's favorite images. In: Mobile HCI 2003, pp. 347–351 (2003)
9. Wiedenbeck, S., Waters, J., Birget, J., Brodskiy, A., Memon, N.: Pass-points: Design and longitudinal evaluation of a graphical password system. Int. J. Human-Computer Studies 63, 102–127 (2005)
10. Sakaguchi, S., Huang, H.-H., Kawagoe, K.: Tag Association Based Graphical Password with Image Search Web Services. In: 3rd EDB, pp. 152–163 (2011)
11. Lowe, D.G.: Object recognition from local scale-invariant features. In: 7th ICCV, pp. 1150–1157 (1999)
12. Bay, H., Tuytelaars, T., Van Gool, L.: SURF: Speeded Up Robust Features. In: Leonardis, A., Bischof, H., Pinz, A. (eds.) ECCV 2006. LNCS, vol. 3951, pp. 404–417. Springer, Heidelberg (2006)

ACARP: Author-Centric Analysis of Research Papers

Weiming Zhang, Xueqing Gong, Weining Qian, and Aoying Zhou

Institute of Massive Computing,
Software Engineering Institute,
East China Normal University
51091500016@ecnu.cn,
{xqgong,wnqian,ayzhou}@sei.ecnu.edu.cn

Abstract. Scientific publications are an important kind of user-generated content, which contains not only high-quality content but also structures, e.g. citations. Scientific publication analysis has attracted much attention in both database and data mining research community.

In this demonstration, we present a system, named as ACARP, for analyzing research papers in database community. The relationship between a research paper and the authors is analyzed based on a *learning to rank* model. Not only the content of the paper, but also the citation graph is used in the analysis. ACARP can not only *guess* authors for papers under double-blind reviewing, but also analyze the researchers' continuity and diversity of research. This author-centric analysis could be interesting to researchers and be useful for further studying on double-blind reviewing process.

1 Introduction

Research papers are important resource for research activities, since they are formal reports on research progress and results. However, there are so many research papers, so that a single researcher is usually not capable of reading and remembering too many papers. A natural solution in terms of data and content management to this problem is that we may build a system to manage all research papers and provide services for analyzing them to assist researchers.

Research paper management is not new in content management. Existing services include directory-based systems, search-based ones, and mining-based ones. An excellent example of directory-based systems is DBLP [3]. It provides list of papers with author names, paper titles, booktitles indexed. Search-based systems include CiteSeer [4] and Google Scholar [5]. They both provide search services for not only the basic information of research papers but also their content. Mining-based systems, such as ArnetMiner [6] and Dolphin [7], go one step further. ArnetMiner analyzes the co-author graph and determines the roles of authors, while Dolphin analyzes the content of papers and is capable of generating a survey for a specific topic.

S.-g. Lee et al. (Eds.): DASFAA 2012, Part II, LNCS 7239, pp. 287–294, 2012.

ACARP is a mining-based system. It differs from other ones in that: 1) both content and citation graph are used to extract the features, so that the subsequent analysis is more accurate; 2) authors are treated as the major object for analysis; 3) various analysis tasks can be conducted based on ACARP.

1.1 System Overview

The architecture of ACARP is shown in Figure 1. The offline part of the system is essentially a learning to rank trainer. It accepts papers with author information as input. The papers are parsed, and features, such as terms, references, are extracted. Then, a RankSVM [2] model is trained, in which the target is the author's list.

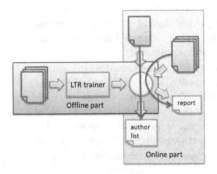

Fig. 1. The architecture of ACARP   **Fig. 2.** The modeling of a research paper

The trained model is used in the online part of the system. The online part provides two services. The first one accepts a paper as input. The author information is not necessarily provided. ACARP parses the paper, along with all papers in its references, extracts features, and tests the paper against the model trained by the offline part. Then, a list of predicted authors, each of which is associated with a score, is returned.

The second service of the online part accepts user queries on authors or papers. ACARP would provide reports on continuity and diversity of research of authors. ACARP also provides information on why it predicts a researcher is the author of a paper.

1.2 Paper Organization

The rest part of the paper is organized as follows. In next section, the method for author prediction of a paper is introduced. Then, in Section 3, the method for analyzing research papers is introduced. Finally, in Section 4, the demonstration outline is presented.

2   Author Prediction

Author prediction is the major part of the ACARP system. The details of the author prediction algorithm is introduced in [1]. An outline of the method is introduced here.

2.1   Modeling of a Research Paper

A research paper can be modeled as it is illustrated in Figure 2. Each paper has a *title*. It may have several *authors*. It should be noted that a paper may not contain any author information in case that the paper is under review satisifying the double-blind-review guideline.

A paper contains *content*. The content is organized in tree structure, in which leaf nodes are plain text. They can also be treated as lists of terms. Furthermore, usually, a paper contains references. Each reference is a point to another paper.

2.2   Features

Features are extracted from papers. They come from both content of the paper itself, and content in references. Each paper is represented by a vector, in which each dimension corresponds to a feature. The features that are used in ACARP are listed as follows.

- **Occurrence** Occurrence means number of times the author appears in the reference. Each author corresponds to a dimension in the vector, while the value is the frequency of the author appears in the reference list. The intuition is that the more an author's paper is cited, the more likely that she is the author.
- **Fame** Fame here means whether the author is famous, which is defined in [8]. If an author has published more than 20 papers in both SIGMOD and VLDB, then the author is marked as a famous one. Otherwise, she is not famous. Thus, the value of the dimension corresponding to the fame of an author is binary: 1 for famous, and 0 for unfamous. It should be noted that famous authors are more likely to appear in references than other ones. Therefore, this feature should be used to prevent the model always predict the researcher who have most papers as author.
- **Title rate** The feature is simply the number of the same term in both reference's title and paper's title. It is quite possible that the continually works share same terms in titles.
- **Time gap** This feature is the difference between the year the paper published and the year of the paper appearing in reference. Each author of the cited paper has one dimension for this feature. This is a feature that measures the relationships between a paper and its reference.
- **Term rate** Similar to the title rate feature, the term rate feature measures the similarity between a paper to its references. It should be noted that not all keywords are used as terms. Terms are identified as those words that are

written in capital, or not appearing in dictionaries. Thus, abbreviations of specific algorithms, names of data sets, etc. can be identified.

– **Publication type** This feature is based on the place the paper in refernce was published. There are six types, which are listed in Table 1. Intuitively, papers in prestigious conferences or journals are more likely to be cited, while technical reports are more likely to be self-citations.

Table 1. Types of publications

Type	Publication
1	Prestigious conferences and journals: SIGMOD, VLDB, SIGKDD, ICDE, VLDBJ, TKDE, ...
2	Other Conferences and journals of ACM and IEEE etc.
3	Published Books
4	Technique report
5	Ph.D. or Master Thesis
6	Others

2.3 Learning to Rank

Let S denotes the training data sets which contains several feature vector sets $X = \{x_1, x_2, \ldots, x_n\}$, and $Y = \{y_1, y_2, \ldots, y_n\}$ means the rank space, $\{r_1, r_2, \ldots, r_n\}$ is a set of ranks, and n is number of ranks. P denotes the preference relationship of a pair of features (x_i, x_j). It equals $+1$ denotes x_i ranks a head of x_j. On contrast, -1 denotes the later one ranks high. For example: $P_{i,j} = +1$, if $y_i > y_j$, and $P_{i,j} = -1$, if $y_i > y_j$. There exists an order among all the ranks: $r_n > r_{(n-1)}, \ldots > r_1$, where ">" denotes the previous one has priority than the later one.

[?] proposed the learning problem by using above pairs of instances. So we assume there is a ranking function f which is linear:

$$f(x) = (\omega, x) \tag{1}$$

$$f(x_i) - f(x_j) = (\omega, x_i - x_j) \tag{2}$$

So plugging the pair-wise result, we obtain:

$$f(x_i) > f(x_j) \longleftrightarrow (\omega, x_i - x_j) > 0 \longleftrightarrow y_i > y_j \tag{3}$$

Based on the given data sets S, there will generate a new data sets S' of each document generated from $x_i - x_j$, which all have the label $p = +1$ or $p = -1$. Then we can train a SVM model on the new sets X' which is equivalent to solving the optimization problem:

$$min\frac{1}{2}\omega^T\omega + C\sum_{i=1}^{n}\xi_i \tag{4}$$

s.t. $p_i(\omega, x_i - x_j) \geq 1 - \xi_i, \xi_i \geq 0, \forall i \in \{1, \ldots, n\}$

where ω denotes the weight of ranking function and ξ is a slack variable. As using SVM to generate a model, we can get a rank function f_{w^*}:

$$f_{w^*}(x) = (\omega^*, x) \tag{5}$$

where ω^* is the weights in the SVM solution for ranking function. And than f_{w^*} can be used to identify potential authors.

3   Extensions and Summary of Results

ACARP provides a set of services based on the author prediction model, which are called *extensions*. They are introduced as follows.

3.1   Are Double-Blind Reviewers Really Blind?

Previous studies, e.g. Tung 2006 [8], show that double-blind reviewing may improve the review process in that reviewers are less biased to *famous* authors. However, experienced reviewers may argue that it is not difficult to know the authors even the the authors' names are hidden.

ACARP can be used directly to guess authors. Figure 3 shows that while features on both content and citation graph are used, the precision can be around 80%[1]. It should be noted that this precision is quite high, since only part of the papers are self-citation ones. The rates of self-citation papers from 1994 to 2010 on SIGMOD are shown in Table 2.

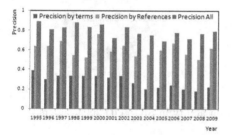
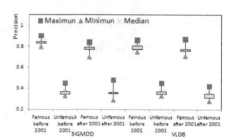

Fig. 3. Precision on author prediction   **Fig. 4.** Comparison of precision in double- and non-double-blind reviewing

In Figure 4, it is shown that though the precision of famous author prediction decreases a little in double-blind reviewing (SIGMOD after 2001), the difference of double-blind reviewing and non-double-blind reviewing is not significant.

---

[1] Here, the precision is defined as $\|P\|/N$, in which N is the number of all papers in testing, and P is the set of papers that have at least one real author is returned by ACARP in the first n results, where n is the number of authors listed in the paper.

Table 2. The rate of self-citation papers

Year	Self--cited	Total	Rate	Year	Self--cited	Total	Rate
1994	37	40	92.5%	2003	43	53	81.13%
1995	33	36	91.67%	2004	59	69	85.51%
1996	40	47	85.11%	2005	53	65	81.54%
1997	36	42	85.71%	2006	50	58	86.21%
1998	38	42	90.48%	2007	56	70	80%
1999	38	41	92.68%	2008	66	78	84.62%
2000	37	42	88.10%	2009	55	63	87.30%
2001	35	44	79.55%	2010	64	80	80%
2002	38	42	90.48%				

ACARP analyzes the disambiguation ability, which means *information gain*, of features for each paper. Thus, while the authors of a paper are guessed, a series of features is returned. Users of ACARP may understand why a specific author is predicted better based on these features. Figure 5 shows the percentage of features that returned for SIGMOD 2009 papers.

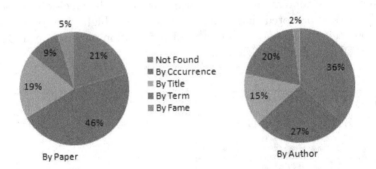

Fig. 5. Features with high disambiguation ability in SIGMOD 2009 papers

Based on statistics returned by ACARP, an interesting conclusion can be made that: *double-blind reviewers are usually not blind*. However, sometimes they may pretend to be blind and be less biased to famous authors.

3.2   Continuity Analysis

The second service ACARP provides is the continuity analysis on research interests of researchers. Since author prediction is based on papers previously written, an author with high rank is actually highly related to his previous work appearing in references. Thus, by listing one researcher's papers in timeline along with her ranks of each paper returned by ACARP, we can see her continuity of research.

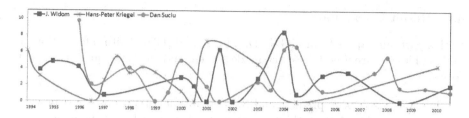

Fig. 6. Three researchers' continuity curves

Table 3. The top and bottom 5 researchers on diversity of interests in 50 most famous researchers in SIGMOD and VLDB

Rank of fame	Top ranked author	Precision	Rank of fame	Bottom ranked author	Precision
32	Hans-Peter Kriegel	0.83	17	Beng Chin Ooi	0.12
21	Rajeev Rastogi	0.81	19	Kian-Lee Tan	0.19
14	Minos N. Garofalakis	0.76	18	Philip S. Yu	0.22
36	Kevin Chen-Chuan Chang	0.73	43	S. Sudarshan	0.30
22	David J. DeWitt	0.71	6	Raghu Ramakrishnan	0.32

Figure 6 shows continuity curves of three researchers, namely Jennifer Widom, Hans-Peter Kriegel, and Dan Suciu. There are several *hills* in each curve. A hill with several points whose ranks are high usually corresponds to a research projects. For Jennifer Widom, for example, The first three hills correspond to the WHIPS (1995-1999), LORE (1996-1999) and TRAPP (2000-2003) projects. The fourth and fifth hill correspond to the STREAM (2001-2005) and TRIO (2005-2009) projects. Papers whose authors that are not correctly returned are usually the first paper of the project, overview papers without references to previous work, or papers that do not appear in list of any project.

3.3 Diversity of Interests

Researchers have different styles. Some researchers' papers are highly related, while some others' papers focus on different topics. ACARP measures the *diversity of interests* of a researcher by percentage that it accurately predict her as the author of all her papers. Table 3 shows researchers whose diversity of interests are ranked highest and lowest of the fifty authors who have most SIGMOD and VLDB papers.

4 Demonstration Outline

Initially, ACARP will be setup on a public service with a learning to rank model trained by papers from SIGMOD, VLDB, and ICDE in three years. The demonstration contains two parts.

4.1   Result Query Part

In this part, all papers from SIGMOD, VLDB, and ICDE (with information on authors hidden) are used to test the model. Then, audience may query the result to see:

- The authors ACARP predicts and the real authors of any specific paper.
- The reason that ACARP thinks a researcher is the author of the paper.
- The continuity of any specific author.
- The diversity of research interests of any specific author.

4.2   Interactive Part

In this part, we encourage audience to provide electronic copy of their own papers, and/or even papers in references, to test the model.

Acknowledgement. This work is partially supported by National Science Foundation of China under grant numbers 60833003, 61170086, and 60803022, National Basic Research (973 program) under grant number 2010CB731402, and National Major Projects on Science and Technology under grant number 2010ZX01042-002-001-01.

References

1. Qian, W., Zhang, W., Zhou, A.: Are Reviewers Really Blind in Double Blind Review, Or They Just Pretend to Be? East China Normal University, Technical Report (2011)
2. Herbrich, R., Graepel, T., Obermayer, K.: Large Margin Rank Boundaries for Ordinal Regression. In: Advances in Large Margin Classifiers, pp. 115–132. MIT Press (2000)
3. Ley, M.: Subharmonic solutions with prescribed minimal period for nonautonomous Hamiltonian systems (2011), http://www.informatik.uni-trier.de/~ley/db/
4. The College of Information Sciences and Technology at Penn State: Scientific Literature Digital Library and Search Engine (2011), http://citeseerx.ist.psu.edu/
5. Google Inc.: Google Scholar (2011), http://scholar.google.com
6. KEG at Tsinghua University: Academic Researcher Social Network Search (2011), http://arnetminer.org/
7. Wang, Y., Geng, Z., Huang, S., Wang, X., Zhou, A.: Academic web search engine: generating a survey automatically. In: WWW, pp. 1161–1162 (2007), http://doi.acm.org/10.1145/1242572.1242745
8. Tung, A.K.H.: Impact of double blind reviewing on SIGMOD publication: a more detail analysis. SIGMOD Record 35(3), 6–7 (2006), http://doi.acm.org/10.1145/1168092.1168093

iParticipate: Automatic Tweet Generation from Local Government Data

Christoph Lofi[1] and Ralf Krestel[2]

[1] IFIS, Carolo-Wilhelmina Universität Braunschweig, Germany
lofi@ifis.cs.tu-bs.de
[2] L3S Research Center, Leibniz Universität Hannover, Germany
krestel@L3S.de

Abstract. With the recent rise of Open Government Data, innovative technologies are required to leverage this new wealth of information. Therefore, we present a system combining several information processing techniques with micro-blogging services to demonstrate how this data can be put to use in order to increase transparency in political processes, and encourage internet users to participate in local politics. Our system uses publicly available documents from city councils which are processed automtically to generate highly informative tweets.

1 Introduction

More and more public institutions and governments provide access to official documents via the Internet. This fosters *transparency* [2] and the possibility for common citizens to participate more actively in political and governmental processes. But the huge amount of information available and the diversity of documents (meeting minutes, requests, petitions, proposals, ...) makes it difficult for non-experts to detect interesting information snippets. Especially the younger generations want to stay informed but are not willing to spent much time digging for related information.

With e-Democracy and e-Government on the rise [6], the use of information technologies for governmental processes becomes a regular means of communication between the government and its citizens. Leveraging all the public information available today could not only yield *social value* but also *commercial value*. Important proposals and eventually decisions are made public by officials but are not easy to find by a broader audience. In particular local politics are not covered by mass media and thus remain obscure and non-transparent despite the goal of Open Government Data.

Besides the informational aspect, open public data can also encourage active *participation* in political discussions and decisions if combined with modern information systems techniques. Politicians can get quick feedback and capture the overall mood towards a question or argument. Communication technology nowadays offers various ways of interaction and participation [3]. Web 2.0 applications enable everyone to share their thoughts, e.g. via blogs, in social networks such as Facebook, or using microblogging (e.g. Twitter).

S.-g. Lee et al. (Eds.): DASFAA 2012, Part II, LNCS 7239, pp. 295–298, 2012.
© Springer-Verlag Berlin Heidelberg 2012

In this paper we present an innovative approach to combine Web 2.0 applications and Open Government Data, directly delivering relevant information to support citizens. We connect and link data offering a space for discussion and sharing of ideas. In detail, we present a system which automatically generates tweets in Twitter based on open local government data.

2   Automatic Tweet Generator

In our demonstrator, we use publicly available data from the city of Hannover, Germany[1]. This data contains various types of documents, such as meeting minutes, proposals, and petitions. The documents are part of official administration processes. The city council, for example, governs the citywide affairs, while for each city district an additional district council manages local issues, such as traffic planning, school and kindergarten organization, maintenance and reorganization of parks and recreational areas, and local town planning. Furthermore, districts may suggest measures to the city council. Each of the district councils meets monthly. For each meeting, members of the council, but also every citizen, may submit petitions to be added to the agenda of the upcoming meeting. These petitions are published online as individual documents as soon as they are received. Briefly before the meeting, the full agenda is published. After the meeting, all decisions are documented in meeting minutes, while the initial petitions are updated to include the passed resolutions.

Currently, although all documents are openly published, the whole process is quite opaque for the average citizen: documents are hard to find and usually only identified by a numeric id. The missing link structure makes it difficult to follow a certain process or stay informed about a particular issue. Here, our prototype systems steps in to increase transparency and convenience of Open Governmental Data. In particular, the following issues are addressed:

- Obtaining relevant documents from local open-data enabled governmental websites and archives
- Automatic topic classification of documents for better accessibility
- Extraction of hash-tags for intuitive navigation between interrelated tweets
- Automated Twitter tweet generation for always up-to-date push-style notifications of notable local political developments

2.1   System Architecture

Figure 1 gives an overview of our system. The first step in generating tweets is to poll the city archives for new documents and store them in an internal database. Each document is then analyzed, classified, and linked to previously obtained documents. Then, the expected "interestingness" of each document is established based on its type, content, and relation to other documents. For each city district,

---

[1] https://e-government.hannover-stadt.de/lhhsimwebre.nsf   (german).   For demonstration purposes, we will use documents machine-translated to English.

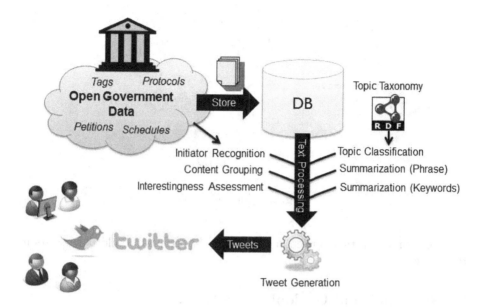

Fig. 1. System Architecture Overview

a twitter account is available allowing users to follow the local politics in this district. Documents which are considered interesting by the system are then transformed into a Twitter tweet, and published via the respective district's account. Each tweet also contains a link to the full document. This task raises two major challenges: a) selection of suitable hash-tags for finding, grouping, and organizing tweets and b) summarization of the document respecting Twitter's highly limited text length (140 characters).

Hash-tags are generated by several different means: Topic hash-tags (like 'schools', 'traffic', 'parks', etc.) briefly summarize the document's content. This classification with respect to a pre-defined topic taxonomy can be achieved by using a machine learning algorithm like support vector machines (SVM) [4]. By training the SVM with a small sample set, all following documents can easily be classified with high accuracy. Hash-tags can then be used to group documents together belonging to a single process or issue. For example, all petitions, agendas, and protocols of one particular district meeting use the same, unique hash-tag, thus allowing for easy navigation between related tweets. Source tags encode the documents political source, i.e. usually containing the shortcut for the responsible political party or spokesperson. These tags help to better explore the political activity within a city. City tags group all tweets of a city's districts together for broader overviews. Finally, different techniques for automatic text summarization [5] and key phrase extraction [1] can be used to summarize and tag the content of a document.

An example tweet illustrating the functionality of our automatic tweet generator is shown in Figure 2. Hannover uses its Twitter account (iPart Hannover)

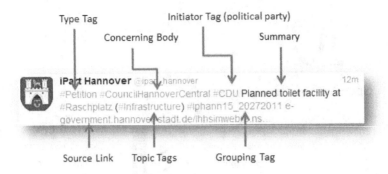

Fig. 2. Example Tweet

to publish news, in this particular case about a petition to build new toilets in the city center.

3 Summary and Outlook

In this paper we presented a system to automatically generate tweets from Open Government Data. Our system can help citizens to stay informed about local affairs. Further, it allows internet users to discuss, follow, and get involved in local politics adding transparency and encouraging participation. For future work, we plan a long-term user study to evaluate the system and estimate the possible impact of new communication technologies like Twitter on local politics.

References

1. Frank, E., Paynter, G.W., Witten, I.H., Gutwin, C., Nevill-Manning, C.G.: Domain-specific keyphrase extraction. In: Proceedings of the Sixteenth International Joint Conference on Artificial Intelligence, IJCAI 1999, pp. 668–673. Morgan Kaufmann Publishers Inc., San Francisco (1999)
2. George, J.M.: Can it help government to restore public trust?: Declining public trust and potential prospects of it in the public sector. In: Sprague, R. (ed.) Proceedings of the 36th Hawaii International Conference on Systems Sciences (HICSS 2003). IEEE Computer Society (2003)
3. Gibson, A.: Local by Social: How Local Authorities Can Use Social Media to Achieve More for Less. NESTA (2010)
4. Joachims, T.: Learning to Classify Text Using Support Vector Machines – Methods, Theory, and Algorithms. Kluwer/Springer (2002)
5. Mani, I.: Advances in Automatic Text Summarization. MIT Press, Cambridge (1999)
6. Parycek, P., Kripp, M.J., Edelmann, N. (eds.): CeDEM11 Proceedings of the International Conference for E-Democracy and Open Government. Edition Donau-Universität Krems (May 2011)

gRecs: A Group Recommendation System Based on User Clustering

Irene Ntoutsi[1], Kostas Stefanidis[2], Kjetil Norvag[2], and Hans-Peter Kriegel[1]

[1] Institute for Informatics, Ludwig Maximilian University, Munich
{ntoutsi,kriegel}@dbs.ifi.lmu.de
[2] Department of Computer and Information Science, Norwegian University of Science and Technology, Trondheim
{kstef,Kjetil.Norvag}@idi.ntnu.no

Abstract. In this demonstration paper, we present *gRecs*, a system for group recommendations that follows a collaborative strategy. We enhance recommendations with the notion of support to model the confidence of the recommendations. Moreover, we propose partitioning users into clusters of similar ones. This way, recommendations for users are produced with respect to the preferences of their cluster members without extensively searching for similar users in the whole user base. Finally, we leverage the power of a top-k algorithm for locating the top-k group recommendations.

1   Introduction

Recommendation systems provide suggestions to users about movies, videos, restaurants, hotels and other items. The large majority of recommendation systems are designed to make personal recommendations, i.e., recommendations for individual users. However, there are cases in which the items to be suggested are not intended for personal usage but for a group of users. For example, a group of friends is planning to watch a movie or to visit a restaurant. For this reason some recent works have addressed the problem of identifying recommendations for a group of users, trying to satisfy the preferences of all the group members.

Our work falls into the collaborative filtering approach, i.e., we offer user recommendations for items that similar users liked in the past. We introduce the notion of support in recommendations to model how confident the recommendation of an item for a user is. We also apply user clustering for organizing users into clusters of users with similar preferences. We propose the use of these clusters to efficiently locate similar users to a given one; this way, searching for similar users is restricted within his/her corresponding cluster instead of the whole database. Moreover, we exploit a top-k algorithm to efficiently identify the k most prominent items for the group.

2   The *gRecs* Group Recommendations Framework

Assume a set of items $\mathcal{I}$ and a set of users $\mathcal{U}$ interacting with a recommendation application. Each user $u \in \mathcal{U}$ may express a preference for an item $i \in \mathcal{I}$, which

S.-g. Lee et al. (Eds.): DASFAA 2012, Part II, LNCS 7239, pp. 299–303, 2012.

is denoted by $preference(u, i)$ and lies in the range $[0.0 - 1.0]$. For the items unrated by the users, we estimate a *relevance score*, denoted as $relevance(u, i)$, where $u \in \mathcal{U}$, $i \in \mathcal{I}$. To do this, a recommendation strategy is invoked.

We distinguish between personal recommendations referring to a single user and group recommendations referring to a set of users.

Personal Recommendations. There are different ways to estimate the relevance of an item for a user. Our work follows the collaborative filtering approach, such as [5] and [3]. To produce relevance scores for unrated items for a user, we employ preferences of users that exhibit similar behavior to the given user. Similar users are located via a *similarity function* $simU(u, u')$, that evaluates the proximity between u and u'. We use $\mathcal{F}_u$ to denote the set of the most similar users to u. We refer to such users as the *friends* of u. Several methods can be employed for selecting $\mathcal{F}_u$. A straightforward method is to locate the users u' with $simU(u, u')$ greater than a threshold δ. This is the method used here.

Given a user u and his friends $\mathcal{F}_u$, if u has not expressed any preference for an item i, the relevance of i for u is commonly estimated as follows:

$$relevance(u, i) = \frac{\sum_{u' \in f_u \wedge \exists preference(u', i)} simU(u, u') \times preference(u', i)}{\sum_{u' \in f_u \wedge \exists preference(u', i)} simU(u, u')}.$$

Typically, users rate only a few items in a recommendation application because of the huge amount of the available items. This is our motivation for introducing the notion of *support* for each suggested item i, for user u. Support defines the percentage of friends of u that have expressed preferences for i, that is,

$$support(u, i) = \frac{|S, S \subseteq \mathcal{F}_u, s.t. \forall u' \in S, \exists preference(u', i)|}{|\mathcal{F}_u|}.$$

To estimate the worthiness of an item recommendation for a user, we propose to combine the *relevance* and *support* scores in terms of a *value* function. Formally, the *personal value* of an item $i \in \mathcal{I}$ for a user $u \in \mathcal{U}$, such that, $\nexists preference(u, i)$, is defined as: $value(u, i) = w_1 \times relevance(u, i) + w_2 \times support(u, i)$, $w_1 + w_2 = 1$.

Group Recommendations. In addition to personal recommendations, there are contexts in which people operate in groups, and so, a model for group recommendations should be defined. Some approaches have been recently proposed towards this direction (e.g., [2]).

In general, collaborative filtering combines the preferences of the single users to predict the preferences for the group as a whole. To this end, in our approach, we first compute the *personal value* scores for the unrated items for each user of the group. Based on these predictions, we then produce the aggregated value scores for the group. Formally, given a group of users $\mathcal{G}$, $\mathcal{G} \subseteq \mathcal{U}$, the group value of an item $i \in \mathcal{I}$ for $\mathcal{G}$, such that, $\forall u \in \mathcal{G}$, $\nexists preference(u, i)$, is: $value(\mathcal{G}, i) = Aggr_{u \in \mathcal{G}}(value(u, i))$.

We employ three different designs regarding the aggregation method $Aggr$, each one carrying different semantics: (i) the *least misery design*, capturing cases where strong user preferences act as a veto, (ii) the *fair design*, capturing more democratic cases where the majority of the group members is satisfied, and (iii) the *most optimistic design*, capturing cases where the most satisfied member of the group acts as the most influential member. In the least misery (resp., most optimistic) design the predicted value score for the group is equal to the minimum (resp., maximum) value score of the scores of the members of the group, while the fair design returns the average score.

Given a group of users and a restriction k on the number of the recommended items, our goal is to provide the group with k suggestions for items that are highly valued, without computing the group value scores of all database items.

3 *gRecs* System Overview

In this section, we describe the main components of the architecture of our system. A high level representation is depicted in Figure 1. Given a group of users, we first locate the friends of each user in the group. Friends preferences are employed for estimating personal recommendations, while in turn, personal recommendations are aggregated into recommendations for the whole group.

Friends Generator. This component takes as input a group of users $\mathcal{G}$ and returns the friends $\mathcal{F}_u$ of each user u in the group. The naive approach for finding the friends of all users in $\mathcal{G}$ requires the online computation of all similarity values between each user in $\mathcal{G}$ and each user in $\mathcal{U}$. We compute the similarity between two users with regard to their Euclidean distance. This however, is too expensive for a real-time recommendation application where the response time is an important aspect for the end users. To speed up the recommendation process, we perform preprocessing steps offline. More specifically, we organize users into clusters of users with similar preferences. For partitioning users into clusters, we use a hierarchical agglomerative clustering algorithm that follows a bottom-up strategy. Initially, the algorithm places each user in a cluster of his own. Then, at each step, it merges the two clusters with the greatest similarity. The similarity between two clusters is defined as the minimum similarity between any two users that belong to these clusters. The algorithm terminates when the clusters with the greatest similarity, have similarity smaller than δ. In this clustering approach, we consider as friends of each user u the members of the cluster that u belongs to. This set of users is a subset of $\mathcal{F}_u$.

Personal Recommendations Generator. In this step, we estimate the personal value scores of each item for each user in $\mathcal{G}$. To perform this operation, we employ the outputs of the previous step, i.e., the friends of the users in $\mathcal{G}$. Given a user $u \in \mathcal{G}$ and his friends $\mathcal{F}_u$, the procedure for estimating the $value(u, i)$ of each item i in $\mathcal{I}$ requires the computation of $relevance(u, i)$ and $support(u, i)$. Pairs of the form $(i, value(u, i))$ are maintained in a set $\mathcal{V}_u$. This component is also responsible for ranking, in descending order, all pairs in $\mathcal{V}_u$ on the basis of their personal value score.

Group Recommendations Generator. This component generates the k highest group-valued item recommendations for the group of users $\mathcal{G}$. To do this, we combine the personal value scores computed from the previous step by using either the *least misery*, the *fair* or the *most optimistic* design.

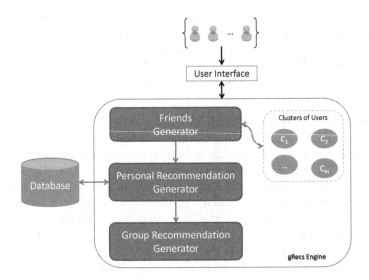

Fig. 1. *gRecs* system architecture

Instead of following the common way of computing the group value scores of all items and ranking the items based on these scores, we employ the TA algorithm [4] for efficient top-k computation. Note that TA is correct when the group value scores of the items are obtained by combining their individual scores using a monotone function. In our approach, aggregations are performed in a monotonic fashion, hence the applicability of the algorithm is straightforward.

4   Demonstration

The *gRecs* system for group recommendations has been implemented in JAVA on top of MySQL. We demonstrate our method using a movie ratings database [1]. In particular, we form groups of users with different semantics and choose an aggregation design from the available ones. After estimating the top-k group value scores, users are presented with the recommended movies. An explanation is also provided along with each movie, i.e., why this specific recommendation appears in the top-k list. For the *least misery* (resp., *most optimistic*) design, we report with each movie its group value score and the member of the group with the minimum (resp., maximum) personal value score for the movie, i.e., the member that is responsible for this selection. Similarly, for the *fair* design, we report with each movie the members of the group with personal value scores for the movie close to the group value score of the movie, i.e., the members that are highly satisfied, and hence, direct towards this selection.

References

1. Movielens data sets, `http://www.grouplens.org/node/12` (visited on October 2011)
2. Amer-Yahia, S., Roy, S.B., Chawla, A., Das, G., Yu, C.: Group recommendation: Semantics and efficiency. PVLDB 2(1), 754–765 (2009)
3. Breese, J.S., Heckerman, D., Kadie, C.M.: Empirical analysis of predictive algorithms for collaborative filtering. In: UAI, pp. 43–52 (1998)
4. Fagin, R., Lotem, A., Naor, M.: Optimal aggregation algorithms for middleware. In: PODS (2001)
5. Konstan, J.A., Miller, B.N., Maltz, D., Herlocker, J.L., Gordon, L.R., Riedl, J.: Grouplens: Applying collaborative filtering to usenet news. Commun. ACM 40(3), 77–87 (1997)

PEACOD: A Platform for Evaluation and Comparison of Database Partitioning Schemes

Xiaoyan Guo[1], Jidong Chen[1], Yu Cao[1], and Mengdong Yang[2]

[1] EMC Labs China
{xiaoyan.guo,jidong.chen,yu.cao}@emc.com
[2] Southeast University
mdyang@seu.edu.cn

Abstract. Database partitioning is a common technique adopted by database systems running on single or multiple physical machines. It is always crucial yet challenging for a DBA to choose an appropriate partitioning scheme for a database according to a specific query workload. In this paper, we present PEACOD, a platform that aims to ease the burden of DBAs for database partitioning design. By automating the processing of database partitioning scheme evaluation and comparison, PEACOD provides the DBAs with a conventional way to choose a suitable scheme.

1 Introduction

Database partitioning (or sharding) has been used for providing scalability for both transactional and analytical applications [1]. There already exist many kinds of general-purpose partitioning algorithms, among which round-robin, range-based, hashing are the most widely used. In the meantime, more ad-hoc and flexible partitioning schemes tailored for specific-purpose applications were also developed, such as the consistent hashing of Dynamo [2] and Schism [3]. As such, choosing the most suitable scheme for an application puts heavy burden on the database administrator (DBA), who needs to consider the following issues: partitioning key selection, data partition algorithms, data placement strategies, load balance, re-partitioning, implementation complexity, etc. The challenges for the DBA lie in the large number of alternative partitioning schemes to choose from, the uncertainty on the practical performance of each scheme, and the difficulty in efficiently comparing all feasible schemes and fast picking up the appropriate one. Without a partitioning advisory tool, usually the DBA has to make a choice based on his past experience and/or some heuristic rules, which very possibly may result in significantly sub-optimal performance of partitioning against the target application.

In this paper, we demonstrate PEACOD, a platform for evaluation and comparison of database partitioning schemes in an automatic and extendible way. In PEACOD, various partitioning schemes can be easily configured and measured against a database and a query workload via either simulated or actual execution, and the evaluation results and statistic logs will be vividly visualized so that the users can fully understand the practical effects of different schemes according to certain performance metrics. PEACOD can also recommend a suitable scheme according to the performance metrics. With the

S.-g. Lee et al. (Eds.): DASFAA 2012, Part II, LNCS 7239, pp. 304–308, 2012.

help of PEACOD, it becomes more convenient for DBAs to choose the right or proper partitioning schemes according to their own requirements. PEACOD is designed to be a stand-alone application which is compatible with multiple database systems. However, it can also be tightly integrated into a certain database system as a plug-in after customizations by the database vendor. In the rest of this paper, we first introduce the system components of PEACOD in Section 2, then describe the demonstration scenarios that we plan to arrange in Section 3, and finally conclude in Section 4.

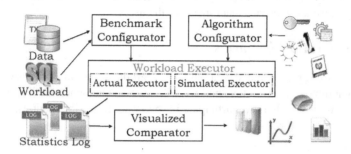

Fig. 1. System Architecture of PEACOD

2   System Overview

PEACOD runs in two phases: a partitioning scheme execution phase and an evaluation result visualization phase. In the first phase, dataset, workload and partitioning schemes are loaded and then evaluated to generate evaluation results and statistic logs, which in turn will be used in the second phase to draw figures for the exhibition and comparison of the effects of partitioning schemes. Figure 1 provides the overview system architecture, which consists of the following four components:

1) *Benchmark Configurator* loads the user-indicated dataset and workload. The dataset source can be a database instance or just plain text files, both of which should include the database schema information. The workload is a trace of SQL statements. PEACOD can handle both analytical and transactional workloads.
2) *Algorithm Configurator* imports the partitioning schemes under evaluation. PEACOD is embedded with several common partitioning schemes, and also supports user-defined schemes. For convenience, we define the following interfaces for each user-defined scheme being plugged to instantiate: how to choose partitioning key, how to partition data, how to route queries and how to re-partition the database when the number of nodes of a non-centralized system changes.
3) *Workload Executor* generates evaluation results and statistic logs that will be visualized later, via either the actual execution of the workload or a simulated execution. The actual execution takes place in the real database system environment, and will be invoked when the original input dataset is small enough or a randomly sampled subset of the dataset can be easily derived by PEACOD. In this case, the actual execution is able to generate the accurate statistics in a short period. Otherwise,

PEACOD will run a lightweight execution simulator which doesn't interact with the underlying database system and thus can run faster without significantly scarifying the quality of derived evaluation results and statistics.

4) *Visualized Comparator* parses and visualizes the execution results and statistic logs in the following three forms: figures that describe the performance of a scheme with a particular setting, figures that compare the performance of a specific scheme with different settings, and figures that compare the performance of different schemes according to the same setting. It also recommends a good candidate scheme based on the performance metrics that ranked by the users.

3 Demonstration Scenarios

We will first use a poster and several slides to introduce the motivation of PEACOD and highlight how it works. After that, we will show the audience the live demonstration system and invite them to participate in the system-user interactions. The whole procedure is described below.

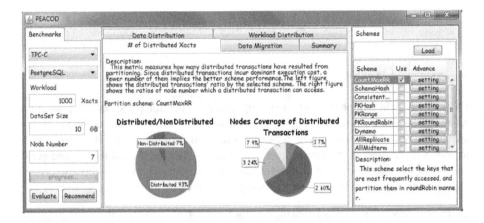

Fig. 2. Demonstration System of PEACOD

Benchmark Configuration. In this demonstration, we will focus on online transactional processing (OLTP) applications. The audience first specify a benchmark to be evaluated. The benchmark setting panel in the left of Figure 2 shows all the pre-defined workloads to be chosen from, which are TPC-C, TPC-W, TPC-E[1] and TATP[2]. The audience can configure the size of the dataset used in the benchmark and the underlying database system running the workload. Moreover, if the database system in assumption is distributed, the number of nodes in the system also can be configured.

Partitioning Scheme Configuration. The audience specify several partitioning schemes to be evaluated and compared. They also configure each scheme with different settings. The scheme setting panel in the right of Figure 2 shows totally eleven

---

[1] http://www.tpc.org

[2] http://tatpbenchmark.sourceforge.net

pre-defined schemes, including round-robin, range, hashing, consistent hashing, etc. We will prepare both text description and illustrating figures in order for the audience to better understand each partitioning scheme.

Simulated Workload Execution. Next, the system will run to generate useful statistic logs, by invoking a lightweight database partitioning simulator which generate required results without accessing the real database system. The simulator samples the dataset and executes the workload in main memory. As the demonstration focuses on OLTP workloads, we define four performance metrics of a partitioning scheme: 1) *Data Distribution*, which shows how uniformly data are distributed across partitions (nodes). 2) *Workload Distribution*, which evaluates how uniformly the data accesses by the workload are distributed across partitions. 3) *Number of Distributed Transactions*, which measures how many distributed transactions have resulted from partitioning. Since distributed transactions incur dominant execution cost, a fewer number of them implies the better scheme performance. 4) *Re-Partitioned Data Migration*, which indicates the amount of data to be migrated among nodes in case of database re-partitioning. Here, we will assume the scenario where database re-partitioning is triggered for the load re-balance of the system with one new node added. During this workload execution step, the simulator collects four types of statistic logs corresponding to the above four performance metrics.

Evaluation Result Visualization. In this step, the audience will review the figures in all three forms mentioned in Section 2 for scheme evaluation and comparison, as depicted on the main panel of Figure 2. The audience can see descriptive evaluation representations of a single scheme. They can modify the settings of a scheme and view the effect of this change - this functionality can facilitate the partitioning scheme parameter tuning. The audience can also compare different partitioning schemes based on the present figures so as to rank the partitioning schemes on their own. In addition, PEACOD can serve as a partitioning scheme advisor by recommending a good candidate scheme when the aduience press the recommend button.

4   Conclusion

In this demonstration, we present a tool, PEACOD, in order to help the DBAs to more efficiently choose the optimal partitioning scheme for a specific application scenario. PEACOD only requires the users to provide the input dataset and workload, and indicate the partitioning schemes that they intend to evaluate and compare. After that, PEACOD will automatically derive the performance results of schemes in comparison and show them to the users with illustrative figures of different forms. We demonstrate PEACOD on different scenarios using well-known benchmarks and present the functionality using its interface.

References

1. DeWitt, D., Gray, J.: Parallel database systems: the future of high performance database systems. Comm. ACM (1992)
2. DeCandia, G., Hastorun, D., Jampani, M., et al.: Dynamo: Amazon's highly available key-value store. In: SOSP (2007)
3. Curino, C., Jones, E., Zhang, Y., Madden, S.: Schism: a Workload Driven Approach to Database Replication and Partitioning. In: VLDB (2010)

Stream Data Mining Using the MOA Framework

Philipp Kranen[1], Hardy Kremer[1], Timm Jansen[1], Thomas Seidl[1],
Albert Bifet[2], Geoff Holmes[2], Bernhard Pfahringer[2], and Jesse Read[2]

[1] Data Management and Exploration Group, RWTH Aachen University, Germany
{kranen,kremer,jansen,seidl}@cs.rwth-aachen.de
[2] Department of Computer Science, University of Waikato, Hamilton, New Zealand
{abifet,geoff,bernhard,jmr30}@cs.waikato.ac.nz

Abstract. **M**assive **O**nline **A**nalysis (MOA) is a software framework
that provides algorithms and evaluation methods for mining tasks on
evolving data streams. In addition to supervised and unsupervised
learning, MOA has recently been extended to support multi-label classification and graph mining. In this demonstrator we describe the main
features of MOA and present the newly added methods for outlier detection on streaming data. Algorithms can be compared to established
baseline methods such as LOF and ABOD using standard ranking measures including Spearman rank coefficient and the AUC measure. MOA
is an open source project and videos as well as tutorials are publicly
available on the MOA homepage.

1 Introduction

Data streams are ubiquitous, ranging from sensor data to web content and click
stream data. Consequently there is a rich and growing body of literature on
stream data mining. For traditional mining tasks on static data, several established software frameworks such as WEKA (Waikato Environment for Knowledge
Analysis) provide mining algorithms and evaluation methods from the research
literature. Such environments allow for both choosing an algorithm for a given
application as well as comparing new approaches against the state of the art.

MOA [4] is a software environment for implementing algorithms and running
experiments for online learning from evolving data streams. MOA is designed to
deal with the challenging problems of scaling up the implementation of state of
the art algorithms to real world dataset sizes and of making algorithms comparable in benchmark streaming settings. MOA initially contained algorithms for
stream classification and was extended over the last years to support clustering,
multi-label classification and graph mining on evolving data streams. In this paper we describe the newly added methods for outlier detection on data streams
in comparison to established baseline methods.

Only two other open-source data streaming packages exist: VFML and a
RapidMiner plugin. The VFML (Very Fast Machine Learning) [5] toolkit was
the first open-source framework for mining high-speed data streams and very
large data sets. It was developed until 2003. VFML is written mainly in standard C, and contains tools for learning decision trees (VFDT and CVFDT),

S.-g. Lee et al. (Eds.): DASFAA 2012, Part II, LNCS 7239, pp. 309–313, 2012.

for learning Bayesian networks, and for clustering. The data stream plugin (formerly: concept drift plugin) [6] for RapidMiner (formerly: YALE (Yet Another Learning Environment)), is an extension to RapidMiner implementing operators for handling real and simulated concept drift in evolving streams.

MOA is built on experience with both WEKA and VFML. The main advantage of MOA is that it provides many of the recently developed data stream algorithms, including learners for multi-label classification, graph mining and outlier detection. Generally, it is straightforward to use or to extend MOA.

2   The MOA Framework

In the following we first describe the general architecture, usage and features of MOA. In Section 2.1 we discuss the new components for outlier detection and Section 2.2 provides information on documentation and tutorials.

Architecture, extension points and workflow follow the design depicted in Figure 1. The three components *data feed*, *algorithm* and *evaluation method* have to be chosen and parameterized in order to run an experiment. For each component a simple interface is available in MOA which can be used to include new components. MOA is written in Java, allowing portability to many platforms and usage of the well-developed support libraries. MOA can be used from the command line, as a library, or via the graphical user interface (cf. Figure 2).

Data Streams. MOA streams can be build using generators, reading ARFF files, joining several streams, or filtering streams. Most of the data generators commonly found in the literature are provided: Random Tree Generator, SEA Concepts Generator, STAGGER Concepts Generator, Rotating Hyperplane, Random RBF Generator, LED Generator, Waveform Generator, and Function Generator. Settings can be stored to generate benchmark data sets.

Classification. MOA contains a range of classification methods such as: Naive Bayes, Stochastic Gradient Descent, Perceptron, Hoeffding Tree, Adaptive Hoeffding Tree, Boosting, Bagging, and Leveraging Bagging.

Clustering. For clustering MOA contains several stream clustering methods, such as StreamKM++, CluStream, ClusTree, Den-Stream, CobWeb, as well as a large set of evaluation methods including the recent Cluster Mapping Measure (CMM) [7]. Dynamic visualization of cluster evolution is available, as depicted in Figure 2.

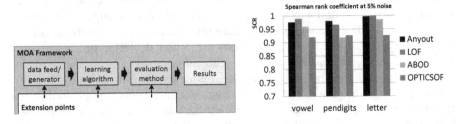

Fig. 1. Left: Architecture and workflow of MOA. Right: outlier detection results.

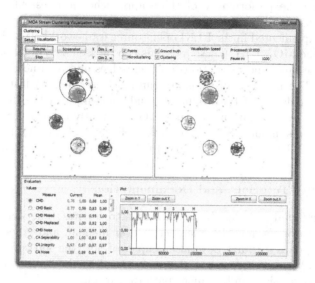

Fig. 2. Visualization tab of the clustering MOA graphical user interface

Multi-label Classification. Multi-label classification has seen considerable development in recent years, but so far most of the work has been carried out in the context of batch learning. MOA implements multi-label stream generators and several state of the art methods: ECC Ensembles of classifier-chains, EPS Ensembles of Pruning Sets, Multi-label Hoeffding Trees, and multi-label adaptive bagging methods.

Graph Mining. MOA also contains a framework for studying graph pattern mining on time-varying streams [3]. All methods work on coresets of closed subgraphs, compressed representations of graph sets. The methods maintain these sets in a batch-incremental manner, but use different approaches to address potential concept drift. MOA implements INCGRAPHMINER, WINGRAPHMINER and ADAGRAPHMINER.

2.1   Outlier Detection on Data Streams Using MOA

Outlier detection is an important task in stream data mining. Applications range from fault detection in network or transaction data to event or error detection in sensor networks and remote monitoring. Several approaches have been proposed in the literature, which make use of different paradigms. In [2] a solution using a hierarchy of clusterings is proposed, other solutions follow density based or distance based approaches. We added an OutlierDetector interface to MOA, which allows for easy inclusion of new or additional methods.

To evaluate the performance of these approaches it is essential to have a comparison to established methods such as LOF or ABOD. We use the outlier algorithms implemented in the ELKI open source framework[1] [1] and make them available in the MOA GUI. These methods can be seen as a baseline since they do not impose any time restrictions on themselves and assume random access to the data. The ELKI algorithms are run on a user defined history of point, where the granularity of the evaluation and the length of the history can be parameterized. As evaluation measures MOA currently provides three standard ranking measures, namely Spearman rank coefficient (cf. Fig. 1), Kendall's Tau and AUC (Area Under the ROC Curve) for outlier detection on data streams.

2.2   Website, Tutorials, and Documentation

MOA can be found at `http://moa.cs.waikato.ac.nz/`. The website includes a video and tutorials, an API reference, a user manual, and a general manual about mining data streams. Several examples of how the software can be used are available. We are currently working on extending the framework to include data stream regression, and frequent pattern learning.

3   Demo Plan and Conclusions

In this demonstrator we focus on presenting the newly added algorithms for outlier detection and the corresponding evaluation on data streams. For researchers MOA yields insights into advantages and disadvantages of different approaches and allows for the creation of benchmark streaming data sets through stored, shared and repeatable settings for the data feeds. Practitioners can use the framework to easily compare algorithms and apply them to real world data sets and settings. Besides providing algorithms and measures for evaluation and comparison, MOA is easily extensible with new contributions and allows for the creation of benchmark scenarios.

Acknowledgments. This work has been supported by the UMIC Research Centre, RWTH Aachen University, Germany.

References

1. Achtert, E., Kriegel, H.-P., Reichert, L., Schubert, E., Wojdanowski, R., Zimek, A.: Visual Evaluation of Outlier Detection Models. In: Kitagawa, H., Ishikawa, Y., Li, Q., Watanabe, C. (eds.) DASFAA 2010. LNCS, vol. 5982, pp. 396–399. Springer, Heidelberg (2010)
2. Assent, I., Kranen, P., Baldauf, C., Seidl, T.: Anyout: Anytime Outlier Detection on Streaming Data. In: Lee, S.-G., et al. (eds.) DASFAA 2012, Part I. LNCS, vol. 7238, pp. 228–242. Springer, Heidelberg (2012)

---

[1] ELKI: `http://elki.dbs.ifi.lmu.de/` - MOA: `http://moa.cs.waikato.ac.nz/`

3. Bifet, A., Holmes, G., Pfahringer, B., Gavaldà, R.: Mining frequent closed graphs on evolving data streams. In: 17th ACM SIGKDD, pp. 591–599 (2011)
4. Bifet, A., Holmes, G., Pfahringer, B., Kranen, P., Kremer, H., Jansen, T., Seidl, T.: Moa: Massive online analysis, a framework for stream classification and clustering. Journal of Machine Learning Research - Proceedings Track 11, 44–50 (2010)
5. Hulten, G., Domingos, P.: VFML – a toolkit for mining high-speed time-changing data streams (2003)
6. Klinkenberg, R.: Rapidminer data stream plugin. RapidMiner (2010), http://www-ai.cs.uni-dortmund.de/auto?self=eit184kc
7. Kremer, H., Kranen, P., Jansen, T., Seidl, T., Bifet, A., Holmes, G., Pfahringer, B.: An effective evaluation measure for clustering on evolving data stream. In: 17th ACM SIGKDD, pp. 868–876 (2011)

Shot Classification Using Domain Specific Features for Movie Management*

Muhammad Abul Hasan[1], Min Xu[1], Xiangjian He[1], and Ling Chen[2]

[1] Centre for Innovation in IT Services and Applications (iNEXT)
[2] Centre for Quantum Computation and Intelligent Systems (QCIS)
University of Technology, Sydney
PO Box 123 Broadway, NSW 2007, Australia
{Muhammad.Hasan,Min.Xu,Xiangjian.He,Ling.Chen}@uts.edu.au

Abstract. Among many video types, movie content indexing and retrieval is a significantly challenging task because of the wide variety of shooting techniques and the broad range of genres. A movie consists of a series of video shots. Managing a movie at shot level provides a feasible way for movie understanding and summarization. Consequently, an effective shot classification is greatly desired for advanced movie management. In this demo, we explore novel domain specific features for effective shot classification. Experimental results show that the proposed method classifies movie shots from wide range of movie genres with improved accuracy compared to existing work.

Keywords: movie management, context saliency, SVM classification.

1 Introduction

Since movie occupies a big portion of entertainment videos, effective movie data management is greatly desired for multimedia database systems. A video shot is a group of video frames with high similarity taken by a single camera at a time. Video shots can be used as basic units for movie understanding and summarization. According to [4], movie shots are determined by the characters' size in a frame. Primarily, movie shot types include: *close shot* (consists of extreme close-up, medium close-up, full close-up, wide close-up and close shot), *medium close shot, medium shot, medium full shot, full shot, over the shoulder shot, cut shot* and *establishing shot*. Accurately classifying movie shots into these types are difficulty because of the wide variety of shooting techniques as well as the movie genres.

In this demo, we classify movie shots using keyframes. A keyframe of a movie shot summarizes the contents of a shot. We consider the first frame of each shot as the keyframe. There are two steps in our proposed method. The first step is feature extraction. The second step is shot classification. Fig. 1 shows the demo interface of our proposed method.

---

* This research was supported by National Natural Science Foundation of China No. 61003161 and UTS ECR grant.

S.-g. Lee et al. (Eds.): DASFAA 2012, Part II, LNCS 7239, pp. 314–318, 2012.

Fig. 1. Demo interface of our proposed method

The main contribution of this work is that we extracted a set of novel movie domain specific features to classify movie shots. Especially, the *Weighted Hue Histogram* and the *Regions with Skin Colour* are proposed as highly distinctive features in movie shot classification.

2   Feature Extraction

A 48 dimensional feature vector is used in movie shot classification. We first introduced the features which were employed by existing approaches before. Next, we introduce the two novel domain specific features: Weighted Hue Histogram and Regions with Skin Color.

Context Saliency Features. Context saliency maps [1] considers background information and statistical information in computing saliency maps. Moreover, it considers geometric context of the keyframe in producing effective saliency maps. In our method, context saliency is firstly computed. Then, the context saliency map is adopted. At the beginning, context saliency is computed. At last, each local features are extracted. To do that, the context saliency map is divided into 9 (3×3) local regions. From each local region, total magnitude of the context saliencies is computed. Then, 9 local context saliency magnitude is normalized by $\hat{c}_{r_{x,y}} = \frac{c_{r_{xy}}}{\sum_x \sum_y c_{r_{x,y}}}$. Here $\hat{c}_{r_{x,y}}$ is the normalized local context saliency and $c_{r_{x,y}}$ is local context saliency. x, y are local indexes and $x \in [1:3]$ and $y \in [1:3]$.

Edge Feature. In close shots, objects are always in focus and backgrounds are mostly out of focus. However, as the distance between camera and the focused object increases, the degree of blurriness decreases. Therefore, amount of edge energy varies greatly in different type of shots. We use Haar wavelet coefficients to measure edge energy. There are two steps in measuring edge energy. 1) Haar wavelet transform is applied on the keyframes and decomposed upto level 3. 2) Edge energy map is computed by using $E_i = \sqrt{lh_i^2 + hl_i^2 + hh_i^2}$, where $i \in [1:3]$

represents the level of the edge map. After calculating edge energy of each pixel, the edge energy map is segmented into 9 (3 × 3) local regions. We extract 9 features from each of these 9 local regions. From each local region, total magnitude of the edge energy is computed. Then, each local edge energy magnitude is normalized by using total edge energy of the corresponding keyframe.

Entropy. For a given keyframe, 9 entropy features are computed from its grayscale values. The keyframe is segmented into 3 × 3 local regions and each region gray pixel histogram is computed. Using the histogram bins, local entropy is computed using $E_{r_{x,y}} = -\sum_{i=1}^{N} h_i \log(h_i)$. Here i is index, N is number of bins and h_i is total number of counts in the histogram bin at index i. Finally, local entropies are normalized using total entropy of the keyframe.

Weighted Hue Histogram. In this subsection, we propose a novel weighted hue histogram scheme. In this scheme, the context saliency is combined with hue histogram to assign importance to the hue bins belonging to salient regions. Firstly, hue histogram of keyframe is computed. Then, the context saliency map is thresholded using entropy thresholding [3] method. The thresholded context saliency map is defined by $S_{x,y} \in \{0, 1\}$, where, x, and y represent the spatial index of a pixel, 1 indicates that the pixel saliencies are higher than the threshold value and 0 indicates the otherwise. Using $S_{x,y}$, we eliminate the hue value of the keyframe as follows: $\theta_{x,y} = \emptyset, \quad \forall S_{x,y} = 0$. Next, we compute hue mean μ_θ and hue standard deviation σ_θ. Using these parameters, a weighted hue histogram is obtained using $H_w(i) = H(i) \times exp\left(-\frac{i - \mu_\theta}{\sigma_\theta^2}\right)$, where H is hue histogram and i is hue index. Hue histogram is a global feature of each keyframe. In our classification method, we consider $i = 12$ bins in computing 12 features from weighted hue histogram.

Regions with Skin Colour. In movies, skin colour varies significantly due to the indoor, outdoor shootings and use of different light sources. In this work, we use a simple and robust threshold based technique to segment regions with skin colour. First, a multiview face detector (MVFD) [2] is used to identify the face regions within the keyframes. Then, the mean $\boldsymbol{\mu}_s = \{\mu_r, \mu_g, \mu_b\}$ and covariance matrix $\boldsymbol{\Sigma}_s$ of the facial region are computed from red, green and blue channels. Then, each pixel of the corresponding keyframe is labeled as skin or non-skin pixels as follows:

$$S(I_{x,y}) = \begin{cases} 1 & \text{if } dist(I_{x,y}, \boldsymbol{\mu}_s, \boldsymbol{\Sigma}_s) < t \\ 0 & \text{otherwise} \end{cases}$$

where $I_{x,y}$ is pixel colour in r, g, b, colour space. $dist(I_{x,y}, \boldsymbol{\mu}_s, \boldsymbol{\Sigma}_s)$ is Mahalanobis distance and t is a predefined threshold. If a keyframe does not contain any face then the previous keyframe's parameters are used in determining pixels belonging to skin. After that, 9 local features are extracted by dividing the frame into 9 (3 × 3) local regions. Then, total of skin pixels of each each local region is computed and normalized using total number of pixels in the keyframe.

3   Shot Classification

For shot classification, Support Vector Machine (SVM) is used. The training and validation data collected form movies almost all kinds of movie genres. We use radial basis function (RBF) as kernel function, $K(x_i, x_j) = exp(-r\|x_i - x_j\|^2)$, $r > 0$, for SVM classification, and one against-all multi-class approach is used. The training data are $(\mathbf{x}_i, y_i), i = 1, ..., l$ where $\mathbf{x}_i \in \mathbb{R}^n$, $y \in \{1, +1\}^l$. SVM finds an optimal separating hyperplane that classifies the one class against all by the minimum expected test error. The experiments are conducted using a dataset of 5134 movie keyframes representing almost equal amount of frames for each class of shots. We segment our dataset into two equal size training and testing sets. The dataset is collected from all types of movie genres (*e.g.* action, horror, romantic, drama, science fiction, comedy, thriller, adventure etc). The ground truth of shot classes was labeled manually. Fig. 2 shows shot classification accuracy rate and a comparison with an existing work. The experimental results

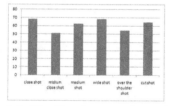
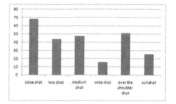

(a) Classification results of proposed method

(b) Classification results proposed in [5]

Fig. 2. Comparison of experimental results

are compared with [5]. The overall accuracy of the proposed method improves significantly. The accuracy for close shot remains almost the same while a slight improvement is noticed in medium and over the shoulder shots. However, a great improvement is achieved in wide shots and cut shots.

4   Conclusion

Movie shot classification is very important in CBVIR application with many challenges. In this demo, we have proposed a domain knowledge based movie shot classification method. We represent the foreground region using context saliency map. Optical effect of camera lens is modelled by edge energy. Background and foreground characteristics are modeled by using entropy and skin features. Using weighted hue histogram, we put emphasis on the foreground region. Experiments demonstrate that the proposed classification method performs better than existing works.

References

1. Shi, L., Wang, J., Xu, L., Lu, H., Xu, C.: Context saliency based image summarization. In: Proceedings of the IEEE ICME, pp. 270–273 (2009)
2. Huang, C., Ai, H., Li, Y., Lao, S.: Vector boosting for rotation invariant multi-view face detection. In: Proceedings of the IEEE ICCV, vol. 1 (2005)
3. Kapur, J.N., Sahoo, P.K., Wong, A.K.C.: A New Method for Gray-Level Picture Thresholding Using the Entropy of the Histogram. Computer Vision, Graphics, and Image Processing 29(3), 273–285 (1985)
4. Cantine, J., Lewis, B., Howard, S.: Shot by Shot; A Practical Guide to Filmmaking, Pittsburgh Filmmakers (1995)
5. Xu, M., Wang, J., Hasan, M.A., He, X., Xu, C., Lu, H., Jin, J.S.: Using Context Saliency for Movie Shot Classification. In: Proceedings of the IEEE ICIP (2011)

PA-Miner: Process Analysis Using Retrieval, Modeling, and Prediction

Anca Maria Ivanescu, Philipp Kranen,
Manfred Smieschek, Philip Driessen, and Thomas Seidl

Data Management and Data Exploration Group
RWTH Aachen University, Germany
lastname@cs.rwth-aachen.de

Abstract. Handling experimental measurements is an essential part of research and development in a multitude of disciplines, since these contain information about the underlying process. Besides an efficient and effective way of exploring multiple results, researchers strive to discover correlations within the measured data. Moreover, model-based prediction of expected measurements can be highly beneficial for designing further experiments. In this demonstrator we present PA-Miner, a framework which incorporates advanced database techniques to allow for efficient retrieval, modeling and prediction of measurement data. We showcase the components of our framework using the fuel injection process as an example application and discuss the benefits of the framework for researchers and practitioners.

1 Introduction

Experiments are an essential part of research and development in a multitude of disciplines. In biology the reactions of organisms to different stimuli or environmental conditions are investigated. In mechanical engineering and related areas potentially complex test-benches are set up and observations are recorded for various input settings. The general scenario consists of setting a number of input variables, running the experiment and recording a number of output variables. Testing all possible input settings is hardly ever possible due to limited resources in terms of time, money, space or machinery hours. Having performed a series of experiments, the following tasks are generally of interest: **Retrieval -** browse your results efficiently to explore and compare different settings and outcomes; **Modeling -** derive models from the experimental data to mathematically describe the relationship between inputs and outputs; **Prediction -** use models to estimate output values for new input settings to plan and steer further series of experiments.

Retrieval and exploration of results is in practice often done by browsing folders of result files and manually opening and comparing results, be it images or numbers. Modeling and prediction are often replaced by guessing and intuition to set up new series of experiments. In this paper we describe PA-Miner, an extensible software framework that combines database and data mining techniques to provide a tool for efficient retrieval, modeling and prediction of experimental measurement data. We detail the components of PA-Miner in Section 2, describe

S.-g. Lee et al. (Eds.): DASFAA 2012, Part II, LNCS 7239, pp. 319–322, 2012.

the GUI along with an application example in Section 3 and state the objectives of our demonstrator in Section 4.

2   PA-Miner Components

In order to use PA-Miner the experimental results have to be initially loaded into the database. To this end an experiment has to be described in the XML format shown in Figure 1(c). We refer to an experiment as an input setting with the corresponding measurements of output variables. The main tag is the **experiment**, which may contain several **input** tags with the input variable **name** and **unit** and a **measurement** tag. Within the measurement tag the **output** tags

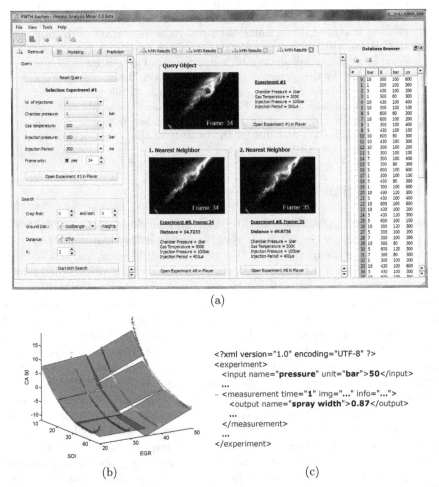

(a)

(b)   (c)

Fig. 1. PA-Miner components: (a) GUI - retrieval tab. (b) Modeling an output variable as a piecewise linear function of two input variables using 6 models. (c) XML format for the description of the experiments.

must be placed. These have a **name** attribute and optionally a **unit**. If for one input setting the output variables are measured at several time steps, then several measurement tags can be used (one for each time step) within the experiment tag. PA-Miner loads all XML files from a user specified folder into the database.

The retrieval component allows the user to either load an external experiment from an XML file or specify an input setting and load the corresponding experiment from the database. In the case of a time series of measurements, the user can then either select a single time step (single result) or the entire time series and search for the most similar results in the data base. This is especially useful for comparing simulation results to results from real experiments (cf. Section 3). For efficient retrieval and similarity search based on attribute values from a single time step we use a regular R*-tree [3]. For similarity search on time series of entire experiments PA-Miner allows for different distance functions, such as dynamic time warping, and uses specialized indexes [2] as well as filter-and-refine approaches [1] to speed up query processing.

Modeling the output variables as a linear function of the input variables based on the experimental measurements can provide a useful insight into the process and give hints for the design of further experiments [7]. To generate a model the user selects an output to be modeled and can optionally restrict the inputs to be fixed or varied individually. Nonlinear behavior can be approximated by allowing a larger number of linear submodels. For modeling the PA-Miner currently uses multiple regression for building linear models, and a simple model regression tree [5] for the approximation of nonlinear models through several linear submodels. Figure 1(b) exemplary shows the output modeled as a piecewise linear function of a 2-dimensional input. The resulting models are visualized and additionally displayed as parameterized equations.

The prediction component finally uses generated models to predict output values for user specified input settings. Prediction can thereby play a significant role in optimizing the design and setup for further experiments. This is especially helpful when experiments are expensive or resources are limited (cf. Section 1). As in the retrieval component the user can either perform a prediction for a specified time step or let PA-Miner predict the entire time series.

3   GUI and Application Example

In this section we show an example application of PA-Miner in the heat and mass transfer domain. Being responsible for the air-fuel mixture formation, the fuel injection process sets the initial conditions for the subsequent combustion process, and is thus a major research topic. The fuel injection process is examined by carrying out experiments in a pressure chamber under different conditions and employing optical measurement techniques [6]. A huge amount of data in form of spray images is generated in this manner and used to analyze, control and stabilize the fuel injection process. The goal is to investigate the impact of the pressure conditions, number of injections as well as the injection time on the overall spray shape (penetration length, spray width, vortex positions), from which conclusions about the fuel injection strategies can be drawn.

PA-Miner offers support for the mechanical engineers to mine the experimental data. The PA-Miner retrieval component for the fuel injection scenario, illustrated in Figure 1(a), allows the user to specify a single spray image or an entire experiment and look for the input values which deliver similar results [4]. In the same way synthetically generated experiments can be loaded and the most similar results from real experiments can be retrieved from the database. Generally, if no images are available, the features are shown in a table instead. The PA-Miner modeling component aims at offering support in understanding the influence of the chamber pressure, and/or that of the injection time, on the spray shape. The nonlinear process of fuel injection is visualized by a piecewise linear function. The PA-Miner prediction component focuses on assisting mechanical engineers in approximating the spray shape or the entire time series of spray shapes for some (not yet measured) input settings to effectively design new series of experiments.

4   Demo Objectives and Conclusion

In our demonstrator we will showcase the three components of PA-Miner using application data from the fuel injection process. The demonstrator will be interesting for researchers developing advanced applications for database systems and practitioners dealing with the analysis of experimental data. We hope to discuss with the audience aspects both w.r.t using the framework for own experimental data as well as integrating new methods for retrieval, modeling or prediction.

Acknowledgments. The authors gratefully acknowledge the financial support of the Deutsche Forschungsgemeinschaft (DFG) within the Collaborative Research Center SFB-686 "Model-Based Control of Homogenized Low-Temperature Combustion".

References

1. Assent, I., Kremer, H., Seidl, T.: Speeding Up Complex Video Copy Detection Queries. In: Kitagawa, H., Ishikawa, Y., Li, Q., Watanabe, C. (eds.) DASFAA 2010. LNCS, vol. 5981, pp. 307–321. Springer, Heidelberg (2010)
2. Assent, I., Krieger, R., Afschari, F., Seidl, T.: The ts-tree: Efficient time series search and retrieval. In: EDBT, pp. 252–263. ACM, New York (2008)
3. Beckmann, N., Kriegel, H.-P., Schneider, R., Seeger, B.: The r*-tree: an efficient and robust access method for points and rectangles. In: SIGMOD, vol. 19, pp. 322–331 (1990)
4. Beecks, C., Ivanescu, A.M., Seidl, T., Martin, D., Pischke, P., Kneer, R.: Applying similarity search for the investigation of the fuel injection process. In: Similarity Search and Applications (SISAP), pp. 117–118 (2011)
5. Karalic, A.: Linear regression in reegression tree leaves. In: International School for Synthesis of Expert Knowledge (1992)
6. Martin, D., Stratmann, J., Pischke, P., Kneer, R.: Experimental investigation of the interaction of multiple GDI injections using laser diagnostics. SAE Journal of Engines 3, 372–388 (2010)
7. Paoletti, S., Juloski, A.L., Ferrari-trecate, G., Vidal, R.: Identification of hybrid systems: a tutorial. European Journal of Control 513(2-3), 242–260 (2007)

Data Management Challenges and Opportunities in Cloud Computing

Moderator: Kyuseok Shim[1],
Panelists: Sang Kyun Cha[1], Lei Chen[2],
Wook-Shin Han[3], Divesh Srivastava[4],
Katsumi Tanaka[5], Hwanjo Yu[6], and Xiaofang Zhou[7]

[1] Seoul National University, Korea
[2] Hong Kong University of Science and Technology, Hong Kong
[3] Kyungpook National University, Korea
[4] AT&T Labs-Research, USA
[5] Kyoyo University, Japan
[6] POSTECH, Korea
[7] University of Queensland, Australia

Abstract. Analyzing large data is a challenging problem today, as there is an increasing trend of applications being expected to deal with vast amounts of data that usually do not fit in the main memory of a single machine. For such data-intensive applications, database research community has started to investigate cloud computing as a cost effective option to build scalable parallel data management systems which are capable of serving petabytes of data for millions of users. The goal of this panel is to initiate an open discussion within the community on data management challenges and opportunities in cloud computing. Potential topics to be discussed in the panel include: MapReduce framework, shared-nothing architecture, parallel query processing, security, analytical data management, transactional data management and fault tolerance.

S.-g. Lee et al. (Eds.): DASFAA 2012, Part II, LNCS 7239, p. 323, 2012.
© Springer-Verlag Berlin Heidelberg 2012

Detecting Clones, Copying and Reuse on the Web (DASFAA 2012 Tutorial)

Xin Luna Dong and Divesh Srivastava

AT&T Labs-Research, Florham Park NJ 07932, USA
{lunadong,divesh}@research.att.com

1 Introduction

The Web has enabled the availability of a vast amount of useful information in recent years. However, the Web technologies that have enabled sources to share their information have also made it easy for sources to copy from each other and often publish without proper attribution. Understanding the copying relationships between sources has many benefits, including helping data providers protect their own rights, improving various aspects of data integration, and facilitating in-depth analysis of information flow.

The importance of copy detection has led to a substantial amount of research in many disciplines of Computer Science, based on the type of information considered. The Information Retrieval community has devoted considerable effort to finding plagiarism, near-duplicate web pages and text reuse. The Multimedia community has considered techniques for copy detection of images and video, especially in the presence of distortion. The Software Engineering community has examined techniques to detect clones of software code. Finally, the Database community has focused on mining and making use of overlapping information between structured sources across multiple databases and more recently on copy detection of structured data across sources.

In this seminar, we explore the similarities and differences between the techniques proposed for copy detection across the different types of information. We do this with illustrative examples that would be of interest to data management researchers and practitioners. We also examine the computational challenges associated with large-scale copy detection, indicating how they could be detected efficiently, and identify a range of open problems for the community.

2 Target Audience

The target audience for this seminar is anyone with an interest in understanding information management across a huge number of data sources. In particular, this includes the attendees at database conferences like DASFAA. The assumed level of mathematical sophistication will be that of the typical conference attendees.

S.-g. Lee et al. (Eds.): DASFAA 2012, Part II, LNCS 7239, pp. 324–325, 2012.

3   Seminar Outline

Our seminar is example driven, and organized as follows.

Information Copying Examples (10 minutes): The seminar will start with a variety of real-world examples illustrating the prevalence of information copying on the Web.

Common Themes in Copy Detection (10 minutes): Next, we overview the common themes underlying copy detection techniques for various types of data. The first common theme is to detect unexpected sharing of data fragments under the no-copy assumption. The second common theme is to be tolerant to distortion or modification of copied information.

Copy Detection for Unstructured Data (25 minutes): In this unit, we present a variety of techniques proposed for detection of reuse in information represented as text, images and video. At the heart of these techniques are scalable algorithms for similarity detection, and we identify common techniques explored across the different types of information.

Copy Detection for Structured Data (25 minutes): In this unit, we present a variety of techniques proposed for copy detection when the information has a richer structure than simple text. We consider approaches for both software code and relational databases. In particular, for software code, we highlight the use of tree structure and dependency graph for copy detection. For relational databases, we differentiate between techniques that simply find overlapping information between structured sources and those that are able to detect evidence of copying. We will highlight the role of source quality metrics like accuracy and coverage in copy detection.

Open Problems (20 minutes): We will present many open problems in copy and plagiarism detection. In general, it is a challenging problem to perform Web-scale copy detection, and exploit evidence from various types of information for detecting copying between structured and unstructured sources. More specifically, for relational data, such open problems include improving scalability of copying detection, detecting copying in an open world where there can be hidden sources, and combining copy detection with other integration techniques such as schema mapping and record linkage for better detection results.

4   Conclusions

Copying of information is prevalent on the Web, and understanding the copying relationships between sources is very important. We expect two main learning outcomes from this seminar. In the short term, we expect that this seminar, by comparing and contrasting the techniques used by different communities for copy detection, will enable the audience to gain a unified understanding of the topic. Taking a more long-term view, we hope that it will foster interactions between researchers across these multiple disciplines to investigate and develop more comprehensive and scalable techniques for copy detection on the Web.

Query Processing over Uncertain
and Probabilistic Databases

Lei Chen[1] and Xiang Lian[2]

[1] Department of Computer Science and Engineering, Hong Kong University
of Science and Technology, Hong Kong, China
leichen@cs.ust.hk
[2] Department of Computer Science, University of Texas,
Pan American, Texas, U.S.A.
xlian@cs.utpa.edu

1 Introduction

Recently, query processing over uncertain data has become increasingly important in many real applications like location-based services (LBS), sensor network monitoring, object identification, and moving object search. In many of these applications, data are inherently uncertain and imprecise, thus, we can either assign a probability to each data object or model each object as an uncertainty region. Based on these models, we have to re-define and study queries over uncertain data. In this tutorial, we will first introduce data models that are used to model uncertain and probabilistic data. Then, we will discuss various types of queries together with their query processing techniques. After that, we will introduce recent trends on query processing over uncertain non-traditional databases, such as sets and graphs. Finally, we will highlight some future research directions. The tutorial aims to introduce the state-of-the-art query processing techniques over uncertain and probabilistic data and discuss the potential research problems.

2 Tutorial Outline

The topics covered in this tutorial include:

1. Data models for uncertain and probabilistic data;
2. Query processing techniques over uncertain and probabilistic data;
3. Query processing techniques over uncertain non-traditional data;
4. Open Challenges and Future Directions.

2.1 Data Models for Uncertain and Probabilistic Data

Based on the correlations existed among uncertain data (objects), we can classify the data uncertainty into two categories, independent uncertainty and correlated uncertainty. For the independent uncertainty, we assume that all objects in uncertain databases are independent of each other. That is, the existence of one

S.-g. Lee et al. (Eds.): DASFAA 2012, Part II, LNCS 7239, pp. 326–327, 2012.

object (or the attribute value of one object) would not affect that of another object. In contrast, the correlated uncertainty indicates that the existence of an object (or the attribute value of an object) may be correlated with that of another object. As an example, the sensors deployed in a spatially close area may report correlated sensory data such as temperature or humidity. Based on the different types of uncertainty, we can propose different data models to model uncertain and probabilistic data.

2.2 Query Processing Techniques over Uncertain and Probabilistic Data

Due to the unique characteristic of uncertain and probabilistic data, efficient query processing is quit challenging due to high computation cost on exponential number combinations of possible data instances. In order to reduce the processing cost, the current works often employ a filter-and-refine frame work. Specifically, before directly computing the query results, we first conduct effective pruning methods, including spatial pruning and probabilistic pruning, then we refine the left candidates by direct computation or sampling solutions.

2.3 Query Processing Techniques over Uncertain Non-traditional Data

In addition to probabilistic relational data and uncertain spatial data, recently, many works have been conducted on queries over uncertain non-traditional data, such as XML, graphs, strings, and time series data. Compared to their counterparts, query processing over uncertain versions of these data is much more challenges. We will present the related work in this area, mainly on pruning and sampling methods.

2.4 Open Challenges and Future Directions

Though many works having been proposed to address the challenges of query processing over uncertain or probabilistic data, there are still many opportunities to explore. For example, query processing on distributed uncertain data, keyword search over uncertain data, and quality measures of uncertain query results.

To summarize, this tutorial discusses the state-of-art query processing techniques on various of uncertain and probabilistic data, we hope the tutorial can give audience an overview of the research in this area and motivate more interesting works in the future.

Acknowledgement. This work is supported in part by Hong Kong RGC grants N HKU-ST612/09, National Grand Fundamental Research 973 Program of China under Grant 2012CB316200.

Tutorial: Data Stream Mining and Its Applications

Latifur Khan[1] and Wei Fan[2]

[1] University of Texas at Dallas
lkhan@utdallas.edu
[2] IBM T.J. Watson Research
weifan@us.ibm.com

Abstract. Data streams are continuous flows of data. Examples of data streams include network traffic, sensor data, call center records and so on. Their sheer volume and speed pose a great challenge for the data mining community to mine them. Data streams demonstrate several unique properties: infinite length, concept-drift, concept-evolution, feature-evolution and limited labeled data. Concept-drift occurs in data streams when the underlying concept of data changes over time. Concept-evolution occurs when new classes evolve in streams. Feature-evolution occurs when feature set varies with time in data streams. Data streams also suffer from scarcity of labeled data since it is not possible to manually label all the data points in the stream. Each of these properties adds a challenge to data stream mining.

Multi-step methodologies and techniques, and multi-scan algorithms, suitable for knowledge discovery and data mining, cannot be readily applied to data streams. This is due to well-known limitations such as bounded memory, high speed data arrival, online/timely data processing, and need for one-pass techniques (i.e., forgotten raw data) issues etc. In spite of the success and extensive studies of stream mining techniques, there is no single tutorial dedicated to a unified study of the new challenges introduced by evolving stream data like change detection, novelty detection, and feature evolution. This tutorial presents an organized picture on how to handle various data mining techniques in data streams: in particular, how to handle classification and clustering in evolving data streams by addressing these challenges. The importance and significance of research in data stream mining has been manifested in most recent launch of large scale stream processing prototype in many important application areas. In the same time, commercialization of streams (e.g., IBM InfoSphere streams, etc.) brings new challenge and research opportunities to the Data Mining (DM) community. In this tutorial a number of applications of stream mining will be presented such as adaptive malicious code detection, on-line malicious URL detection, evolving insider threat detection and textual stream classification.

A Biographical Sketch of the Presenter(s):

- Latifur R. Khan is currently an Associate Professor in the Computer Science department at the University of Texas at Dallas (UTD), where he has taught and conducted research since September 2000. He received his Ph.D. and M.S. degrees in Computer Science from the University of Southern California (USC), USA in

S.-g. Lee et al. (Eds.): DASFAA 2012, Part II, LNCS 7239, pp. 328–329, 2012.

August of 2000, and December of 1996 respectively. His research work is supported by grants from NASA, the Air Force Office of Scientific Research (AFOSR), National Science Foundation (NSF), the Nokia Research Center, and Raytheon. In addition, Dr. Khan's research areas cover data mining, multimedia information management, and semantic web. He has published more than 160 papers in data mining, and database conferences, such as ICDM, ECML/PKDD, PAKDD, AAAI, ACM Multimedia, and journals such as VLDB, TKDE, Bio Informatics, KAIS etc. Dr. Khan has served a PC member of several conferences such as KDD, ICDM, SDM, and PAKDD. Dr. Khan is currently serving on the editorial boards of a number of journals including *IEEE Transactions on Knowledge and Data Engineering (TKDE)*.

- Wei Fan received his PhD in Computer Science from Columbia University in 2001 and has been working in IBM T.J.Watson Research since 2000. He published more than 90 papers in top data mining, machine learning and database conferences, such as KDD, SDM, ICDM, ECML/PKDD, SIGMOD, VLDB, ICDE, AAAI, ICML, IJCAI, TKDE, KAIS etc. Dr. Fan has served as Associate Editor of TKDD, Area Chair, Senior PC of SIGKDD'06/10, SDM'08/10/11/12, ECML/PKDD/11/12 and ICDM'08/09/10, sponsorship co-chair of SDM'09, award committee member of ICDM'09/11, His main research interests and experiences are in various areas of data mining and database systems, such as, risk analysis, high performance computing, extremely skewed distribution, cost-sensitive learning, data streams, ensemble methods, easy-to-use nonparametric methods, graph mining, predictive feature discovery, feature selection, sample selection bias, transfer learning, novel applications and commercial data mining systems. His thesis work on intrusion detection has been licensed by a start-up company since 2001. His co-teamed submission that uses Random Decision Tree (www.dice4dm.com) has won the ICDM'08 Championship. His co-authored paper in ICDM'06 that uses "Randomized Decision Tree" to predict skewed ozone days won the best application paper award. His co-authored paper in KDD'97 on distributed learning system "JAM" won the runner-up best application paper award. He received IBM Outstanding Technical Achievement Awards in 2010 for his contribution in building Infosphere Streams.

URLs of the Slides/Notes:

```
http://www.utdallas.edu/~lkhan/
DASFAATutorial2012Final.pptx
```

Storing, Querying, Summarizing, and Comparing Molecular Networks: The State-of-the-Art

Sourav S. Bhowmick and Boon-Siew Seah

School of Computer Engineering, Nanyang Technological University, Singapore
Singapore-MIT Alliance, Nanyang Technological University, Singapore
{assourav,seah0097}@ntu.edu.sg

Abstract. A grand challenge of systems biology is to model the cell. The cell can be viewed as an integrated network of cellular functions. Each cellular function is defined by an interconnected ensemble of molecular networks and represent the backbone of molecular activity within the cell. The critical role played by these networks along with rapid advancement in high-throughput techniques has led to explosion in molecular interaction data. In this tutorial we explore the data management and mining techniques that have been proposed in the literature for storing, querying, summarizing, and comparing molecular networks and pathways. It offers an introduction to these issues and a synopsis of the state-of-the-art.

1   Tutorial Overview

In recent times, data on molecular interactions are increasing exponentially due to advances in technologies such as mass spectrometry, genome-wide chromatin immuno-precipation, yeast two-hybrid assays, combinatorial reverse genetic screens, and rapid literature mining techniques. Data on thousands of interactions in humans and most model species have become available. This deluge of information presents exciting new opportunities for comprehending cellular functions and disease in the future.

In this tutorial we explore the field of data management and mining techniques that have been proposed in the literature for understanding molecular networks and pathways. We begin by introducing fundamental concepts related to molecular networks data and the critical role interaction networks play in functioning of cells. Next, our discussion focuses on the following five main components: (a) techniques for modeling and representing molecular network data, (b) techniques for storing and querying molecular network data, (c) techniques for summarizing molecular networks, and (d) molecular network comparison techniques. We conclude by identifying potential research directions in the context of data management in this area. To the best of my knowledge, the content of this tutorial has not been presented in any major conference.

2   Full Description of the Tutorial

The tutorial consists of the following topics.

Section 1: Introduction and Motivation. This section includes a brief overview on the fundamentals of molecular networks. It serves as a tutorial for the audience to get familiarize with molecular networks and related issues.

S.-g. Lee et al. (Eds.): DASFAA 2012, Part II, LNCS 7239, pp. 330–331, 2012.

Section 2: Modeling and Representing Molecular Network Data. In this section, we discuss various graph data structures used to model molecular network data. We also introduce to the audience XML-based formats such as SBML, PSI-MI, CellML, and BioPAX that are widely used to represent and exchange molecular pathway information. We compare these models and representations and highlight their differences.

Section 3: Storing and Querying Molecular Pathways. Centre to any endeavor of managing molecular network or pathway data is the creation and maintenance of interaction and pathway databases. Unfortunately, as is the case with many biological data resource, interaction and pathway databases have unique data models, distinct access methods, different file formats, and subtle semantic differences. This diversity of implementation makes it extremely difficult to collect data from multiple sources, and therefore slows down scientific research involving pathways. This has given rise to increasing research activities to create integration, storage and querying techniques to manage large volumes of heterogeneous biological network data. In this section, we discuss a wide variety of such data management systems. We highlight the key features of these systems as well as critically compare their advantages and disadvantages. We also discuss query languages designed primarily for querying molecular pathways.

Section 4: Molecular Network Summarization. The amount of information contained within large biological networks can often overwhelm researchers, making systems level analysis of these networks a daunting task. As majority of function annotation and high throughput or curated interaction data are encoded at protein or gene level, higher-order abstraction (summary) maps such as complex-complex or process-process functional landscapes, are crucial. Such summary information is invaluable as it not only allows one to ask questions about the relationships among high-level modules, such as processes and complexes, but also allows one to visualize higher order patterns froma bird's eye perspective. In this section, we discuss state-of-the-art approaches for summarizing biological networks.

Section 5: Molecular Network Comparison. Molecular network comparison is the process of aligning and contrasting two or more interaction networks, representing different species, conditions, interaction types or time points. This process aims to answer a number of fundamental biological questions: which proteins, protein interactions and groups of interactions are likely to have equivalent functions across species? Based on these similarities, can we predict new functional information about proteins and interactions that are poorly characterized? In this section, we introduce the notion of molecular network comparison and discuss various state-of-the-art techniques and algorithms that have been proposed recently in literature to compare a set of molecular networks.

3   Speakers

Sourav S Bhowmick and Boon-Siew Seah (doctoral student) have published several papers in the area of biological network analytics. One of their papers received the best paper award at the ACM BCB 2011. They have also developed a state-of-the-art biological network summarization system called FUSE which was recently demonstrated in ACM SIGHIT 2012. Biography of Sourav can be found at www.ntu.edu.sg/home/assourav.

Author Index